TCP/IP COMPLETE

COMPLETE SERIES

TCP/IP
Complete

Ed Taylor

McGraw-Hill

New York · San Francisco · Washington, D.C. · Auckland · Bogotá
Caracas · Lisbon · London · Madrid · Mexico City · Milan
Montreal · New Delhi · San Juan · Singapore
Sydney · Tokyo · Toronto

Library of Congress Cataloging-in-Publication Data

Taylor, Ed, 1958—
   TCP/IP complete / D. Edgar Taylor.
      p.   cm.
   Includes index.
   ISBN 0-07-063400-9
   1. TCP/IP (Computer network protocol) I. Title.
TK5105.585.T33 1998
004.6′2—dc21
                                       97-43395
                                       CIP

McGraw-Hill

A Division of The **McGraw·Hill** *Companies*

P/N 0-07-063411-4
Part of
ISBN 0-07-063400-9

The sponsoring editor for this book was Steven Elliot, the editing supervisor was Frank Kotowski, Jr., and the production supervisor was Sherri Souffrance. It was set in Vendome ICG by Michele Pridmore and Paul Scozzari of McGraw-Hill's Professional Book Group composition unit, Hightstown, N.J.

Printed and bound by R. R. Donnelley & Sons Company.

McGraw-Hill books are available at special quantity discounts to use as premiums and sales promotions, or for use in corporate training programs. For more information, please write to the Director of Special Sales, McGraw-Hill, 11 West 19th Street, New York, NY 10011. Or contact your local bookstore.

 This book is printed on recycled, acid-free paper containing a minimum of 50% recycled, de-inked fiber.

To:
Ginny and Maegan

From:
Ed Taylor

CONTENTS

Contents

Contents

ACKNOWLEDGMENTS

The following people and companies are appreciatively acknowledged for having contributed to the book.

Charlie Casale
Steve Elliot
Judy Kessler
Frank Kotowski, Jr.
Iwao Matsushita
Donna Namoratto
Jane Palmieri

Airborne
Altec Lansing
Bud Industries
Creative Labs
DHL
Emery Airfreight
Federal Express
Fluke Instruments
Hewlett Packard
Hubbell
IBM
Information World, Inc.
McAfee
Microsoft
MJH
Roadway
SMC
SnapOn Tools
Sony Corporation
3ComUSRobotics
Tripp Lite
United Parcel Service (UPS)
United States Post Office
Wagner Edstrom

PURPOSE OF THIS BOOK

The reason I wrote this book is because a need existed to put certain information together in a single source. I have included some fundamental data communication information, a real network example using TCP/IP, TCP/IP details that have been beneficial to me along the way. I can be reached at the following address:

Internet: IWIinc@aol.com
                IWIinc@ibm.net
                IWIinc@msn.com
                Edtaylor@aol.com
                zac0002@ibm.net

AOL: IWIinc
        Edtaylor

Compuserve: 72714,1417

HOW TO USE THIS BOOK

You can read it from front to back. You can use it as as reference.

Transmission Control Protocol/Internet Protocol (TCP/IP)

Transmission Control Protocol/Internet Protocol, generally referred to as *TCP/IP*, is an upper-layer network protocol in widespread use around the world today. This chapter presents the core components and issues related to TCP/IP, beginning with a historical perspective.

1.1   A Historical Perspective

A good place to begin is in the late 1960s. An entity in the U.S. government called the Advanced Research Projects Agency, ARPA for short, was exploring technologies of all sorts. One of those technologies led to a need to create a network based on packet-switching technology to help members of the agency experiment with what they built. It was also seen as a means whereby telephone lines could be used to connect scientists and personnel who were in physically different locations, so they could work together on this network.

By late in 1969, the necessary components had come together to create the ARPAnet. In short order, a few individuals had put together a network that was capable of exchanging data. As time passed, additions and refinements were made to ARPAnet.

In the 1970s

In 1971, the Defense Advanced Research Projects Agency (DARPA) succeeded ARPA. As a result, the ARPAnet came under the control of DARPA. DARPA's forté was satellite, radio, and packet-switching technology.

During this same time period, the ARPAnet was using what was called a Network Control Program (NCP). Since the NCP was so closely tied to the characteristics of ARPAnet, it had limitations for coping with the areas of research, capabilities, and other requirements. The protocols ARPAnet utilized (namely the NCP) were characteristically slow and had periods where the network was not stable. Since ARPAnet was now officially under DARPA's umbrella and a new approach to ARPAnet was needed, a different direction was taken.

Beginning around 1974, DARPA sponsored development of a new set of protocols to replace the ones in use at that time. This endeavor led to the development of protocols that were the basis for TCP/IP.

The first TCP/IP began to appear in the 1974-1975 timeframe. While these technical matters were in full force, another phenomenon was occurring.

In 1975, the U.S. Department of Defense (DoD) put ARPAnet under the control of the Defense Communication Agency (DCA); the DCA was responsible for operational aspects of the network. It was then that the ARPAnet became the foundation for the Defense Data Network (DDN).

Time passed, and TCP/IP continued to be enhanced. Many networks emerged, working with and connecting to the ARPAnet with TCP/IP protocols. In 1978, TCP/IP was stable enough for a public demonstration from a mobile location connecting to a remote location via a satellite. It was a success.

In the 1980s

From 1978 until 1982, TCP/IP gained momentum and was continually refined. In 1982, multiple strides were made. First, DoD made a policy statement adopting TCP/IP protocols and making it the overseeing entity for uniting distributed networks. The next year, 1983, DoD formally adopted TCP/IP as the standard for the protocol to use when connecting to the ARPAnet.

Early 1983, when DoD formally discontinued support for the Network Control Program (NCP) and adopted TCP/IP protocol, marks the birth of the Internet. The term Internet was an outgrowth of the term internetworking, which was a technical term used to refer to the interconnection of networks. Nevertheless, the term Internet has maintained its association, reflecting the multiple networks around the world today.

In the 1990s

The Internet today consists of numerous interconnected networks. The National Research and Education Network (NREN) is a dominant part of the Internet today. Other networks that are part of the Internet include the National Science Foundation (NSF), NASA, the Department of Education, and many others, including educational institutions.

Organizations of all types are connected to the Internet. An industry of service providers for the Internet seems to be emerging.

1.2   Forces Contributing to Its Growth

The history reviewed in the previous section sheds some light on the technology surrounding TCP/IP and the Internet, but does not explain certain aspects of the Internet that may aid in understanding the technological impact it had on TCP/IP.

Technology

The Internet is based upon TCP/IP, as the U.S. government made it the standard. The Internet is worldwide, and all sorts of entities are connected to it. We can deduce that these entities are using TCP/IP. This alone counts for a tremendous amount of TCP/IP in the marketplace—and at the current rate, it is increasing rapidly.

The 1980s can be characterized as a decade of rapid technological growth. Many companies capitalized on the U.S. government endorsement of TCP/IP being the standard for the Internet and began producing products to meet this need.

This influx of TCP/IP products nursed the need for additional products. For example, in the 1980s, two technologies dominated: PCs and LANs. With the proliferation of PCs and LANs, an entirely new industry began emerging. These technological forces seemed to propel TCP/IP forward because TCP/IP and PCs made for a good match when implementing LANs. TCP/IP implemented on an individual basis is referred to as an *internet* (note the lowercase *i*).

Market Forces

A factor that contributed to the growth of TCP/IP in the market was corporate downsizing. This might seem strange, but during the 1980s, I witnessed many cases where TCP/IP-based networks grew while others were shrinking. This was not the only reason for TCP/IP's healthy market share, but it did contribute.

For example, I experienced a situation in which a corporation (which I will not name) had its corporate offices in the northeast, with over 50 satellite offices around the nation. This corporation needed these satellite offices to have independence for daily operations and at the same time be connected to the corporate data center. They achieved this by implementing TCP/IP-based LANs in their satellite offices, then connecting them to the data center. This example is one of many I have seen.

Availability

TCP/IP could be purchased off the shelf from many computer stores by the end of the 1980s. This degree of availability says a lot for a product that, at the beginning of the decade, was not readily available to end-users.

Another factor played a role in the availability of TCP/IP. The DoD not only encouraged the use of TCP/IP, they funded a company called Bolt, Beranek, and Newman to port TCP/IP to UNIX. In addition, DoD encouraged UC Berkeley to include TCP/IP in their BSD UNIX operating system. By acquiring Berkeley UNIX, therefore, users got TCP/IP free. It was not long before TCP/IP was also added to AT&T's System V UNIX operating system.

Individual Knowledge

By the late 1970s and into the 1980s, TCP/IP was in most colleges and many other educational institutions. Since by the mid-1980s it was shipping free with Berkeley UNIX, it became dominant in learning institutions. The obvious occurred: Individuals everywhere began graduating from educational institutions and, if their background included computer science, odds were they had been exposed to TCP/IP.

These individuals entered the workplace and began penetrating the technical and managerial departments. When it came to contributing to a decision about a network protocol, which would be the likely choice in many cases?

In the 1980s, the marketplace paid a premium for those who understood TCP/IP. Now, in the mid-1990s, the market has a considerable number of individuals who have varying degrees of TCP/IP knowledge.

All these factors weave together to make TCP/IP as dominant as it is today. TCP/IP has become a prevalent upper-layer protocol worldwide.

1.3 Layer Analysis

In the early days of the Internet, the term *gateway* became commonplace. It generally meant a connection from a specific location into the Internet. This was adequate at the time, but now confusion abounds with the use of this term.

According to the American Heritage Dictionary, the term *gateway* is defined as: "1. An opening, as in a wall or a fence, that may be closed by a gate. 2. A means of access." I believe the original intent of the term's meaning was "a means of access." This is fine, and you are probably wondering why it is even mentioned. Well, today, an entire industry called *internetworking and integration* has appeared, and with it specialized devices exist. One such device is a gateway.

Integrators and those who integrate heterogeneous networks have agreed upon the definition of the term *gateway* as a device that, at a minimum, converts upper-layer protocols from one type to another. It can, however, convert all seven layers of protocols.

The purpose of explaining this here is simple. Throughout this book, the term *gateway* may appear. The term has such a foothold in the TCP/IP community that it is still used. However, when the term *gateway* is used in many instances with TCP/IP and the Internet, technically, the term should be `router`.

1.4 Overview and Correlation to the OSI Model

TCP/IP is an upper-layer protocol. TCP/IP is implemented in software, although some specific implementations have abbreviated TCP/IP protocol stacks implemented in firmware. TCP/IP can operate on different hardware and software platforms, and it supports more than one data link layer protocol.

The OSI model is a representation of the layers that should exist in a network. Figure 1.1 shows TCP/IP compared to the OSI model. Notice that

Figure 1-1
TCP/IP correlated to
OSI

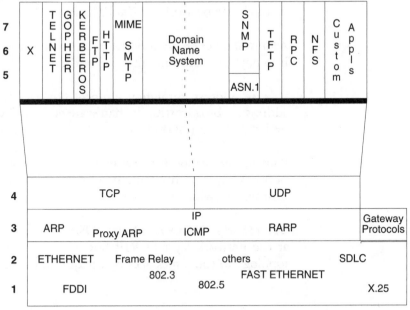

TCP/IP has three layers: network, transport, and the upper three layers combined to function as the application layer. TCP/IP is flexible when it comes to the lower two layers. It can be implemented in a variety of ways.

TCP/IP can operate with a number of data link layer protocols. Some are listed in Fig. 1-1. The remainder of this chapter highlights popular components at each layer.

1.5   Network Layer Components and Functions

The third layer of the OSI model is the network layer. In TCP/IP, it is the lowest layer in the TCP/IP protocol suite. TCP/IP network layer components include the following:

Internet Protocol (IP)   IP has an addressing scheme used to identify the host in which it resides. IP is involved in routing functions.

Internet Control Message Protocol (ICMP) ICMP is a required component in each TCP/IP implementation. It is responsible for sending messages through the network via the IP header.

Address Resolution Protocol (ARP) ARP dynamically translates IP addresses into physical (hardware interface card) addresses.

Reverse Address Resolution Protocol (RARP) RARP requests its host IP address by broadcasting its hardware address. Typically, a RARP server is designated and responds.

Routing Information Protocol (RIP) RIP is a routing protocol used at the network layer. If implemented, it performs routing of packets in the host within which it resides.

Open Shortest Path First (OSPF) OSPF is a routing protocol implemented at the network layer as RIP, but it utilizes knowledge of the internet topology to route messages by the quickest route.

1.6   Transport Layer Components and Functions

Layer four of the OSI model is the transport layer. In TCP/IP, it is the same. Transport layer components include the following:

TCP This transport-layer protocol is considered reliable and performs retransmissions if necessary.

UDP This transport-layer protocol is considered unreliable and does not perform retransmissions. Instead, this is left up to the application using its services.

1.7   Popular Application-Layer Offerings

Above the transport layer in TCP/IP, a number of popular applications exist:

X   This is a windowing system that can be implemented in a multi-vendor environment.

TELNET   This application provides remote logon services.

File Transfer Protocol (FTP)   This application provides file transfer capabilities among heterogeneous systems.

Simple Mail Transfer Protocol (SMTP)   This application provides electronic mail services for TCP/IP-based users.

Domain Name Service (DNS)   This application is designed to resolve destination addresses in a TCP/IP network. This application is an automated method of providing network addresses without having to update host tables manually.

Trivial File Transfer Protocol (TFTP)   This UDP application is used best in initialization of network devices where software must be downloaded to a device. Since TFTP is a simple file transfer protocol, it meets this need well.

Simple Network Management Protocol (SNMP)   This is the way most TCP/IP networks are managed. SNMP is based on an agent-and-manager arrangement. The agent collects information about a host, and the manager maintains status information about hosts participating with agents.

Network File Server (NFS)   NFS is an application that causes remote directories to appear to be part of the directory system of the user's host system.

Remote Procedure Call (RPC)   This is an application protocol that enables a routine to be called and executed on a server.

Custom Applications   Custom applications can be written using UDP as a transport-layer protocol. By doing so, peer communications can be achieved between applications.

1.8   TCP/IP Network Requirements

Before exploring the details of TCP/IP, you need to know the basic requirements for a TCP/IP network to function. For example, TCP/IP

Figure 1-2
Hosts use TCP/IP as
their networking
standard

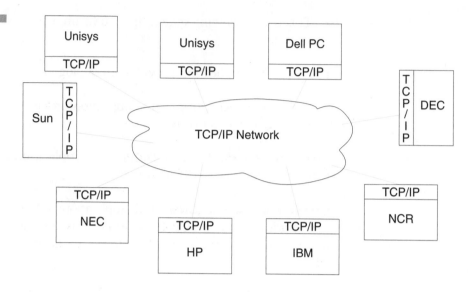

networks require that all participating hosts have TCP/IP operating on them, and they must be connected directly or indirectly to a common link. This may require some gateway functionality for some systems, but Fig. 1-2 is an example of a typical TCP/IP network with different vendor computers.

Figure 1-2 includes vendors whose operating systems are different. They also have different hardware platforms. However, if the link is established to the TCP/IP network, these different computers can communicate effectively.

1.9 Summary

TCP/IP has evolved over the past few decades. Its origins were in development, education, and research entities. Many technologies and market forces converged in the 1980s that caused TCP/IP to gain market momentum.

TCP/IP is technically a four-layer protocol. It can be compared and contrasted to the OSI model. TCP/IP can operate with numerous lower-layer protocols, as this chapter revealed.

Many of the applications that operate on top of TCP and UDP have become popular today. Their functionality has become common in many software packages that incorporate them.

TCP/IP rose to popularity for many reasons. One of those reasons is the unifying factor the protocol offers. It is possible to use TCP/IP with every major operating system in the marketplace today.

CHAPTER 2

Internet Protocol (IP)

IP resides at network layer three. IP routes *packets* (units of data) from source to destination. Some people refer to a packet in the sense of IP as a *datagram*. An IP datagram is a basic unit moved through a TCP/IP network.

IP is connectionless. It implements two basic functions: fragmentation and addressing. Fragmentation (and reassembly) is accomplished by a field in the IP header. Fragmentation is required when datagrams need to be smaller for passing through a small packet-oriented network.

2.1 IP Header Format

The addressing function is also implemented in the IP header. The header includes the source and destination address, as well as additional information. Figure 2-1 is an example of an IP header.

The components in the IP header, and their meanings, are as follows:

VERSION The VERSION field is used to indicate the format of the IP header.

IHL IHL stands for *internet header length*. It is the length of the internet header in 32-bit words, and points to the beginning of data.

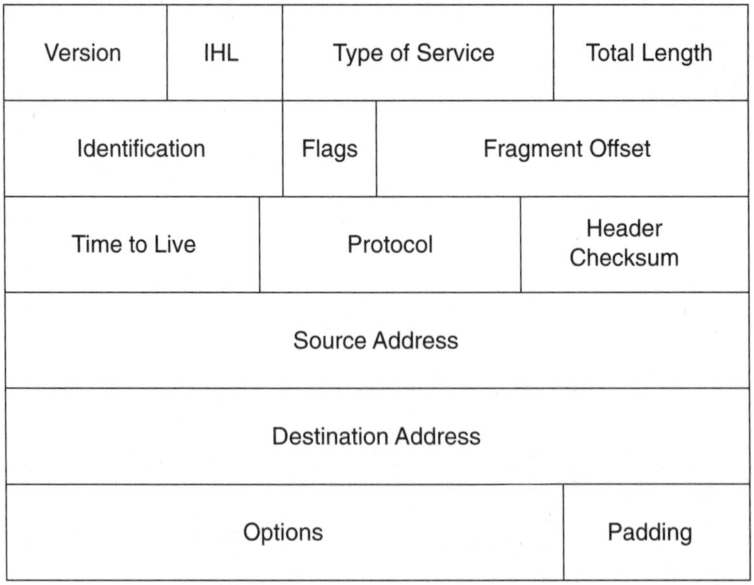

Figure 2-1
The contents of an IP header.

Version	IHL	Type of Service	Total Length
Identification		Flags	Fragment Offset
Time to Live		Protocol	Header Checksum
Source Address			
Destination Address			
Options			Padding

TYPE OF SERVICE   The TYPE OF SERVICE field specifies how the datagram is treated during its transmission through the network.

TOTAL LENGTH   This field indicates the total length of the datagram; this includes the IP header and data.

FLAGS   The FLAG field has three bits. If fragmentation is supported, they are used to indicate not to fragment, more fragments, and last fragment.

FRAGMENT OFFSET   This indicates where in the datagram the fragment belongs, assuming fragmentation has occurred.

TIME TO LIVE   This indicates the maximum time a datagram is permitted to stay in the internet system (whether this is a local internet or the Internet). When the value equals zero, the datagram is destroyed. Time is measured in units per second, and each entity that processes the datagram must decrease the value by one, even if the process time is less than one second.

PROTOCOL   This field determines whether the data should be sent to TCP or UDP in the next layer in the network.

HEADER CHECKSUM   This is a header checksum only. Some header fields change, and the header checksum is recomputed and verified every time the header is processed.

SOURCE ADDRESS   This is the originator of the datagram. It consists of 32 bits.

DESTINATION ADDRESS   This is the target for the header and data. Like the source address, it too is 32 bits.

OPTIONS   Options may or may not appear in datagrams. Options must be implemented in IP modules; however, they might not be used in any given transmission.

A number of variables exist for the OPTIONS field. The following is a list of those variables, including brief explanations:

NO OPTION   This option can be used between options to correlate the beginning of a following option on a 32-bit boundary.

SECURITY   Security is a mechanism used by DoD. It provides hosts a way to use security by means of compartmentation, handling restrictions, and transmission control codes (TCC). The compartmentation value is used when information transmitted is not compartmented. Handling restrictions are defined by the Defense Intelligence Agency. TCC permits segregation of data traffic.

LOOSE SOURCE and RECORD ROUTE   This provides a way for a source of a datagram to supply routing information to aid in forwarding the datagram. It also serves to record the route information.

STRICT SOURCE and RECORD ROUTE   This option permits the source of a datagram to supply information used by routers and record the route information.

RECORD ROUTE   This is simply a way to record the route of a datagram as it traverses the network.

STREAM IDENTIFIER   This provides a way for a stream identifier to be carried through networks that do not support this stream concept.

TIMESTAMP   This option includes a pointer, overflow, flag field, and internet address. Simply put, this provides the time and date when a router handles the datagram.

PADDING   The PADDING option is used to ensure the header ends on the 32-bit boundary.

2.2 IP Version 6: A Perspective

IP version 6 (IPv6) is a new version of the Internet Protocol, designed as a successor to IP version 4 (IPv4) [RFC-791]. The changes from IPv4 to IPv6 fall primarily into the following categories:

Expanded Addressing Capabilities   IPv6 increases the IP address size from 32 bits to 128 bits, to support more levels of addressing hierarchy, a much greater number of addressable nodes, and simpler auto-configuration of addresses. The scalability of multicast routing is improved by adding a "scope" field to multicast addresses. A new type of address called an "anycast address" is also defined, used to send a packet to any one of a group of nodes.

Header Format Simplification   Some IPv4 header fields have been dropped or made optional, to reduce the common-case processing cost of packet handling and to limit the bandwidth cost of the IPv6 header.

Improved Support for Extensions and Options   Changes in the way IP header options are encoded allow for more efficient forwarding, less stringent limits on the length of options, and greater flexibility for introducing new options in the future.

Flow Labeling Capability   A new capability is added to enable the labeling of packets belonging to particular traffic flows for which the sender requests special handling, such as non-default quality of service or real-time service.

Authentication and Privacy Capabilities   Extensions to support authentication, data integrity, and data confidentiality are specified for IPv6. This document specifies the basic IPv6 header and the initially defined IPv6 extension headers and options. It also discusses packet-size issues, the semantics of flow labels and priority, and the effects of IPv6 on upper-layer protocols. The format and semantics of IPv6 addresses are specified separately in [RFC-1884]. The IPv6 version of ICMP, which all IPv6 implementations are required to include, is specified in [RFC-1885].

2.3   Internet Control Message Protocol (ICMP)

ICMP works with IP; it is also located at layer three with IP. Since IP is connectionless oriented, it has no way to relay messages or errors to the originating host. ICMP performs these functions on behalf of IP. ICMP sends status messages and error messages to the sending host.

ICMP utilizes IP to carry the ICMP data within it through a network. Just because ICMP uses IP as a vehicle does not make IP reliable; it simply means that IP carries the ICMP message. The structure of an ICMP message is shown in Fig. 2-2.

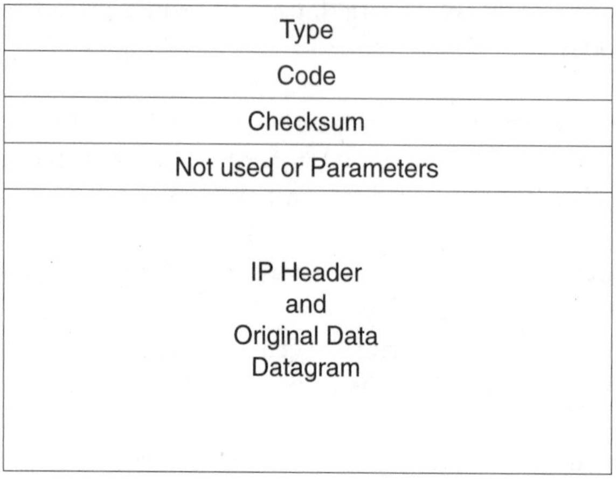

Figure 2-2
The ICMP message
format.

Type
Code
Checksum
Not used or Parameters
IP Header and Original Data Datagram

The first part of the ICMP message is the TYPE field. This field has a numeric value reflecting its meaning; this field identifies its format as well. The numeric values that can appear in the type field, and their meanings, are shown in Fig. 2-3.

The next field in the ICMP message is the CODE field. It, too, has a numeric value assigned to it. Different numeric values have different associated meanings, as shown in Fig. 2-4.

The CHECKSUM is computed from the ICMP message starting with the ICMP type.

The NOT USED field means that; I referenced RFC 792. The next field is the IP HEADER AND DATA DATAGRAM.

ICMP is the source for many messages a user sees on their display. For example, if a user attempts a remote logon and the host is not reachable, then the user will see the message "host unreachable" on the screen. This message comes from ICMP.

ICMP detects errors, reports problems, and generates messages as well. For IP to be implemented, ICMP must be part of it because of the design of IP.

2.4 Address Resolution Protocol (ARP)

ARP is located at layer three along with IP and ICMP. ARP maps IP addresses to the underlying hardware address. Actually, ARP dynamically binds these addresses.

Figure 2-3
ICMP message types
and their meanings.

Type	Meaning of Message
0	Echo reply
3	Destination unreachable
4	Source quench
5	Redirect
8	Echo request
11	Time exceeded for a Datagram
12	Parameter problem on a Datagram
13	Timestamp request
14	Timestamp reply
17	Address mask request
18	Address mask reply

Figure 2-4
ICMP codes and their
meanings.

0	Network unreachable
1	Host unreachable
2	Protocol unreachable
3	Port unreachable
4	Fragmentation needed
5	Source route failed
6	Destination Network unknown
7	Destination Host unknown
8	Source Host isolated
9	Administrative restrictions to destination Network. Communication prohibited
10	Communication with destination Host prohibited by Administration
11	Network unreachable for service type
12	Host unreachable for service type

Perspective

Since TCP/IP works at layer three and above, it must have a mechanism to function with interface boards. When TCP/IP is implemented, it is done so in software. Each host participating on a TCP/IP network must have TCP/IP and have a unique IP address. This IP address is considered a software address, since it is implemented at layer three in software.

Because any one of many data link protocols could be used, IP requires a way to correlate the IP address and the data link address. Data link addresses are generally considered hardware addresses. For example, if TCP/IP is implemented with ETHERNET, there is a 48-bit ETHERNET address that must be mapped to the 32-bit IP address. If Token Ring is used, a 12-digit hexadecimal address is used as the hardware address. Neither of these data link protocol addresses match the 32-bit IP address of TCP/IP. This is the reason for ARP.

ARP Theory of Operation

Using ETHERNET for a data link, ARP can be explained this way: Assume five hosts reside on an ETHERNET network. Assume a user on host A wants to connect to host E. Host A uses ARP to broadcast a packet that includes A's IP address and ETHERNET address, and host E's IP address.

All five hosts on the network "hear" the ARP broadcast for host E. However, only host E recognizes its IP address inside the ARP request. Figure 2-5 depicts this. When host E recognizes its hardware address, it replies back to host A with its IP address. Figure 2.6 is an example of this process.

It is obvious that all hosts shown in Fig. 2-6 must examine the ARP request. This is expensive in regards to network utilization. To avoid this constant barrage, an *ARP cache* is maintained. This is a list of network hosts' physical and IP addresses, which curbs the number of ARP packets on the network.

When a host receives an ARP reply, that host keeps the IP address of the other host in this ARP table. Then, when a host wants to communicate with another host on the network, it first examines its ARP cache for the IP address. If the desired IP address is found in the cache, there is no need to perform an ARP broadcast. The way the communication occurs is via hardware communication; that is for example, ETHERNET boards communicating with one another.

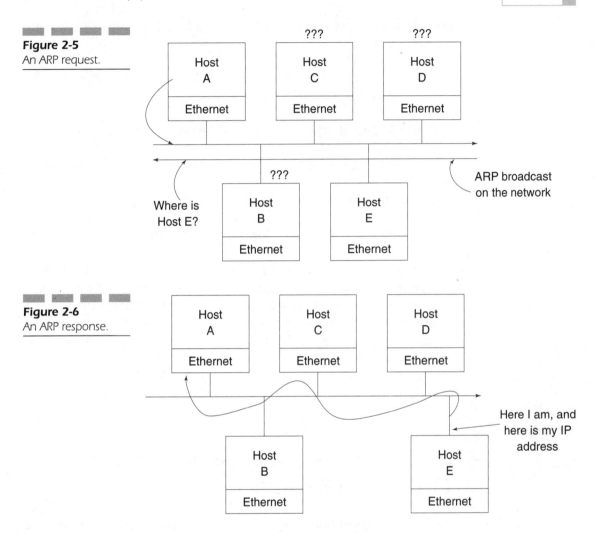

Figure 2-5
An ARP request.

Figure 2-6
An ARP response.

ARP Message Format

Figure 2-7 is an example of ARP message format. The following lists the fields in the ARP packet and briefly explains their meanings:

HARDWARE TYPE   Indicates the hardware interface type.

PROTOCOL TYPE   Specifies the upper-level protocol address the originator sent.

HARDWARE ADDRESS LENGTH   Specifies the length of the bytes in the packet.

PROTOCOL ADDRESS LENGTH Specifies the length in bytes of the high-level protocol.

OPERATION CODE Specifies one of the following: ARP request, ARP response, RARP request, or RARP response.

SENDER HARDWARE ADDRESS Supplied by the sender, if known.

SENDER PROTOCOL ADDRESS Like the hardware address, sent if known.

TARGET HARDWARE ADDRESS Specifies the destination address.

TARGET PROTOCOL ADDRESS Contains the IP address of the destination host.

Since ARP functions at the lowest layers within a network, the ARP request itself must be encapsulated within the hardware protocol frame because the frame itself is what physically moves through the network at this level. Conceptually, the frame carrying the ARP message and the frame appears like Fig. 2-8.

ARP's dynamic address translation provides a robust method for obtaining an unknown address. The efficiency of ARP is in the utilization of the caching mechanism.

2.5 Reverse Address Resolution Protocol (RARP)

RARP is the reverse of ARP. It is used commonly where diskless workstations are implemented. When a diskless workstation boots, it knows its hardware address because it is in the interface card connecting it to the network. However, it does not know its IP address.

RARP Request and Server Operation

Devices using RARP require that a RARP server be present on the network to answer an RARP request. RARP requests ask the question, "What is my IP address?" This is broadcast on the network, and a designated RARP server replies by examining the physical address received in

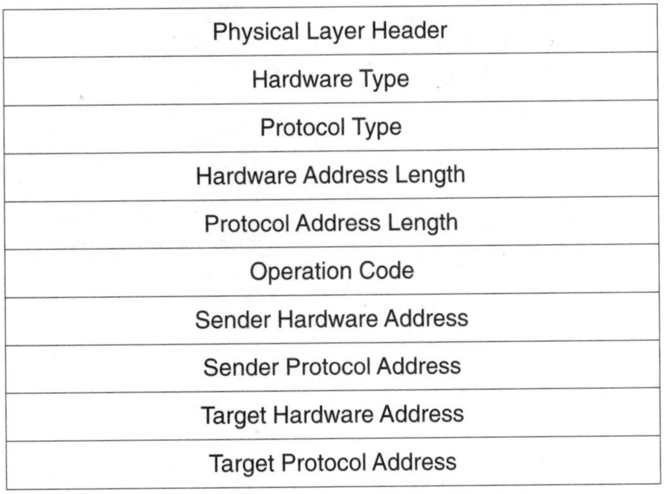

Figure 2-7
An example of an
ARP packet.

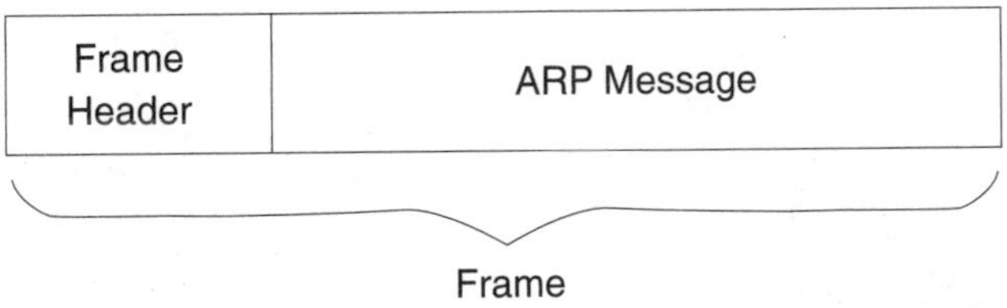

Physical Layer Header
Hardware Type
Protocol Type
Hardware Address Length
Protocol Address Length
Operation Code
Sender Hardware Address
Sender Protocol Address
Target Hardware Address
Target Protocol Address

Figure 2-8
A conceptual view of
the frame and the
ARP message.

Frame Header	ARP Message

Frame

the RARP packet, comparing it against its tables of IP addresses, and sending the response back to the requesting host. Figure 2-9 is an example of an RARP broadcast.

Figure 2-9 shows the RARP request going to all hosts on the network. It also shows a RARP server. In Fig. 2-10, the RARP server answers the RARP request. For RARP to be used in a network, a RARP server must exist. In most implementations where RARP is used, multiple RARP servers are used. One is designated as a primary server and another as a secondary server.

2.6 Router Protocols

Normally, this section would be called "Gateway Protocols," but as mentioned previously, this is incorrect if defined according to functionality. For clarity's sake, this section covers the gateway protocols, but here they are called *routers* because that is what they do—route.

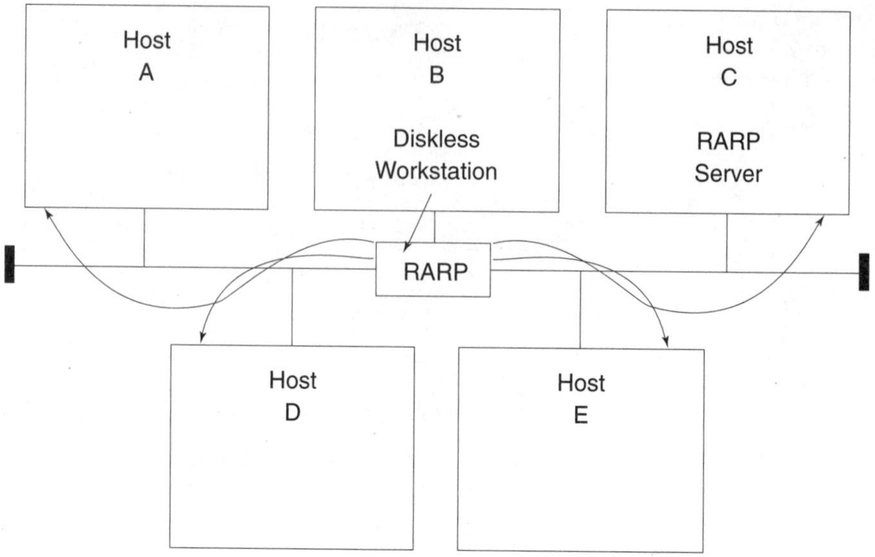

Figure 2-9
A RARP broadcast.

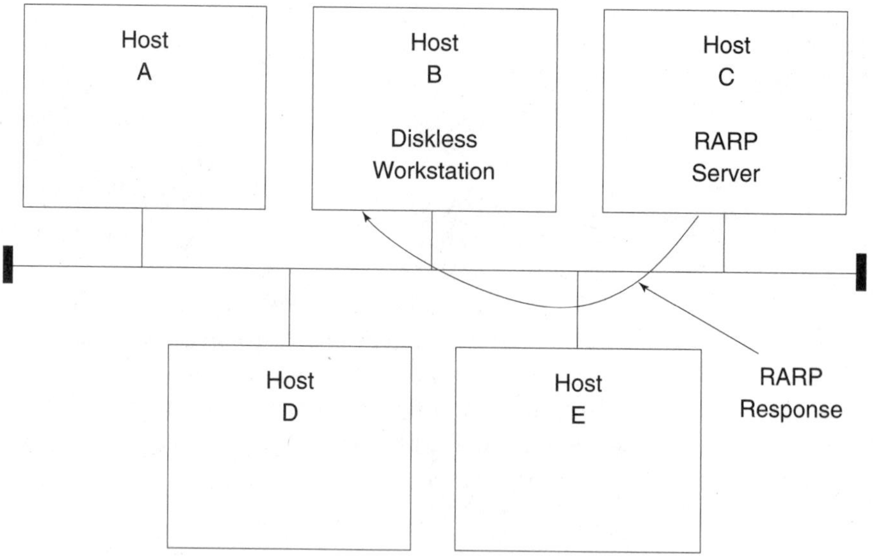

Figure 2-10
A RARP response.

This section focuses upon what are called Interior Gateway Protocols. An Interior Gateway Protocol is defined as routing in an autonomous system. An autonomous system is a collection of routers controlled by one administrative authority and using a common interior gateway protocol. Two popular routing protocols exist in this category: Router Information Protocol (RIP) and Open Shortest Path First (OSPF). These protocols are used by network devices such as routers, hosts, and other devices normally implemented in TCP/IP software. However, it is feasible that these protocols can be implemented by firmware.

RIP

RIP has its origins in Xerox's network systems protocol. It was at one time included in the software distribution of TCP/IP with Berkeley UNIX.

RIP is an example of a *de facto* protocol. It was implemented before a standard RFC existed. It was part of TCP/IP, it worked, and it was needed; all the ingredients to make a product popular! RFC 1058 brought RIP into a formal standard.

RIP Header Analysis

Consider Fig. 2-11, which is an example of the RIP message format. The following is a brief description of each field in the RIP message format:

COMMAND This specifies an operation, which could be a request or response.

VERSION This identifies the protocol version.

ZERO A blank field.

ADDRESS FAMILY IDENTIFIER This is used to identify the protocol family under which the message should be interpreted.

ZERO A blank field.

IP ADDRESS This field usually has a default route attached to it.

ZERO A blank field.

ZERO A blank field.

DISTANCE TO NET This is a value indicating the distance to the target network.

RIP messages either convey routing information or request routing information. RIP is based on broadcast technology. Periodically, a router (or designated device) broadcasts the entire RIP routing table throughout the network, be that a LAN or otherwise. This aspect alone has become a problem in some environments because of the lack of efficiency. In addition to broadcasts and updates from that process, RIP also gets updates due to changes in the network configuration. These updates are referred to as *responses*.

Another characteristic of RIP is that it relies on other devices (adjacent nodes) for routing information for targets that are more than one "hop" away. RIP also calculates its distances by costs per hop. One hop is defined as a metric. The maximum number of hops RIP can make along one path is 15.

RIP maintains a table with entries. This table is the one referred to previously that is broadcast throughout the network. Information contained in each entry in this table includes the following:

- The destination IP address
- The number of hops required to reach the destination
- The IP address for the next router in the path
- Any indication if the route has recently changed
- Timers along the route

RIP is still used today. Many vendors support it. In certain environments, it might be a good gateway protocol to use. However, many vendors support Open Shortest Path First (OSPF).

Open Shortest Path First (OSPF)

The philosophy of OSPF differs from RIP. It was recommended as a standard in 1990. By late 1991, version 2 of OSPF was available, giving an

Figure 2-11
The RIP message format.

Command
Version
Zero
Address Family Identifier
Zero
IP Address
Zero
Zero
Distance to Net A

indication of its popularity. Some of the tenets OSPF maintains include the following:

- Provides type-of-service routing
- Allows virtual networks to be defined
- Offers route distribution
- Allows minimized broadcasts
- Supports a method for trusted routers

Other tenets support OSPF. Depending on the vendor, a variety of them may be implemented.

OSPF Advertisements

OSPF uses what are called *advertisements.* These advertisements are ways that routers can inform other routers about paths. There are four distinct types of advertisements:

AUTONOMOUS   Has information about routes in other autonomous systems.

NETWORK   Contains a list of routers connected to the network.

ROUTER   Contains information about given router interfaces in certain areas.

SUMMARY   Maintains route information outside a given area.

These advertisements enable a more focused approach to spreading information throughout a network. Besides the advertisements, OSPF uses a number of messages for communication. Some of these messages include the following:

- Hello
- Database Description
- Link State Request
- Link State Update
- Link State Acknowledgement

Two of these are presented and explained in detail in the following sections, to provide insight on the operation of OSPF. You can find additional information about OSPF from RFCs 1245, 1246, and 1247.

OSPF Header Analysis

The OSPF packet header appears as in Fig. 2-12. The OSPF header fields are as follows:

VERSION   Indicates the protocol version.

TYPE   indicates messages as one of the following:

1. Hello
2. Database Description
3. Link Status
4. Link Status Update
5. Link Status Acknowledgement

MESSAGE LENGTH   Indicates the length of the field, including the OSPF header.

SOURCE GATEWAY IP ADDRESS
  Provides the sender's address.

AREA ID   Identifies the area from which the packet was transmitted.

CHECKSUM   Provides the checksum, which is performed on the entire packet.

AUTHENTICATION TYPE   Identifies the authentication type that will be used.

AUTHENTICATION   Includes a value from the authentication type.

The HELLO packet includes messages that are periodically sent on each link to establish if a destination can be reached. The HELLO packet appears as Fig. 2-13.

The following provides a list of fields in the HELLO packet and a brief explanation of each:

OSPF HEADER   A required field.

NETWORK MASK   Contains the network mask for the network where the message originated.

Figure 2-12
The OSPF packet
header.

Version
Type
Packet Length
Router ID
Area ID
Checksum
Authentication type
Authentication

Figure 2-13
The HELLO packet
format.

OSPF Header
Network Mask
Dead Timer
Hello Interval
Router Priority
Designated or Backup Router
Neighbor 1 IP Address
Neighbor 2 IP Address
Neighbor 3 IP Address
etc
.
.
.

DEAD TIMER A value (in seconds) that indicates a neighbor is dead if no response is received.

HELLO INTERVAL A value, in seconds, reflecting the amount of time between a router sending another HELLO packet.

ROUTER PRIORITY Used if a designated router is used as a backup.

DESIGNATED or BACKUP ROUTER Identifies the backup router.

NEIGHBOR ROUTER ID This field, and subsequent ones, indicates the ID of routers that have recently sent HELLO packets within the network.

The Database Description packet message includes an OSPF header and fields of information. These fields include information about messages received. They can be broken into smaller units. Information is also provided to indicate if information is missing. The packet also includes information about the type link and its ID. A checksum is provided to ensure that corruption has not occurred. The Link State packet header includes an OSPF header and fields that provide information such as router, network, and link station type.

The essence of OSPF is that it reduces traffic overhead in the network because it performs individual updates rather than broadcasts across the entire network. OSPF also provides an ability for authentication. Another strength of OSPF is that it can exchange subnet masks as well as subnet addresses.

2.7 Summary

TCP/IP is undergoing a slow evolution. IP itself is at versions 4 and 6 in the marketplace today. Even though version 6 is the direction of IP products and implementations, much equipment still operates at version 4.

IP uses ICMP to operate at layer three in networks. IP addressing is segregated into three different classes in version 4 and is expanded to encompass a large number of hosts in version 6.

Material in this chapter is fundamental. Details on IP versions 4 and 6 are presented in chapters later in this book.

CHAPTER 3

Transmission Control Protocol and User Datagram Protocol

TCP operates at layer four and is a transport protocol. It takes data passed to it from its applications and places it in a send buffer. Then TCP divides the data into what is called a *segment*. A segment is the application data and a TCP/IP header, which is necessary because data is delivered in datagrams.

3.1 Characteristics and Functions

TCP treats data in different ways. For example, if a user enters a simple command and presses the Enter key, TCP uses what it refers to as a *PUSH* function to make the command happen. This push ability is a function of TCP. On the other hand, if a large amount of data is passed from an application to TCP, it segments it and passes it to IP for further processing.

Applications that use TCP pass data to TCP in a stream fashion. This is in contrast to some other protocols, where applications pass data down the protocol stack in byte fashion. TCP is also a connection-oriented protocol. This means TCP keeps the state and status of streams passing into and out of it. It is TCP's responsibility to ensure reliable end-to-end service.

Two other major features that TCP provides are multiplexing and full-duplex transmission. TCP is capable of performing multiplexing user sessions because of the addressing scheme used in TCP/IP. This addressing scheme is covered later in this chapter in a separate section.

TCP also performs a function called resequencing. This function manages segments if they reach the destination out of order. It can perform this function because of sequence numbers used for acknowledgements.

3.2 TCP Header Analysis

Many of the aforementioned functions and others can be understood by examining the TCP header in Fig. 3-1. The following lists the parts of the TCP header segment, with a brief description of each part:

SOURCE PORT This field identifies the upper-layer application using the TCP connection.

DESTINATION PORT This field identifies the upper-layer application using the TCP connection.

SEQUENCE NUMBER The value in this field identifies the transmitting byte stream. This value is used during connection-management operations.

ACKNOWLEDGEMENT NUMBER The value in this field reflects a value acknowledging data previously received.

DATA OFFSET This value determines where the field of data begins.

RESERVED This field is reserved, and the bits within it are set to zero.

URGENT This field indicates if the urgent pointer is used.

ACKNOWLEDGEMENT This indicates whether or not the acknowledgement field is significant.

PSH This field indicates if the PUSH function is used.

RESET The value in this field indicates whether the connection should be reset.

SYNCHRONIZE This is used to indicate that sequence numbers are to be synchronized.

FINISHED This field is used to indicate that the sender has no more data to send.

WINDOW This value indicates how much data the receiving host can accept. The value in this field is contingent on the value in the acknowledgement field.

CHECKSUM This performs a checksum on the 16-bit words in the segment.

URGENT POINTER This field, if used, indicates that urgent data follows.

OPTIONS At the current time, three basic options are implemented in this field: end-of-list, no operation, and maximum segment size.

PADDING This field is used to ensure the header length equals 32 bits.

DATA User data follows in this field.

Figure 3-1
A TCP segment.

Source Port
Destination Port
Sequence Number
Acknowledgment Number
Data Offset
Reserved
Urgent
Acknowledgment
Push
Reset
Synchronizer
Finished
Window
Checksum
Urgent Pointer
Options
Padding
Data

TCP's reliable data transfer, connection-oriented nature, stream support for applications, multiplexing, full duplex transmission, PUSH functions, flow control, and other characteristics make it a popular and reliable protocol. The TCP protocol is defined in RFC 793. Additional information can be obtained from it if details are required.

3.3 User Datagram (UDP) Protocol

User datagram protocol (UDP) resides at transport layer four. In many ways, it is the opposite of TCP. UDP is connectionless-oriented and unreliable. It does little more than provide a transport-layer protocol for applications that reside above it.

UDP Header Analysis

The extent of information about UDP is brief compared to TCP. An example of the UDP datagram is shown in Fig. 3-2.

The following lists the components in the UDP datagram and provides a brief description of each:

SOURCE PORT   The value in this field identifies the origin port. (Ports are used in addressing and will be discussed in detail in the section on addressing.)

DESTINATION PORT   This identifies the recipient port for the data.

LENGTH   The value in this field indicates the length of the data sent, including the header.

CHECKSUM   This algorithm computes the pseudo-IP header, the UDP header, and the data.

DATA   This field holds the data passed from applications using UDP.

3.4   UDP Applications

UDP is a useful protocol in situations where there is a need for a custom application. When this is the case, UDP is a good transport protocol. However, because UDP is unreliable and does not perform the retransmissions and other services that TCP offers, the custom application must perform these functions.

Because of UDP's nature, it leaves work for application programmers. The necessary operations that it cannot perform can be achieved via the application; it merely requires more work on behalf of the one creating the application.

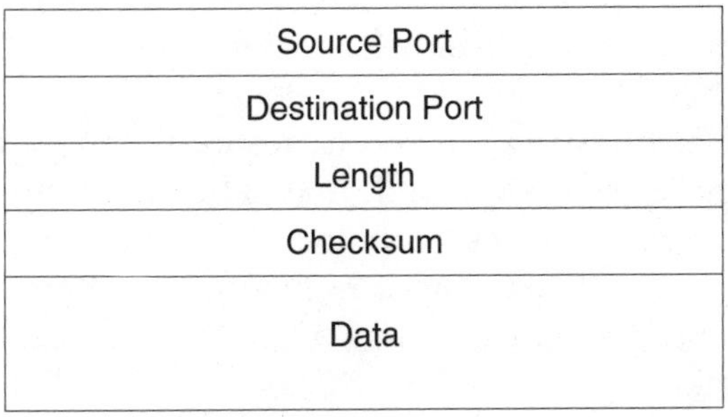

Figure 3-2
A UDP datagram.

Source Port
Destination Port
Length
Checksum
Data

Messages sent to UDP from applications get forwarded to IP for transmission. Some applications that reside on the UDP protocol pass messages directly to IP and ICMP for transmission.

3.5   TCP/IP Addressing

Addressing in TCP/IP consists of a variety of factors that work together to make TCP/IP a functioning upper-layer network protocol. Some of these factors include the following:

- IPv4 and v6 addressing
- Address classifications
- Ports
- Well-known ports
- Port manipulation
- Sockets
- Hardware addresses

Each of these is presented in the following sections, with a synthesis to aid in understanding how they relate. It is important to understand these addressing schemes/mechanisms in order to understand TCP and UDP operation.

IP Addressing v4

The Internet protocol uses a 32-bit addressing scheme. This addressing is implemented in software; however, in some network devices, it is stored in firmware and/or non-volatile RAM.

Each host participating in a TCP/IP network is identified via a 32-bit IP address. This is significant because it is different than the host's hardware address.

The IP addressing scheme structure is shown in Fig. 3-3. Figure 3-3 shows five classes of IP addresses. The IP addressing scheme is *dotted decimal notation*. The class of address indicates how many bits are used for a network address and how many for a host address. Before examining these in detail, a word about how these addresses are assigned is beneficial.

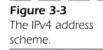

Figure 3-3
The IPv4 address
scheme.

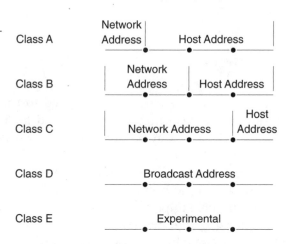

As Fig. 3-3 shows, multiple classes of addresses exists. A reasonable question is, why? The answer relates to the fact that two implementations of TCP/IP networks are possible: the Internet and internets.

The Internet is a worldwide network that has thousands of entities connecting to it. An agency responsible for maintaining Internet addresses assigns IP addresses to entities connecting to the Internet; the entity itself has no say in the matter. On the other hand, if a TCP/IP network is implemented in a corporation, for example, the IP addressing scheme is left up to the implementers responsible for that corporate network. In other words, it is locally administered.

When the latter implementation is the case, it is best someone understands the ramifications of selecting an IP addressing scheme. Multiple issues factor into the equation for selecting an IP addressing scheme.

In Fig. 3-3, five classes of addresses were shown. The following explains the numerical meaning and how this effects hosts implemented with IP addresses:

Address Class	Assigned Numbers
A	0 to 127
B	128 to 191
C	192 to 223

Address Class	Assigned Numbers
D	224 to 239
E	240 to 255

Class A addresses have fewer bits allocated to the network portion (one byte) and more bits (three bytes) dedicated for hosts addressing. In other words, more hosts can be implemented than networks, according to this addressing scheme.

Class B addressing allocates an equal amount of bits for network addressing (two bytes) and host addressing (two bytes). This class is popular in locally administered implementations.

Class C addressing allocates more bits (three bytes) to the network portion and fewer bits (one byte) to the host portion.

Class D is generally used as a broadcast address. The numerical value in each of the four bytes is 255.255.255.255.

Class E networks are for experimental purposes. I know of no class-E networks that are currently implemented.

Implementing an internet uses these addresses in conjunction with aliases. For example, an address assigned to a host would usually have a name associated with it. If a host had a class-B address, such as 137.1.0.99, its alias name might be RISC. This alias and internet address reside in a file on UNIX systems called /etc/hosts. Another file related to this is the /etc/networks file. In a UNIX environment, these two files are normal in the configuration. Additional information is included in later sections in this chapter regarding TCP/IP configuration in a UNIX environment.

IP Addressing v6

IPv6 addresses are 128-bit. There are three types of addresses:

- *Unicast* is an identifier for a single interface. A packet sent to a unicast address is delivered to the interface identified by that address.

- *Anycast* is an identifier for a set of interfaces (typically belonging to different nodes). A packet sent to an anycast address is delivered to one of the interfaces identified by that address (the "nearest" one, according to the routing protocol's measure of distance).

■ *Multicast* is another identifier for a set of interfaces typically belonging to different nodes. A packet sent to a multicast address is delivered to all interfaces identified by that address.

There are no broadcast addresses in IPv6. Broadcast addresses are superseded by multicast addresses. Here, address fields are given a specific name, such as *subscriber*. When this name is used with the ID for identifier after the name *subscriber ID,* it refers to the contents of that field. When it is used with the term *prefix,* it refers to all of the address up to and including this field.

In IPv6, all zeros and ones are legal values for any field unless specifically excluded. Prefixes may contain zero-valued fields or end in zeros.

IPv6 addresses of all types are assigned to interfaces, not nodes. Since each interface belongs to a single node, any of the unicast addresses for that node's interfaces may be used as an identifier for the node.

An IPv6 unicast address refers to a single interface. A single interface may be assigned multiple IPv6 addresses of any type (unicast, anycast, and multicast). There are two exceptions to this model, however:

1. A single address may be assigned to multiple physical interfaces if the implementation treats the multiple physical interfaces as one interface when presenting it to the internet layer.

2. Routers may have unnumbered interfaces on point-to-point links to eliminate the necessity to manually configure and advertise the addresses.

Addresses are not required for point-to-point interfaces on routers if those interfaces are not to be used as the origins or destinations of any IPv6 datagrams.

IPv6 continues the IPv4 subnet model that is associated with one link. Multiple subnets may be assigned to that link. There are three conventional forms for representing IPv6 addresses as text strings:

1. The preferred form is x:x:x:x:x:x:x:x, where the *x*s are the hexadecimal values of the eight 16-bit pieces of the address. Here are two examples:

```
FEDC:BA98:7654:3210:FEDC:BA98:7654:3210

1080:0:0:0:8:800:200C:417A
```

2. Due to the method of allocating certain styles of IPv6 addresses, it is common for addresses to contain long strings of zero bits. In

order to make it easier to write addresses containing zero bits, a special syntax is available to compress the zeros. The use of :: indicates multiple groups of 16 bits of zeros. The :: can only appear once in an address. It can also be used to compress the leading and/or trailing zeros in an address. For example, consider the following addresses:

```
1080:0:0:0:8:800:200C:417A    a unicast address
FF01:0:0:0:0:0:0:43           a multicast address
0:0:0:0:0:0:0:1               the loopback address
0:0:0:0:0:0:0:0               the unspecified addresses
```

These may be represented as follows:

```
1080::8:800:200C:417A    a unicast address

FF01::43                 a multicast address

::1                      the loopback address

::                       the unspecified addresses
```

3. An alternative form that is sometimes more convenient when dealing with a mixed environment of IPv4 and IPv6 nodes is x:x:x:x:x:x:d.d.d.d, where the *x*s are the hexadecimal values of the six high-order 16-bit pieces of the address, and the *d*s are the decimal values of the four low-order 8-bit pieces of the address (in standard IPv4 representation). Examples include the following:

```
0:0:0:0:0:0:13.1.68.3

0:0:0:0:0:FFFF:129.144.52.38
```

In compressed form, these appear like this:

```
::13.1.68.3

::FFFF:129.144.52.38
```

Ports

Ports are the addressable end points of TCP and UDP. This is partially how applications above TCP and UDP are addressed.

TCP and UDP have popular applications that use them for a transport protocol. Without some standardization of ports and relationships to applications, there would be chaos. As a result, TCP and UDP have applications that are assigned to a *well-known port*. Those working in the field of TCP/IP know this, generally. It is a standard to which most people adhere, although some flexibility does exist.

These well-known port names reflect specific applications of wide implementation and usage. Ports are the end points, an addressable entity to create a logical connection. Also known as *service contact ports,* these ports provide services to callers (requesters) of a particular service. The following list includes the port's decimal number as it is known, the name of the reference associated with a specific port, and a brief description of what each port provides. The list is not exhaustive; it is intended to provide the reader with a reference for common ports used in TCP/IP networks.

Decimal	Name	Description
0	Reserved	
1	TCPMUX	TCP Port Service Multiplexer
2-4		Unassigned
5	RJE	Remote Job Entry
7	ECHO	Echo
9	DISCARD	Discard
11	USERS	Active Users
13	DAYTIME	Daytime
15		Unassigned
17	QUOTE	Quote of the Day
19	CHARGEN	Character Generator
20	FTP-DATA	File Transfer (Data)
21	FTP	File Transfer (Control)
23	TELNET	TELNET
25	SMTP	Simple Mail Transfer
27	NSW-FE	NSW User System FE
29	MSG-ICP	MSG-ICP
31	MSG-AUTH	MSG Authentication

Decimal	Name	Description
33	DSP	Display Support Protocol
35		Any Private Printer Server
37	TIME	Time
39	RLP	Resource Location Protocol
41	GRAPHICS	Graphics
42	NAMESERVER	Host Name Server
43	NICNAME	Who Is
49	LOGIN	Login Host Protocol
53	DOMAIN	Domain Name Server
67	BOOTPS	Bootstrap Protocol Server
68	BOOTPC	Bootstrap Protocol Client
69	TFTP	Trivial File Transfer
79	FINGER	Finger
101	HOSTNAME	NIC Host Name Server
102	ISO-TSAP	ISO TSAP
103	X400	X.400
104	X400SND	X.400 SND
105	CSNET-NS	CSNET Mailbox Name Server
109	POP2	Post Office Protocol version 2
110	POP3	Post Office Protocol version 3
111	SUNRPC	SUN RPC Portmap
137	NETBIOS-NS	NETBIOS Name Service
138	NETBIOS-DGM	NETBIOS Datagram Service
139	NETBIOS-SSN	NETBIOS Session Service
146	ISO-TP0	ISO TP0
147	ISO-IP	ISO IP
150	SQL-NET	SQL-NET
153	SGMP	SGMP

156	SQLSRV	SQL Service
160	SGMP-TRAPS	SGMP TRAPS
161	SNMP	SNMP
162	SNMPTRAP	SNMPTRAP
163	CMIP-MANAGE	CMIP/TCP Manager
164	CMIP-AGENT	CMIP/TCP Agent
165	XNS-COURIER	Xerox
179	BGP	Border Gateway Protocol

Port Manipulation

Port numbers can be changed. Usually they are not, but there is good reason for this capability to exist. For example, some port numbers are available for development of custom applications. During the explanation of UDP, the concept of custom applications was presented. This is an example where being able to use a "free" port number would be required.

The downside of changing a port number is that if that application using that port is popular, it could cause problems in the network from a user's perspective.

Sockets

A *socket* is the combination of an IP address and the port number appended to the end. Sockets are used in programming and are not normally of any concern for general users. However, in some instances, it is important to understand this socket concept.

Hardware Address

TCP/IP operates at layers three and above in a network; therefore, it stands to reason that an interface of some type is needed for a TCP/IP host to participate in a network. The question then becomes what is the lower-layer protocol used? If it is ETHERNET, a 48-bit addressing scheme is used. If token ring, a 12-digit hexadecimal address is used. Other lower-layer protocols have their own addressing schemes.

Understanding this addressing scheme is important, especially for those who troubleshoot networks. It is also important for those designing networks and implementers who have to make them work.

Synthesis

Understanding the previous information in this chapter is important for planning a TCP/IP network. The size of the network, the purpose for the network, and other site-specific parameters should be considered when selecting IP address classes. Planning, with the technical implications understood in the beginning, can save time and money in the long run.

3.6 Summary

TCP and UDP are transport-layer protocols. Both have popular applications used in the marketplace today. TCP is connection-oriented, while UDP is connectionless-oriented. Both TCP and UDP map addresses to IP.

TCP and UDP have well-known ports identified. These ports are acknowledged entry points to applications that use TCP and UDP. A combination of sockets, ports, IP addressing, and even data link addresses work together to make TCP and UDP work.

Additional details on TCP and UDP are presented in chapters later in this book.

Popular TCP and UDP Applications

In this chapter, popular applications that use TCP as a transport-layer protocol are explained. Those presented in this chapter include the following:

- X window system
- TELNET
- File Transfer Protocol (FTP)
- Simple Mail Transfer Protocol (SMTP)
- Domain Name System (DNS)

4.1 X Window System

The X window system (or X, as it is known in the marketplace) is a distributed windowing system. It was created at MIT in the early 1980s by developers who were looking for a way to develop applications in a distributed computed environment. This was cutting-edge technology at the time. During their work, they realized that a distributed windowing system would meet their needs very well.

The MIT group met with individuals at Stanford who had performed similar work, and who gave the MIT group a starting point to begin this endeavor. The group at Stanford working with this technology had dubbed it *W*, for *windowing*. The individuals at MIT named their system X, based upon the reason that it was the next letter in the alphabet. The name stuck.

By the late 1980s, X commanded a considerable market share in UNIX-based environments. One of the factors for its growth was its hardware and software independence. Today, X is a dominant user interface in the UNIX environment and has spread into MS-DOS and VMS environments as well.

X is asynchronous and based on a client/server model. It can manipulate two-dimensional graphics on a bitmapped display. Before examining some of the operational aspects of X, consider the layer of X and its relationship to the TCP/IP protocol suite, as shown in Fig. 4-1.

Figure 4-1 shows the TCP/IP protocol suite, but the focus is upon X. The protocol suite is there to help you understand the relationship of X with TCP/IP. X is not a transport-layer protocol; it uses TCP for a transport protocol.

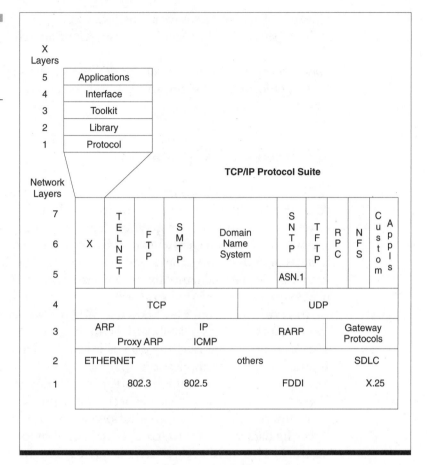

Figure 4-1
A conceptual view of
the example of X lay-
ers in respect to
TCP/IP.

X can be evaluated two ways. From a TCP/IP perspective, it comprises layers 5, 6, and 7. However, X itself has five layers. X's layer names and functions include the following:

Protocol   This is the lowest layer in X. It hooks into TCP. This layer is comprised of actual X protocol components.

Library   The X library consist of a collection of C-language routines based upon the X protocol. X library routines perform functions such as responding to the pressing of a mouse button.

Toolkit   The X toolkit is a higher level of programming tools. Examples of support provided from this layer are functions that provide programming related to scroll bars and menus.

Interface   The interface is what a user sees. Examples of an interface include SUN's OpenLook, HP's OpenView, OSF's Motif, and NeXT's interface.

Applications   X applications can be defined as client applications that use X and conform to X programming standards, and that interact with the X server.

X Theory of Operation

X clients and X servers do not function in the way other clients and servers do in the TCP/IP environment. Normally, a client initiates something and servers serve, or answer the request of, clients. In X, this concept is skewed.

An X display manager exists in the X environment. Its basic function is starting and keeping the X server operating. The X display manager itself can be started manually or automatically. In respect to X, the display manager (also referred to as *Xdm*) is a client application.

An X display server (also known as *Xds*) is a go-between for hardware components (such as a keyboard or mouse) and X client applications. The Xds operates by catching data as it is entered and directing it to the appropriate X client application.

The correlation of Xdm and Xds can be understood by an example. Consider two windows that are active on a physical display. Each window functions as a client application. With this in mind, the idea of directing data to the appropriate X client application takes on a different meaning. This architectural arrangement is required to maintain order because multiple windows may be on the display (say four or five).

The X display manager and X server control the operations on the display, which is what a user sees. Most entities in an X environment function as X client applications. Examples of this include the Xclock, an Xterm that is an emulator, or even a TN3270 emulation software package used to access a 3270 data stream in an SNA environment.

4.2 TELNET

TELNET is a TCP application. It provides the ability to perform remote logons to adjacent hosts. TELNET consist of a client and server. The majority of TCP/IP software implementations have TELNET, simply because it is part of the protocol suite. As previously stated, *clients* initiate something (in this case a remote logon) and *servers* serve requests of clients. Figure 4-2 shows the TCP/IP protocol suite with TELNET highlighted, showing its client and server.

The example of TELNET as shown in Fig. 4-2 is the same on practically all TCP/IP host implementations if the protocol suite has been developed according to the RFCs. Exceptions do occur, however. For example, TCP/IP on a DOS-based PC cannot implement a TELNET server because of the architectural constraints of the PC. In short, the PC cannot truly multitask, and other nuances apply also. Furthermore, on some network devices, this implementation cannot work. However, on most host implementations, such as UNIX, VMS, MVS, VM, and VSE, the TELNET client and server will function.

Figure 4-3 is an example of TELNET client and server interaction on different hosts. Figure 4-3 shows a RISC/6000 user invoking a TELNET client, native to that machine because it is in the TCP/IP protocol suite.

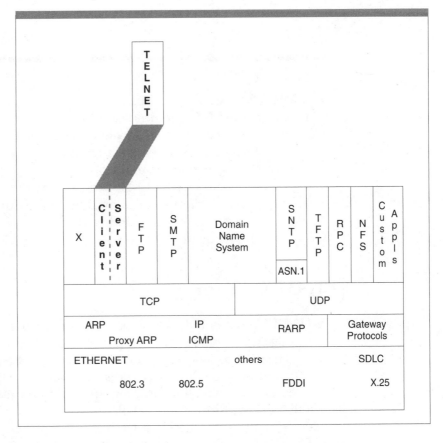

Figure 4-2
A highlighted view of TELNET

Figure 4-3
TELNET client/server
operation

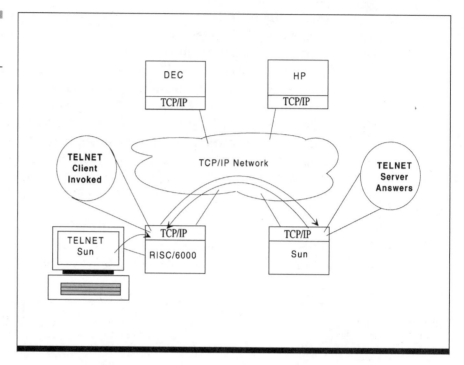

The RISC/6000 user wants to log on to the SUN host. The SUN host has TELNET in its TCP/IP protocol suite. Consequently, the TELNET server answers the client's request and a logical connection is established between the RISC/6000 and the SUN host. It appears to the RISC/6000 user that he or she is physically connected to the SUN.

This functionality of TELNET works with the majority of major vendors in the marketplace today. The key to understanding the client/server concept is to remember that clients initiate and servers serve the client's request.

TN3270 Client

A TELNET protocol is defined, called TN. An individual with the required knowledge who wishes to design a program based on the TN protocol can do so. The most common program written using TN protocol is an emulator application providing data translation services between ASCII and EBCDIC and vice versa. This program (application) is called a *TN3270 client*.

TELNET (within a native TCP/IP protocol stack) has ASCII-based data, so it does not natively fit into SNA, which is dominated by EBCDIC. In the SNA world, EBCDIC goes a step further and defines data streams. A few specific data streams exist, but the dominant one is the 3270 data stream. It is used with terminals interactively.

Because of the difference in ASCII and EBCDIC, converting ASCII into EBCDIC, specifically into a 3270 or a 5250 data stream, is required. But where will this process take place? With the data stream dilemma between TCP/IP networks and SNA networks, this fundamental issue must be resolved. So, how do users on a TCP/IP-based network have ASCII data converted into EBCDIC? Two possible solutions exist:

1. A raw TELNET client can be used to establish a logical connection between a UNIX or other non-EBCDIC host and an EBCDIC-based host. If this is the case, then ASCII-to-EBCDIC translation will occur on the EBCDIC host (with the exception of when a gateway is used between the two and translation services are provided).

2. A TN3270 client application can be used like a raw TELNET to gain entry into the SNA environment, but a TN3270 client application performs data translation. This means it sends an EBCDIC 3270 or 5250 data stream to the destination host. In this case, the TN3270 client application translates ASCII data into EBCDIC.

Figure 4-4 shows two UNIX hosts and a VM host connected to a network. On one UNIX host, a TN3270 client application exists. The TN3270 client is shown establishing a logical connection with the TELNET server native to the TCP/IP protocol on the VM machine. The data stream leaving the ASCII-based host is EBCDIC. This works because data format (be it ASCII or EBCDIC) gets formatted at layer six within a network. By the time the data gets down to the interface card connecting it to the network, the data is represented by voltages or light pulses, whichever the network is based upon.

The net effect of having a TN3270 client is that it does pay for itself over a period of time. Instances do exist where TN3270 applications are not needed and provide little benefit to the end user. Both a raw TELNET and a TN3270 client provide the user with remote logon capability. Both the raw TELNET and TN3270 client are client applications. The difference is merely where data translation is performed.

TELNET Client Usage

As mentioned previously, TELNET consist of a client and server. A client always initiates a logical connection and a server always answers the client's request. To use TELNET, a command must be entered to invoke the TELNET client. The command to invoke the TELNET client from the TCP/IP suite is TELNET Assuming TCP/IP has been installed properly

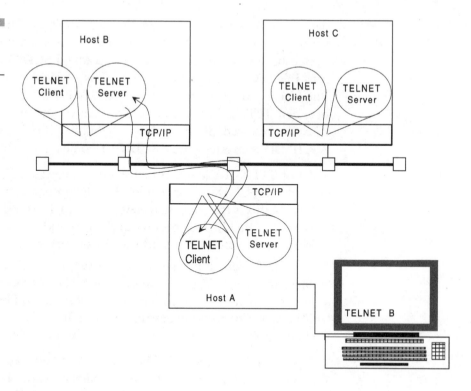

and normal setup occurred, entering the TELNET command invokes the TELNET client from the TCP/IP protocol stack.

If the **TELNET** command is entered without a target host name, alias, or internet address, the following prompt appears:

```
telnet>
```

This command is generated from the TELNET client on that host. When this prompt appears, valid TELNET client commands can be entered against it.

Valid TELNET Client Commands

Valid TELNET client commands can be entered at the TELNET client prompt. If a user does not know valid commands to execute against a TELNET client prompt, a question mark can be entered (?), and a list of valid TELNET commands will be displayed. An abbreviated list of valid TELNET client commands and a brief explanation of each is given here:

- **close**—This closes a current connection if one is established.
- **display**—This command will display the operating parameters in use for TELNET. Because these parameters can be changed, they are site-dependent.
- **mode**—This command indicates whether entry can be made line-by-line or one character at a time.
- **open**—This command is required prior to the target host name in order for the session to be established.
- **quit**—This command is entered to exit the **telnet** prompt, thus exiting TELNET.
- **send**—In certain instances, special characters might need to be transmitted. This provides the means to accommodate some of these characters.
- **set**—This command is used to set certain parameters to be enforced during a TELNET session.
- **status**—This command provides information regarding the connection and any operating parameters in force for the TELNET session.
- **toggle**—This command is used to toggle (change) operating parameters.
- **z**—This command will suspend the **telnet** prompt.
- **?**—This command prints valid TELNET commands that can be entered against the **telnet** prompt.

TELNET Use

Using TELNET is straightforward. Once the newness of the technology is overcome, it is not difficult for users to work with. Learning TELNET is easier when one understands basic TELNET operation, TELNET commands, and how to log on to hosts appropriately.

Since TELNET is part of the TCP/IP protocol suite, it works with other components in the suite. For example, if a user attempts to establish a remote logon with a target host, and after a period of time a response such as "host unreachable" is displayed on the terminal, this indicates problems not necessarily related to TELNET. In this example, the "host unreachable" message comes from the Internet Control Message Protocol (ICMP) component. This is an integral part of the IP layer. It provides messages responding to different conditions. Here, a destination host is not reachable by the TELNET client. The obvious question is, why? With this example, a couple of possible reasons would be viable. It could be the host is unreachable because of a break in the physical cable connecting the hosts together. Or, it could be that the host is located on another segment of the network and for some reason, inaccessible at the moment. Other possibilities exist as well.

When messages such as these appear, they are most often generated from the ICMP portion of the TCP/IP suite. It would be helpful to familiarize yourself with common messages and understand their meaning. It can prove to be a valuable troubleshooting tool.

4.3 File Transfer Protocol (FTP)

FTP is a file transfer application that uses TCP for a transport protocol. FTP has a client and server like TELNET; operationally, they work the same. The difference is that TELNET enables remote logon whereas FTP permits file transfers.

FTP does not actually transfer a file from one host to another—it copies it. After an FTP operation, the original file exists on one machine, and a copy of it has been put on a different machine. Figure 4-5 depicts this scenario.

Figure 4-5 shows a user on a SUN host performing two steps. First, the SUN user executes **FTP HP** and a logon is established. Second, the SUN user issues the FTP command **GET** and designates the filename as FILEABC. The dotted line shows the file is copied from the HP's disk to the SUN's disk.

This illustration shows DELL, IBM, and MVS hosts. The same operation can be performed among any of these as well. An FTP can be performed on a PC and not a TELNET because the nature of FTP uses two ports to function, and because FTP is merely requesting a file transfer, which does not require multitasking on behalf of the host. The DEC host can perform any of the TCP/IP functions just as the SUN or HP.

An interesting twist on this scenario (shown in Fig. 19.22) is that a DEC user could TELNET to the HP, then from the HP execute a FTP against the DELL PC and move a file to or from the DELL PC. This type of networking scenario is quite powerful. Another twist (also shown in Fig. 19.22) involves an HP user executing an FTP against the MVS host, using

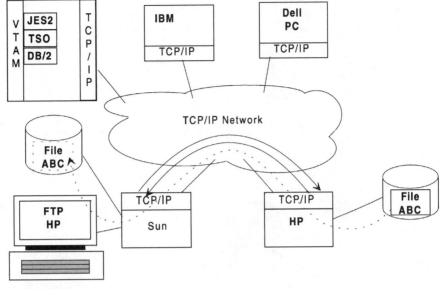

Figure 4-5
FTP client/server
operation.

Step 1: Logon
Step2: Get file ABC

the PUT command on file FILEABC to put it into the MVS JES2 subsystem, then having it print on the printer attached to the MVS server. These few examples convey some of the practically limitless ways FTP operations can be performed.

4.4 Simple Mail Transfer Protocol (SMTP)

SMTP is another TCP application. It does not use a client and server, but the functionality is similar. SMTP utilizes what is called a *user agent* and a *message transfer agent.* Figure 4-6 is a simple example of how SMTP operates.

Sending mail is accomplished by invoking a user agent, which in turn causes an editor to appear on the user's display. After the mail message is created and sent from the user agent, it is transferred to the message transfer agent. The message transfer agent is responsible for establishing communication with the message transfer agent on the destination host.

Figure 4-6
SMTP components

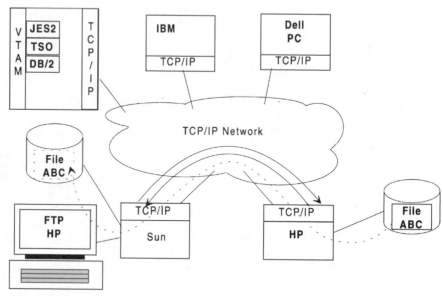

Step 1: Logon
Step2: Get file ABC

Once this is accomplished, the sending message-transfer agent sends the message to the receiving message-transfer agent, then stores it in the appropriate queue for the user. The recipient of the mail only needs to invoke the user agent on that machine to read the mail.

4.5   Domain Name System (DNS)

In its early days, the Internet used a *hosts* file to keep track of hosts on the Internet. This meant when new hosts were added to the Internet, all participating hosts had to have their hosts file updated. As the Internet grew, this task of updating the hosts file became insurmountable. The Domain Name System (DNS) grew from the need to replace such a system.

The philosophy of DNS was to replace the need to FTP updated hosts files throughout the entire network. Thus, the foundation of DNS was built around a distributed database architecture.

DNS Structure

DNS is a hierarchical structure that conceptually appears as an upside-down tree. The root is at the top, and the layers are below. Figure 4-7 is an example of how DNS is implemented in the Internet.

The legend for the DNS structure in Fig. 4-7 is as follows:

- *Root* is the root server, which contains information about itself and the top-level domains immediately beneath it.
- *GOV* refers to government entities.
- *EDU* refers to any educational institutions.
- *ARPA* refers to any ARPAnet (Internet) host ID.
- *COM* refers to any commercial organizations.
- *MIL* refers to military organizations.

Figure 4-7
The DNS structure

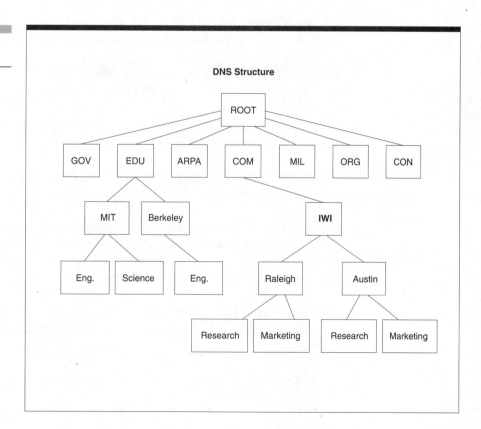

- *ORG* serves as a miscellaneous category for those not formally covered.
- *CON* refers to countries conforming to ISO standards.

Figure 4-7 shows the Internet implementation of DNS. Three examples are shown to aid in understanding the structure. Notice IWI is under COM, which indicates a commercial organization. Beneath IWI is Raleigh and Austin, and beneath them is research and marketing. The other examples are MIT and Berkeley. The example with MIT shows beneath it two *zones*, engineering and science. The Berkeley example has one layer beneath it, entitled engineering.

At a local level, such as in a corporation, most sites continue to follow this naming scheme and structure because it is consistent, and if a connection is ever made to the Internet, restructuring of DNS is not necessary.

DNS Components

To better understand DNS, it is helpful to know the components that make it functional. These components include the following:

Domain The last part in a domain name is considered the domain. For example, in the case of eng.mit.edu, edu is the domain.

Domain Name Defined by the DNS as being the sequence of names and domain. For example, a domain name could be eng.mit.edu.

Label The DNS identifies each part of a domain name as a label. For example, eng.mit.edu has three labels: eng, mit, and edu.

Name Server The name server is a program operating on a host that translates names to addresses. It does this by mapping domain names to IP addresses. Also, the term *name server* may be used to refer to a dedicated processor running name-server software.

Name Resolver This is software that functions as a client regarding its interaction with a name server. It is sometimes referred to simply as the client.

Name Cache This is storage used by the name resolver to store information frequently used.

Zone This is a contiguous part of a domain.

Theory of Operation

Figure 4-8 shows a TCP/IP network with five hosts. Of these five hosts, host B has been designated as the name server. It has a database with a list of aliases and IP addresses of participating hosts in the network. When the user on host A wants to communicate with host C, the name resolver checks its local cache. If no match is found, then the name resolver (client) sends a request (also known as a *query*) to the name server.

The name server, in turn, checks its cache for a match. If no match is found, then the name server checks its database. Though not shown in this figure, if the name server were unable to locate the name in its cache or database, it would forward the request to another name server, then return the response back to host A.

In an internet environment that implements DNS, some assumptions are made. For example, a name resolver is required. Also, a name server, and usually a foreign name server, is part of the network.

Figure 4-8

A conceptual view of DNS

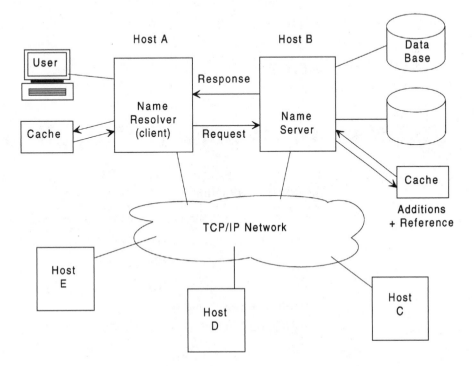

Implementation with UDP

The DNS provides service for both TCP and UDP. This is why figures have shown DNS residing above part of TCP and part of UDP. It serves the same purpose for both transport-layer protocols.

Obtaining Additional Information

Additional information should be consulted on this issue if DNS is implemented. The following RFCs are a good beginning point:

- RFC 882
- RFC 883
- RFC 920
- RFC 973
- RFC 974
- RFC 1034
- RFC 1123
- RFC 1032
- RFC 1033
- RFC 1034
- RFC 1035

4.6   Popular UDP Applications

This section presents popular UDP applications. A list of those covered include the following:

- Simple Network Management Protocol (SNMP)
- Trivial File Transfer Protocol (TFTP)
- Network File System (NFS)
- Remote Procedure Call (RPC)
- Custom applications
- PING and Finger

4.7 Simple Network Management Protocol (SNMP)

SNMP is considered the de facto standard for managing TCP/IP networks as of this writing. SNMP uses agents and application managers (or simply managers). A user agent can reside on any node that supports SNMP, and each agent maintains status information about the node on which it operates. These nodes, which may be a host, gateway, router, or other type network device, are called *network elements* in SNMP. The term *element* is merely a generic reference to a node.

Normally, multiple elements exist in a TCP/IP network, and each has its own agent. Typically, one node is designated as a network management node. Some refer to this node as the *network manager*. This host (network management node) has an application that communicates with each network element to obtain the status of a given element. The network management node and the element communicate via different message types. Some of these messages are given here:

GET REQUEST This type of request is used by the network manager to communicate with an element to request a variable or list about that particular network element.

GET RESPONSE This is a reply to a GET REQUEST, SET REQUEST, or GET NEXT REQUEST.

GET NEXT REQUEST This request is used to sequentially read information about an element.

SET REQUEST This request enables variable values to be set in an element.

TRAP This message is designed to report information such as link status, whether or not a neighbor responds, whether or not a message is received, and the status of the element.

Information stored on elements is maintained in a *Management Information Base* (MIB). This MIB is a database containing information about a particular element; each element has a MIB. Examples of MIB information include statistical information regarding segments transferred to and from the manager application, a community name, an interface type, and other element-specific information.

The MIB information structure is defined by the *Structure of Management Information* (SMI) language. SMI is a language used to define a data structure and methods for identifying an element for the manager application. This information identifies object variables in the MIB. A minimum of object descriptions defined by SMI include the following:

ACCESS Object access control is maintained via this description.

DEFINITION This provides a textual description of an object.

NAMES This term is synonymous with *object identifier*. It refers to a sequence of integers.

OBJECT DESCRIPTOR This is a text name ascribed to the object.

OBJECT IDENTIFIER This is a numeric ID used to identify the object.

STATUS This describes the level of object support for the status.

SNMP implementations use ASN.1 for defining data structures in network elements. Because this language is based on a datatype definition, it can be used to define practically any element on a network.

SNMP itself is event-oriented. An event is generated when a change occurs to an object. SNMP operation is such that approximately every 10 to 15 minutes, the manager application communicates with each network element regarding their individual MIB data.

Additional information about SNMP can be obtained from RFC 1155, RFC 1156, and RFC 1157.

4.8   Trivial File Transfer Protocol (TFTP)

TFTP is an application that uses UDP as a transport mechanism. The program itself is simpler than its counterpart, FTP, which uses TCP as a transport mechanism. TFTP is small enough in size so that it can be part of ROM on diskless workstations.

TFTP's maximum packet size is 512 bytes. Because of this and the nature of operation, TFTP is popular with network devices such as routers and bridges. If implemented, it is normally used upon initial device boot.

TFTP has no provisions for security or authentication. However, it does have some basic timing and retransmission capabilities. TFTP uses five basic types of protocol data units (PDUs):

- Acknowledgement
- Data Error
- Read Request
- Write Request

These PDUs are used by TFTP during file transfer. These five PDUs comprise the operational ability of TFTP. It is straightforward and not as complex as FTP. The first packet of TFTP establishes a session with the target TFTP program. It then requests a file transfer between the two. Next, it identifies a filename and whether or not a file will be read or written.

Additional information on TFTP can be obtained from RFCs 783 and 1068.

4.9   Remote Procedure Call (RPC)

RPC is a protocol. Technically speaking, it can operate over TCP or UDP as a transport mechanism. Applications use RPC to call a routine (thus executing like a client) and make a call against a server on a remote host. This type of application programming is a high-level, peer relationship between an application and an RPC server. Consequently, these applications are portable to the extent that RPC is implemented.

Within RPC is the *External Data Representation* (XDR) protocol. XDR data description language can be used to define datatypes when heterogeneous hosts are integrated. Having the capability to overcome the inherent characteristics of different architectures lends RPC and XDR a robust solution for distributed application communication. This language permits parameter requests to be made against a file of an unlike type. In short, XDR permits datatype definition in the form of parameters and transmission of these encoded parameters.

XDR provides data transparency by way of encoding (or encapsulating) data at the application layer so lower layers and hardware do not have to perform any conversions. A powerful aspect of XDR is automatic data conversion performed via declaration statements and the XDR compiler. The XDR compiler generates required XDR calls, thus making the operation less manual by nature. Figure 4-9 is an example of this type of implementation.

RPC implements what is called a *port mapper*. It starts upon RPC server initialization. When RPC services start, the operating system assigns a port number to each service. These services inform the port mapper of their port numbers, program numbers, and other information required by the port mapper for it to match a service with a requester.

When a client application issues a service request to a port mapper, the port mapper identifies the requested service and returns the appropriate parameters to the requesting client application. In other words, the port mapper is similar in function to a manager, knowing what services are available and their specific addressable locations.

The port mapper can be used in a broadcast scenario. For example, a requesting RPC call broadcasts a call to all hosts on a network. Applicable port mappers report back to the information sought after by the client. From this process comes the term *Remote Procedure Call* (RPC).

Additional information on RPC and related components can be found in RFCs 1057 and 1014.

4.10   Network File System (NFS)

NFS is a product of Sun Microsystems that permits users to execute files without knowing the locations of these files. They may be local or remote

Figure 4-9
A conceptual view of RPC and XDR.

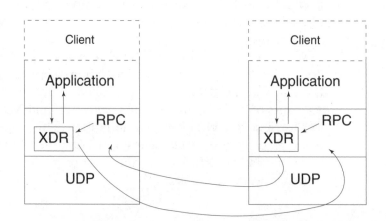

in respect to the user. Users can create, read, or remove a directory. Files themselves can be written to or deleted. NFS provides a distributed file system that permits a user to capitalize on access capabilities beyond their local file system.

NFS uses RPC to make execution of a routine on a remote server possible. Conceptually, NFS, RPC, and UDP (which it typically uses) work together as shown in Fig. 4-10.

The idea behind NFS is having one copy of it on a server that all users on a network can access. The consequence of this is that software (and updates) can be installed on one server and not on multiple hosts in a networked environment. NFS is based on a client/server model. However, with NFS, a single NFS server can function to serve the requests of many clients.

NFS origins are in UNIX, where it is implemented in a hierarchical (tree) structure. However, NFS can operate with IBM's VM and MVS operating systems. It can also operate with Digital Equipment's VMS operating system.

NFS uses a *mount* protocol. This protocol identifies a file system and remote host to a local user's file system. The NFS mount is known by the port mapper of RPC, thus it is capable of being known by requesting client applications.

NFS also uses the NFS protocol; it performs file transfers among systems. NFS uses port number 2049 in many cases, however, this is not a well-known port number (at least at the time of this writing). Consequently, the best approach is to use the NFS port number with the port mapper.

In a sense, an NFS server operates with little information identified to it. A loose analogy of NFS operation is UDP. UDP assumes that a custom application (or other entity operating on top it) will perform requirements such as retransmissions (if required) and other procedures that would otherwise be performed by a connection-oriented transport protocol such as TCP. NFS assumes required services are implemented in other protocols.

Figure 4-10
A conceptual view of
NFS, RPC, and UDP.

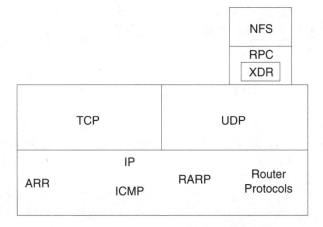

From a user perspective, NFS is transparent. Typical user commands are entered, then passed to the NFS server. In most cases, a user does not know the physical location of a file in a networked environment.

Additional information about the Network File System and related components can be obtained from RFC 1094, RFC 1014, and RFC 1057.

4.11   Custom Applications

Custom applications can be written to use UDP as a transport mechanism. One scenario could be where two hosts need peer program communication through a network. Writing a custom application using UDP can achieve this task, as Fig. 4-11 shows.

4.12   PING and Finger

Packet Internet Groper (PING) is actually a protocol that uses UDP as a transport mechanism to achieve its function. It is used to send a message to a host and wait for that host to respond to the message (if the target host is "alive"). PING uses ICMP echo messages along with the echo reply messages.

PING is a helpful tool on TCP/IP networks, where it is used to determine if a device can be addressed. It is used in a network to determine if a network itself can be addressed. A PING can also be issued against a remote host name. The purpose for this function is name verification. It is generally used by individuals who troubleshoot TCP/IP networks.

Finger is a command issued against a host which will cause the target host to return information about users logged onto that host. Some of the information retrievable via Finger includes user name, user interface, and the job name the user is running. Additional information about Finger can be obtained in RFC 1288.

4.13   Summary

TCP/IP is an upper-layer protocol that has a proven track record. It began around 1975, public demonstration of its capabilities was presented in 1978, and in 1983, the DoD endorsed it as the protocol to use for connection to the Internet.

Many vendors supply TCP/IP products today. TCP/IP can operate on different hardware and software platforms. This flexibility, along with its

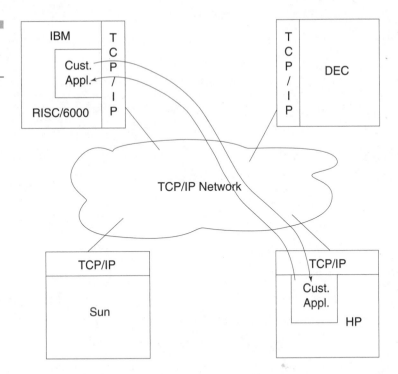

Figure 4-11
Custom applications
using UDP

cost efficient pricing, puts it in a favorable position for those looking for a protocol that provides a variety of services, such as:

- Remote logon
- File transfer
- Electronic mail
- A windowing system
- Programmatic interface support
- Network management capabilities
- Distributed processing support

TCP/IP is dominant throughout the marketplace worldwide today; most major vendors around the world support it to varying degrees. Its flexibility with data link layer protocols makes it attractive. TCP/IP has two transport-layer protocols, which adds to its flexibility. Some users need the reliability of transport provided by TCP, while others need a connectionless transport like UDP. TCP/IP supports both.

TCP/IP has in many ways become a de facto standard. It is used in government, commercial, educational, and nonprofit organizations, as well as by individuals.

Designing a TCP/IP Network

Designing a network goes beyond picking hardware and software, installing it, and hoping it works. Networks that have a lasting ability and quality during the heaviest loads have generally been well thought-out. Success rarely just happens in the data communications and networking industry. As I told a friend, "You can be a few feet off landing a jet and not worry much, but being a few *bits* off in the computing industry can bring down an entire network."

TCP/IP is an excellent protocol around which to build a network. The network designed and presented here uses TCP/IP as its backbone network protocol. A network consists of many components and pieces beyond the upper-layer protocol. This chapter, and the next, explains what you should consider for your TCP/IP-based network.

5.1   Network Requirements

When a company or group of people decide a network is needed, several reasons usually exist. In too many places I have seen networks that emerged in various departments to meet particular needs, only to eventually end up crashing into the higher needs within a company.

Your company size, customer size, anticipated company size, and customer base should all factor into the equation that is the basis for building your network.

Internal

Each company, regardless of size, has internal needs. Most companies usually have shipping and receiving, sales, marketing, operations, support, financial, and personnel services. This is not an all-inclusive list of operations, but it is a place to begin.

Internal needs might be subtle, but take time to examine the inner workings of the company. If technology is involved, then engineering, support, and sales might all need access, in varying depths, to information about the product or services the company offers. What does this mean for a network? At a minimum, it means people in these departments will need to be able to access information, and most likely exchange information. The information might include graphics, internal reports, external feedback from potential or existing customers, and possibly a database of information.

There are other pieces of internal information that are typically required for a given company to maintain some type of order. For example, online calendars are handy. This information is considered generally public within the company, and might well be needed by shipping and receiving. For example, if Ms. Hoover is on vacation this month and receives multiple packages, the receiving department could take advantage of an online calendar to know she is out of the office, and whether or not any special instructions are standing regarding her absence.

Online calendars also provide help in scheduling meetings. A company might be able to maintain fewer conference rooms if everyone in the company knew when a meeting was going to occur and what facility would be used to accommodate it. The needs in the conference rooms themselves might also affect the network. For example, are telephone services needed in the conference rooms? Are data network services needed? If so, how will these services be provided? Will those with portable computers have access to appropriate cabling in the conference rooms? What is the capacity of connections that can be made in the conference rooms at any given time? These and similar questions factor into the equation of network design. Overlook these points and you might well end up hiring a crew to tear part of a wall out to make something available that should have been available in the first place.

Many other types of information might be needed to be exchanged within a company. Regardless of what that might be, you should attempt to plan as far into the future as possible when adding information into the network design equation. Granted, nobody can plan with 100% accuracy five to ten years out, but it is possible to plan for worst-case and best-case scenarios. Do this, and your work with data communication gear will be less painful. However, you would be wise to memorize the seven best words I have ever heard spoken about this industry:

"Comm gear ain't never gone in easy."

Words from experience.

External

Needs external to the company will also affect your network. For example, the way your customers interact with the company is important. If there is a need for them to be able to access certain information, such as download files or read certain information made available to the public, then this should be factored into the network design equation.

Beyond awareness of customer interaction with the company is the number of customers (current and planned) that will need this interactive ability. Circumstances might justify an entirely separate network to meet the needs of the customer base. If so, it is important to consider how this network will or will not interact with the internal network being designed.

The marketplace today is full of companies that cannot meet the demands placed upon them by customers. This is typically due to a company's lack of attention to the need, unawareness that the need exists, or inability to cope with the need in a positive way. Too many times today the philosophy of "less is more." People who have this mentality are either on the way out of business, or they already are. (Feel free to quote me on this.) Less never has been more, and in the data communications industry, more is required to get more results. The point here is simple: Be cognizant of the consumers; know their needs, desires, and requirements.

5.2 Physical Requirements

Because it discusses design, this chapter is most likely the most important one in this entire book. When network design commences, it should be done on paper first. I have a philosophy: If I cannot solve a problem on paper, then I cannot fix the problem in reality. It is easier to work through painstaking areas of design on paper.

Consider the physical part of network design. Before we get to the technical part, we must address fundamentals. Assume you have worked through the needs of the company and customers on paper. You are ready to begin ordering. In this case, assume the physical location you are in is the location where the network will be built. The following should be a minimum checklist you use as a rule of thumb when ordering hardware:

1. Is there enough free space so that when equipment begins to arrive, it will not impede day-to-day operations?

2. Will any of the equipment arrive on pallets? If multiple pallets of equipment arrive, where will they be put initially to unload, and then where will they be stored until the equipment is used?

3. Are stairs used to access the equipment where it is stored? If so, how many?

4. Can a forklift enter the location to unload and load equipment?

5. What is the estimated weight the floor can sustain per square foot?

6. How wide are your doors?

7. If handtrucks are needed, are yours capable of sustaining the weight of the new equipment?

8. Are elevators required to get to a location where the equipment will be stored? How accessible are they? Will using these elevators affect daily operations of your company or another company?

9. Will the location where the equipment is stored until implementation be secure?

10. Does the addition of the new equipment cause any code infringement with fire, police, or any other city, county, state, or federal government entity?

11. If any of the equipment is shipped on pallets or in large boxes, how do you plan to dispose of them?

12. Is any of the equipment sensitive to any environmental conditions in your location, such as heat, cooling, or humidity? Have you verified this with the vendor who manufactured the equipment?

13. Do you have a single point of contact with the company shipping equipment to you?

At a minimum, these questions should have clear, definable answers before you give the okay to receive the first shipment of equipment. Your particular location might require that additional questions be posed and answered. I encourage you to think the matter of your physical location through. It might even be helpful if you have someone else who knows the physical plant well go through this phase with you. Somebody should take responsibility for this task of physical premises evaluation and preparation; if this is not addressed prior to receipt of equipment, you could learn about some matters the hard way. Then someone will take responsibility, like it or not.

5.3   Electrical Requirements

The importance of this section cannot be understated. It is imperative that you, or the appropriate person, know the electrical capability of your physical location. Once you have determined the equipment that is to be deployed, then ascertain the specifications about each piece of equipment. When you create the logical network illustration, this information will be critical.

If you do not have the personnel capable to perform a certified electrical evaluation, contact the following company, which does have the personnel to do so:

Information World, Inc.
Fort Worth, TX

You might think, "We have plenty of electrical outlets, so we will not need additional sources." That is just where the problem begins. Most offices have multiple electrical outlets tied into one circuit breaker. Suppose a given room has six outlets. Three are being used by a computer, light, and radio. To assume you could add a mid-size photocopier to this room and plug it into an available wall outlet is at the least misguided. Even the addition of two laser printers could easily overload a single circuit breaker.

Network design and electrical considerations begin with determining all pieces of equipment that require electricity. List all these pieces on paper. Next, obtain from the manufacturer the amount of watts, amps, and volts the device will consume upon power-up and in idle state. You could be in for a surprise.

Once you have obtained information about all devices, next determine how power conditioning will be included into the equation. Forthcoming chapters explain the network designed and used as an example in this text. This network used Tripp Lite power-protection equipment. Once electrical information was obtained about each device, information about each piece of Tripp Lite equipment was needed. Just because power-protection equipment has multiple outlets on it does not mean you can fill each one and expect power protection. In the case of Tripp Lite equipment, the specifications of each piece of equipment was adhered to during the installation phase of the network. In order for power-protection equipment to operate the way it was designed, you must use it the way the manufacturer designed it.

Many offices have 20-amp circuit breakers for electrical outlets. Exceptions abound, but this is the general rule. My recommendation is do not exceed 70% to 80% of the circuit-breaker ability. For example, if you know a given room has four outlets tied into one 20-amp breaker and none of these outlets is currently in use, then add equipment that will not exceed 15 amps. Why? Because it is safe. Odds are that after you install equipment into this hypothetical office, being careful not to exceed 70% to 80% of the circuit breaker, someone will come along behind you in a few weeks or months and add a few additional items that will use the remaining percentage of the circuit ability.

If no planning went into the electrical part of the location where the network will be installed, you are in for some extra work. Do not assume you can superimpose a network into an existing site without coping with the electrical factor.

Some companies have "computer rooms" where the bulk of computing and network equipment resides. This room has usually been preplanned to handle computing and networking equipment. However, if you do not have such a room or designated area, you must start from scratch. The best place to begin is to contact the facilities manager. Usually, he or she will be able to get you in touch with electricians who can answer questions or assist you in the planning phase.

The network designed, built, and explained in forthcoming chapters has two laser printers. This is a good example because it is typical of many scenarios; it could even be typical of what you intend to implement. Both laser printers are IBM. Both have their power consumption stated explicitly on the appropriate plate as defined by national and international requirements. Additionally, both have power requirements stated clearly in the documentation shipped with the printers. These two printers use a single Tripp Lite spike and surge protector and voltage regulator. You might be surprised to learn these two printers have a single 20-amp dedicated circuit breaker. Figure 5-1 illustrates this.

This figure shows the circuit breaker, wire size, Tripp Lite spike and line noise filter, voltage regulator, and the IBM printers. You might think, "Gee, Ed, this is overkill." Really? What if I told you these two printers absorb 12.5 to 13.5 amps when powered up simultaneously? Now assume you have some additional equipment to be added to this outlet. What if you took this equipment and plugged it into the place you have the network printers connected? Would these additions work without overloading a circuit breaker? This level of detail needs to be obtained and factored into the network design equation.

While plans were formulated for the network, Fluke test instruments were used to determine the existing power supplied to the facility. Consider Figs. 5-2 through 5-6.

Notice the voltage differences among the illustrations. These show how much incoming voltage varies in my location during the week I monitored the AC. Your usage might be different, but I suspect you

Figure 5-1
Multistage power
protection

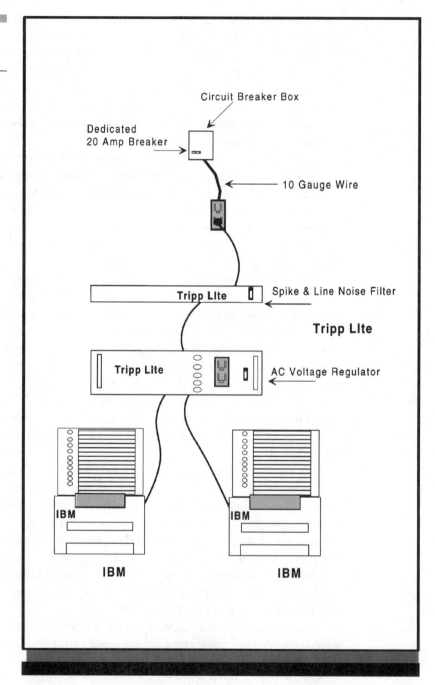

Figure. 5-2
AC voltage on day 1.

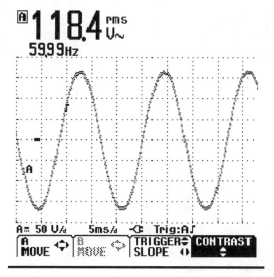

Figure. 5-3
AC voltage on day 2.

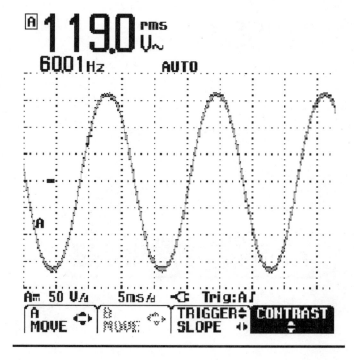

Figure. 5-4
AC voltage on day 3.

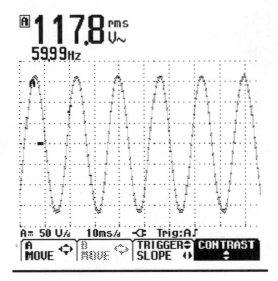

Figure. 5-5
AC voltage on day 4.

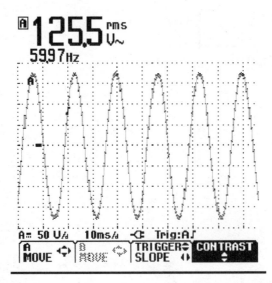

Figure. 5-6
AC voltage on day 5.

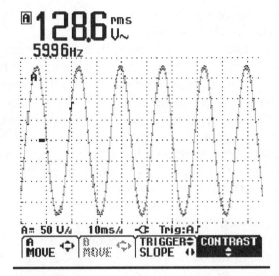

might find similar voltages. These voltages were obtained using a certified, calibrated Fluke 123 scopemeter. Fluke includes software and cabling with a 123 scopemeter kit to interface it into a PC running Windows or Windows NT. This makes capturing real-time line voltages and printing out capabilities easy.

Information such as this is part of the foundation of the network. The better your measurements of what the electrical reality is in your facility, the more solid your network will be.

5.4 Network Personnel

Networks don't get built without people. Neither is it possible to maintain them without people. Some time back, a movement occurred in the data communication/telecommunication industry to cut back on personnel. This mindset was general in many companies of all sizes. It is most unfortunate for many companies that some think money is more important than people. I have seen companies that were cash rich and brain poor when it came to technical people. Companies exist that have incredible amounts of money but cannot keep good technical talent. It seems that managers in some companies think technical people are dispensable. Ironically, the past few years have dealt the management of many companies a great blow. Why is this important? No, I am not just expressing an opinion.

Networks are either planned for, or they are not. In the planning of a network, you must include in the equation the human factor. It takes more than good people with technical skills to design a network. It also takes representatives from every aspect of a company that will use the network to participate in the planning of it. You might think this is trite; think again. Not too long ago, a great move was to get networks and data centers to operate in "lights-out" mode...meaning as few people as possible operating it. Well, this too has passed. I have seen data centers that went from lights-out operation to everything-out operation! The more advanced the networks and technology deployed, the more brains it takes to keep them running.

Well-trained people are not optional when it comes to network design and maintaining daily operations. If you happen to be the one responsible for network design, I encourage you to get with the appropriate people and include (as best as possible) the skill sets required to keep the network operational. At the least, you will have documented on paper estimates of reality; at most, this phase of planning could get the right attention and make life easier for many as network implementation begins to happen.

My implication here is not to have an army of people for the sake of having people. It is simply that I have yet to see any technology that created itself, maintained itself, and phased growth into itself. It is foolish to buy the best equipment available in the market today and not pay to get reasonably skilled people, *and keep them,* to maintain it; it does not matter what a business plan says about this.

An important aspect of the human factor is education. In the late 1980s, Derek Bok, a professor at Harvard University (if memory serves) was on the cover of *Time* magazine. Inside, an article quoted him as saying, "If you think education is expensive, try ignorance." My thoughts exactly.

Go ahead and factor into the network equation continuing education across the board for all technical people who will be involved in it. Since most network implementations vary, I can only endorse a principle here. Train your people or have them trained.

5.5 Network Growth

The network you are about to design, or are in the process of designing, will evolve. Sooner or later someone will make a suggestion that this or that be added to the network, and it probably will. As the network evolves, be sure to document additions and changes to the existing architectural layout of the network.

More than likely, as time goes by, some people will leave the company for a variety of reasons. Try to plan as best as possible for this. This is part of the evolution the network will experience.

It seems most networks tend to reflect the personalities of those who have designed and worked with it, directly and indirectly. As the network changes, do not lose sight of its original design intent. This is critical. I have seen many people argue this or that for or against a network's operation. Typically, these situations were squashed when two questions were asked: What was the original design intent of the network? And is it meeting that intent? Unfortunately, some people maintain the bizarre notion that "if it is not doing this or that, it is not *right.*" These situations can be easily resolved; simply ask, "Why is it the modem will never print laser-quality reports?" Sometimes it takes holding a mirror up to the situation to show the lack of thought behind certain criticisms.

A real-life experience is a good fit here. In the past year I was in a certain location, with about six other professionals who were designing a network. I came onto the scene a little late. Many aspects of the network had been penciled in on paper.

As I was being briefed about the network, its purpose, the size, and plan for implementation, I read through notes of what had been done so far. The network was large. It was fairly costly. As I examined the preliminary design, I could not believe it. My thoughts must have shown on my face as I was reading through the materials because someone asked, "What is the matter? Is there something that needs changing"? That is when I really got concerned.

I explained, "This network design is going to break. It is not a matter of *if*, it's merely a matter of *when.*" You could have heard a pin drop. I will leave it at that.

My point here is that some group of people had made a preliminary design of a network around a single point of failure. Those with this mindset need to be doing something other than designing networks.

So, during your design, factor in evolution. Do not even think about designing a network that has a single point of failure. However, if that is the only way a design can be achieved, then design in redundancy—an alternative architecture (with real equipment) that will kick in when the first fails. You will probably get a raise or promotion if you do.

5.6   Factors of Technology

During the penciling phase of network design, include a detailed examination of the following. These should be the minimum to look into and estimate the impact they will have for the network:

1. Telephone line abilities, or lack thereof
2. Heat considerations once all equipment is in place
3. Cooling considerations once all equipment is in place
4. Backup contingencies for electrical, telephone, and operations
5. Electromagnetic field interference
6. Radio frequency (RF) interference
7. Cable lengths (maximum for AC, DC, voice, and data)
8. Cable location
9. Service accessibility for all equipment
10. Hot site location for back-up operational plans, architectural design plans, emergency procedures, etc.
11. Identified chain of command (technical people) to execute plans in case of emergency
12. Labelling cables: Data, voice, video, AV, DC, etc.
13. Checklist for testing *all* cables prior to use with anything, especially if they are new
14. Ten worst-case technical scenarios and detailed action plans ready to implement for any variation of need
15. Physical and logical security
16. Categorization of technical needs into critical, urgent, and as soon as possible (remember, most things are not urgent.)

Your site might require a variation on this list. However, I have used it in many places with networks of all sizes.

5.7   Summary

This chapter has presented information to assist you in the planning of your network. Network design includes electrical considerations, cooling, heating, weight factors, height, width, storage, and component functions.

Forthcoming chapters present additional information that goes hand-in-hand with this one. Together, these chapters will provide the foundation you need to begin a solid network design. The examples from my experience are real. I hope you profit from them.

TCP/IP Network Components

This chapter starts by presenting the logical network design I created. After the design is explained, further sections in this chapter explain the components that comprise the network.

6.1 Network Design

I began this network design the same way I have all others; that is, at the drawing board, literally. I use a marker board to work out my ideas. It can take days—sometimes weeks—for me to sift through the requirements, my thoughts, and variations of what equipment is needed and how it fits into the overall design scheme.

The network design explained in this chapter and those to come is based more on principle than a given piece of technology. The purpose of the network design here is to meet current needs and sustain growth. The original design intent of this network is to do what I want it to do now and be flexible enough to change and accommodate growth in the near future. It is not designed for anyone else's criteria.

Before examining the components of the network, consider the logical network design shown in Fig. 6-1. Although some necessary network components, such as network interface cards or particular wiring, are not shown here, the figure does show the overall logical design of the network. Close inspection of the figure might seem to imply a single point of failure in a given place, but redundancies have been built into the system.

The reason this is the logical network design is because it was driven by user requirements. This network enables users to exchange files, send e-mail, perform remote logons to systems such as servers, and even use network printers. The remaining sections in this chapter explain the components of the network.

6.2 Component Overview

Before examining each component, consider the following list of what is part of this network:

- Bud Industries rack enclosure
- SMC Ethernet network hubs
- SMC SNMP management board for network hubs

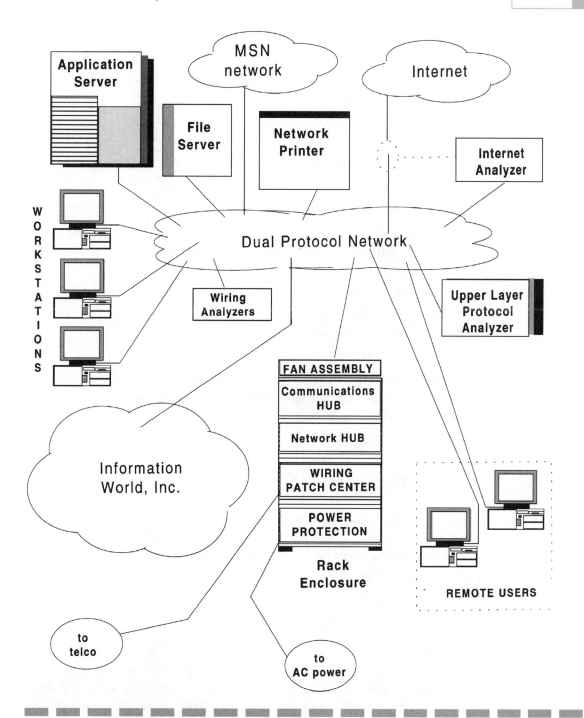

Figure 6-1   A conceptual view of the example network.

- Hubbell Premise Wiring 48-port inline panel
- Hubbell 10BaseT wire for all interface cards
- Hubbell patch cables for patch panel
- Fluke 123 scopemeter
- Fluke Model 41B power analyzer
- Fluke 685 Enterprise LANmeter
- Microsoft software
- McAfee software
- Tripp Lite power protection
- IBM network printer
- Hewlett Packard Internet Advisor and Upper Layer Protocol Analyzer
- U.S. Robotics enterprise network hub
- IBM personal computers
- Creative Labs multimedia support
- Miscellaneous equipment

The components in this list are in no particular order; it is simply a list of the vendor components that have been put together to make the network possible.

6.3 Personal Computers

IBM personal computers used in this network are from the PC350 and PC365 series. Here is a typical example of the general specifications for the base system units (model 350s) used in this network:

- 200 MHz Pentium MMX processor
- 2.6 Gig hard disk
- 16MB non-parity EDO memory
- 3.5-inch floppy disk drive

Units used in this network employ a PCI Busmaster controller and SMART capabilities. These systems also include PCI enhanced IDE hard drives, Universal Serial Bus ports, infrared ports, 64-bit PCI graphics, and wake-on-LAN capability.

The availability of Universal Serial Bus (USB) ports makes peripheral connectivity easier. The hot-connect ability enables peripheral devices to be connected in seconds. Such devices can be added or removed without reconfiguring or rebooting. Each USB port permits up to 127 USB-capable devices.

Some of the PCs used in this network have the capability for symmetrical multi-processing (SMP) when dual processors are used. An L2 external CPU cache of 256KB is included, as well as pipeline burst L3 cache. The BIOS type is 256KB flash, SurePath.

The systems can accommodate up to 192MB RAM at a speed of 60ns deployed by 72-pin SIMMs. The average seek time of their hard disks is 12 ms, with a latency of approximately 5.8 ms. They support RAID and hot-swappable drive bays.

For graphics, these systems employ a S3 Trio64 V+Graphics chipset. The result is SVGA graphics and data width of 64-bit video RAM. Graphic resolution (with the standard video RAM) is 1280×1024, 16 colors. The maximum resolution (with a maximum video RAM) is 1280×1024 with 65,536 colors. The graphics bus interface uses PCI architecture.

The systems have a 200-watt power supply for either 110 or 220 volts, with a universal manual switch. The heat and sound emissions are 48 dB. The typical weight of each cabinet is 28 pounds, height is 6.3 inches, width is 16.5 inches, and depth is 17.6 inches.

Systems used in this network include the following security features:

- Boot sequence control
- Boot without keyboard or mouse
- Cover key lock
- Diskette boot inhibit
- Diskette write protect (switch)
- Diskette I/O control
- Hard disk I/O control
- Parallel I/O control
- Power-on password
- Secure fixed DASD
- Secure removable media
- Serial I/O control
- Setup utility password (administrator password)
- U-bolt tie-down support

The systems specifications used in this network also include the following product approvals and/or certifications, according to IBM:

■ BABT (UK)

■ CE

■ CISPR-22 Class B

■ CSA C22.2 No. 950 (Canada)

■ DEMKO (EN 60950)

■ EIF (SETI) (EN 60950)

■ Energy Saving Law (refer to N-B 1-9174-001)

■ FCC Class B (U.S.)

■ IECEE CB Certificate and report to IEC-950 Second Edition

■ ISO 9241-3 Compliant

■ JATE

■ NEMKO (EN 60950)

■ NS/G/1234/J/100003 (Telecommunications Safety only: no approval mark)

■ OVE (EN 60950)

■ Power Line Harmonics (refer to N-B 2-4700-017)

■ SEMKO (EN 60950)

■ TUV-GS (EN 60950)

■ TUV-GS - ZH1/618

■ UL-1950 First EditionVCCI Class 2 (Japan)

In addition, IBM's current limited warranty is type 3: three year, first year on-site, second and third year carry-in, three years parts and labor.

The IBM desktop systems used in this network came with pre-installed software. Some of these systems were reconfigured to meet the needs of the network, but all legal and ethical consideration was given to the manufacturers of the hardware and software products. Each system used in this network is covered by either a site license or has a dedicated piece of software, and each system has one user. In the case of servers, workstations, or otherwise, each manufacturer's legal guidelines were followed. I strongly recommend these matters be factored into network design of any network. Simply put, using an unpaid piece of software, unless it is clear that it is freeware, is stealing. It is no different than stealing a tangible item. Consider this when you design your network.

For more information on IBM products, contact IBM via the Web at www.ibm.com or by mail:

IBM North America
1133 Westchester Avenue
White Plains, NY 10604

6.4 Rack Enclosure

A Bud Industries rack enclosure is used in this network. This enclosure includes an industry-standard 19-inch rack mount in the front and rear. Figure 6-2 illustrates it.

Figure 6-2 shows a front view of the rack enclosure. The enclosure used has a 42-inch rack capability inside. When the rack enclosure is assembled, with casters on, its total height is 48 inches. Figure 6-3 is a rear view of the Bud rack enclosure.

This figure is similar to Fig. 6-2; however, the chassis-support brackets run the length of the cabinet. Notice the rear view shows that the chassis-support brackets are mounted beneath the fan assembly at the

Figure 6-2
Bud Industries' rack
enclosure, front view.

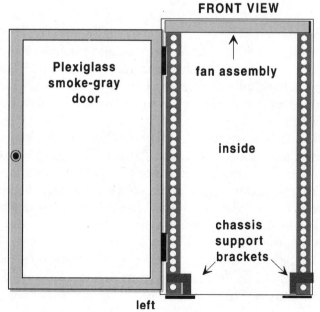

Figure 6-3
Bud Industries' rack
enclosure, rear view.

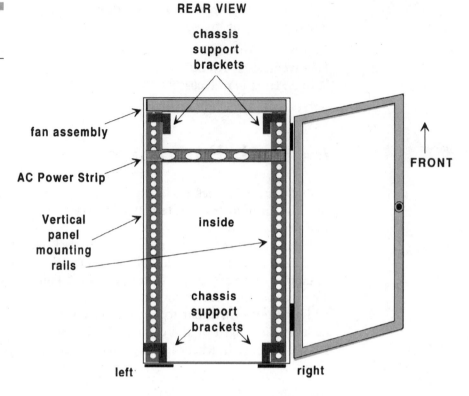

Figure 6-3
Bud Industries' rack enclosure, rear view.

top, as well as on the bottom. As Figs. 6-2 and 6-3 show, there are vertical-panel mounting rails in the front and the rear of the enclosure. It is possible to mount equipment in the enclosure without rear vertical-panel mounting rails and chassis-support brackets. This is a judgment call you need to make. The enclosure is sturdy enough to hold considerable weight.

Bud Industries has a variety of options that can be used in the cabinets they offer. My decision to use front and rear vertical mounting rails was due to weight concerns of the equipment mounted in the enclosure. Your situation could differ. However, having the enclosure with front and rear vertical rails and chassis-support brackets connecting them makes the enclosure even stronger than it already is. It is an added strength.

The enclosure has a unique benefit: Both sides can be removed without any tools. Also, both sides of the enclosure are vented to assist in air circulation. The front and rear door are both keyed and lockable. They are also removable. This flexibility makes the cabinet accessible from every side. The top has a circle of machined holes to aid in circulation. In addition, Bud ships the fan assembly unassembled. This is a great advantage. The orientation of the fans inside the fan assembly determines the direction of air flow. This is very important, since equipment mounted in the cabinet used in this example has fans mounted inside a rack mount cabinet. Because the Bud enclosure fan orientation can be changed, all devices can circulate air in the same direction.

The Bud enclosure arrived at the location where the network was physically assembled. It arrived by way of tractor-trailer, on a pallet. The pallet weighed in at approximately 250 pounds. Bud included all the required parts to assemble the enclosure with the least amount of tools. A Phillips screwdriver, a wrench to tighten the nuts and bolts on each caster, and pliers were required.

If you desire more information about rack enclosures, regardless of your size requirements, contact:

Bud Industries, Inc.
4605 East 355th Street
P. O. Box 998
Willoughby, OH 44094
www.budind.com

  or

Bud West
P. O. Box 41190
Phoenix, AZ 85080

This enclosure is a good example of what I referred to earlier about having a facility capable of storing equipment. This pallet arrived within a day of another pallet of equipment. Together, they weighed over 600 pounds and required approximately 10 square feet of floor space.

6.5   Electrical Test Equipment

Knowing the electrical requirements for all network equipment is not optional. Neither is knowing the electrical ability of the environment to which the equipment will be installed. Someone needs to be well trained to perform electrical testing prior to, during, and after installation of each piece of equipment. In addition, this person should be able to analyze the power of the site, factor in various components to be used, and convey what changes need be made, if any.

My background includes training in electricity and in test equipment. Fluke instruments were used in this network. Actually, Fluke equipment was used prior to any piece of equipment being powered up so that a baseline reading and capacity could be obtained.

The Fluke analyzers were calibrated and tested for mechanical and electrical function. This is important, as you know if you understand analyzers. Calibration puts the analyzer at a known position to begin. A reference point must be obtained for any analyzer or meter to be reliable.

Consider Figs. 6-4 and 6-5. These figures show the electrical outlet prior to any equipment being connected. They are typical readings. These readings were viewed in real-time by a Fluke 41B Power Analyzer and captured with Fluke software on a PC. The PC in this case was an IBM ThinkPad using batteries, so no skewed readings here!

After initial readings were obtained, including information regarding which outlets were wired to which breaker, additional readings were taken. Figure 6-6 is a snapshot of a Tripp Lite voltage regulator with an IBM network printer attached. Notice the figure shows the voltage and amperage. In this case, the amperage reading is 0.81 amps. Notice, as well, the duty cycle of 60.0, displayed in the lower left of the figure.

Figure 6-7 is the same Fluke instrument and the same Tripp Lite voltage regulator, with the same IBM network printer. Examine the figure.

Figure 6-4
Voltage and frequency sampling.

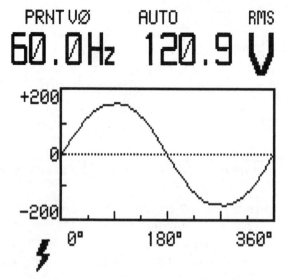

Figure 6-5
A snapshot of a voltage in isolation.

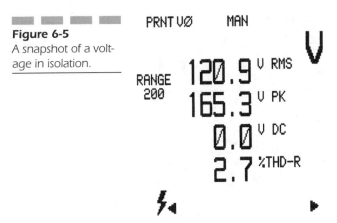

Figure 6-6
A minimum amperage reading.

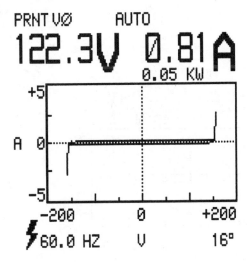

Notice the amperage used is 5.12. The printer is drawing this current for the heating element inside it. Now consider Fig. 6-8. This snapshot was taken seconds after Fig. 6-7. The amperage is back down to 0.39. In Fig. 6-9, the current drawn is 8.11 amps.

Assume you intend to install two network printers. Assume both can draw 8.11 amps when required. That is a total of 16.22 amps. Now, assume these printers were plugged into a 20-amp circuit breaker. This scenario would be fine, but a word of caution is due, since circuit utilization is approximately 65% of capacity.

The significance of this section is to show you that when you are designing your network, you need to get very specific before the equipment is deployed. Then, once the equipment is deployed, you need to

Figure 6-7
Amperage use with
one printer.

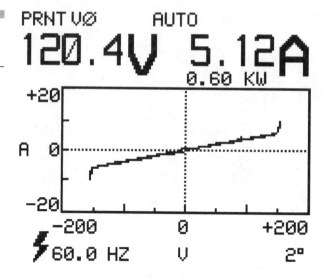

Figure 6-8
Voltage and amper-
age sampling.

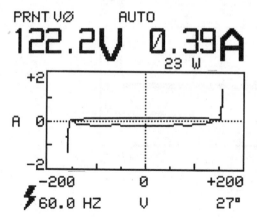

Figure 6-9
Amperage utilization.

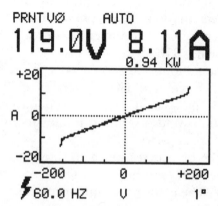

verify your prior estimates. Electricity requires respect; violate that respect and negative consequences are certain, the only question is how severe. Electricity is not anything to play with or guess about. You might think I am pushing the point here; if I am, it is because I understand what electricity can do. Network equipment is sensitive. A considerable amount of equipment can easily overload circuits if you are not careful. Unfortunately, this oversight is common.

When you are addressing the electrical considerations for your network, rely on qualified expertise. If you have that expertise in house and need equipment, Fluke can be reached at

Fluke Corporation
P. O. Box 9090
Everett, WA 98206
www.fluke.com

or

Fluke Europe B. V.
P. O. Box 680
7600 AR, Alemelo
The Netherlands

After you have performed preliminary testing during the installation phase of network design, you will need to continue to monitor electrical considerations day by day. This part of network maintenance works together with power protection, which is addressed later in this chapter.

6.6   Network Hubs

Most networks have some form of *hub*, a device that serves as the vehicle for the lower-layer network protocols. In this case, Standard Micro Systems (SMC) network interface cards (NICs) were selected both for the desktop systems and notebook computers. SMC network hubs were also chosen.

The NICs can accommodate 10BaseT or BNC connectors. The exposed end of the card (where the cables connect) has LEDs showing the link and transmit/receive status of the board. These boards support SNMP, so management at lower-layer interface-board level is possible. SMC PCMCIA cards are used with the notebook computers. They have the same functionality of the NICs used in the desktop systems.

The SMC NICs in this network are 10BaseT 14 port-stackable, rack mount enclosures. This particular hub (the Elite 3812TP) is configurable and stackable to accommodate up to a maximum of 112 ports. Figure 6-10 illustrates how these appear inside the rack enclosure.

This illustration shows two hubs with two SCSI ports on the front of each. A cable is used to daisy chain these together to make them stackable. In this case, two are used; one cable is used to connect them, and each hub has a terminator supplied by SMC in order to "close" the connection.

The rear of the hubs is where network device cabling is connected. Consider Fig. 6-11. The rear of the hubs have 10BaseT connection points. Though not shown in this illustration, the rear of the hubs have LEDs indicating the status of links, sources, and partitions.

6.7   Patch Panel and Wiring

Wiring any network is either planned for, or chaos exists. I speak from experience, having walked through layers of cable behind equipment that was not labeled. With that thought in mind, the design for wiring in this network was examined from multiple angles. First, reliability of the cable maker was considered. Second, the actual implementation was reviewed: Would it be easy to reconfigure equipment once installed?

Figure 6-10
SMC hub mounting, front view.

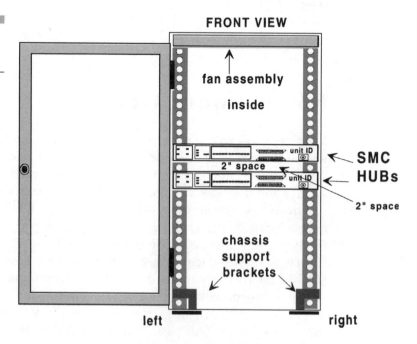

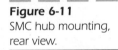

Figure 6-11
SMC hub mounting,
rear view.

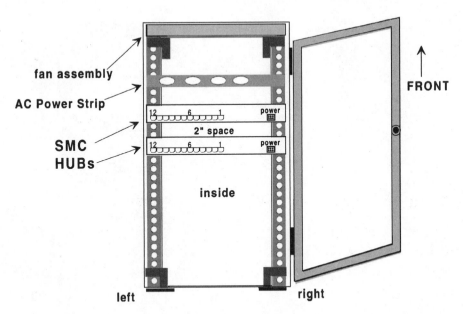

How would one know which cables go where? With this level of thought and multiple designs on the marker board, the determination was made to use Hubbell equipment for the rack enclosure as well as wiring.

Figure 6-12 shows an inside view of the enclosure from the rear. The figure shows that the inline panel has RJ-45 female connectors on what is considered the back side.

Figure 6-13 shows front and rear views of the inline panel. The significance of this inline panel will become very clear forthwith. Rack mount panels are available that have RJ-45 female connectors on the front, but the rear has a 110 connector. This is neither good nor bad. Some environments require this type of rear connection point for each connection. However, in this case, it is different.

Hubbell offers a patch panel with all connection points wired straight through. So what, you say? Well, with a rack enclosure full, would you like to work inside of the rear or be able to work with all the equipment from the front? That's what I thought. Now, consider Fig. 6-14.

Figure 6-14 shows a side view of the rack enclosure. Notice there is a U.S. Robotics enterprise network hub, two SMC network hubs, and the Hubbell inline panel. Look closely at the area with the dotted-line

Figure 6-12
Inline panel mounting, rear view.

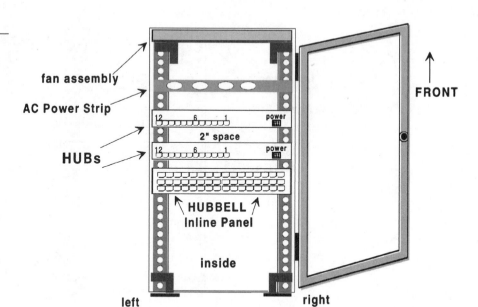

REAR VIEW

fan assembly

AC Power Strip

HUBs

Figure 6-12
Inline panel mounting, rear view.

HUBBELL
Inline Panel

inside

FRONT

left   right

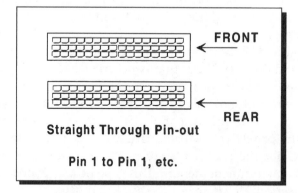

HUBBELL Inline Panel

Figure 6-13
The Hubbell inline panel.

FRONT

REAR

Straight Through Pin-out

Pin 1 to Pin 1, etc.

rectangular box. This illustrates any combination of connections can be achieved from the front of the enclosure. How? All equipment is connected into the rear of the inline panel. Actual physical configuration is then made via patch cables from the front. In this network, a 48-port inline panel is used, but Hubbell has larger panels as well.

This example of component and wiring design illustrates the concept of designing ease-of-use, flexibility, and expandability into the network from

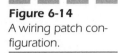

Figure 6-14
A wiring patch con-
figuration.

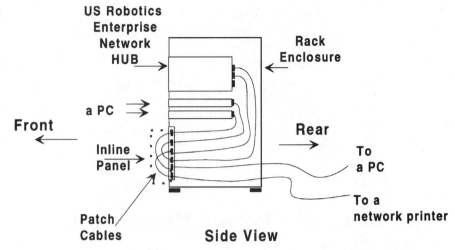

the outset. Go one step further with Fig. 6-14 and consider this: With a Fluke 685 LANmeter, any cables can be tested *from the front* of all equipment with relative ease. The same is true for troubleshooting or monitoring upper-layer protocols. The Hewlett Packard Internet Advisor can be easily connected to the inline panel to monitor network traffic and other operating parameters. To make life even better, Hubbell wiring is color-coded. During the installation of this equipment, colors, numbers, and labels are assigned to all ports, equipment, and other entry points to the network.

There is no reason why a network design should not be documented completely from the very beginning. A well-designed network will enable anyone who understands the technology to perform any applicable work function.

If you have any questions about inline panels, wiring, fiber connections, surface-mount housing, jacks, or other components required to design your network, contact one of the following sources:

Hubbell Corporate Headquarters
14 Lord's Hill Road
Stonington, CN 06378
800-626-0005
www.hubbell-premise.com
www.hubbell-canada.com

Hubbell Ltd
Ronald Close, Woburn Road Industrial Estate
Kempston, Bedford, England MK42 7SH
44-1234-855444

Hubbell-Taiwan Co., Ltd.
12 Floor, 66, Sec. 2, Chien-Kuo North Road
Taipei, Taiwan
886-2-515-0855

Hubbell Canada, Inc.
870 Brock Road South
Pickering, Ontario, Canada L1W 1Z8
905-839-1138

6.8 Power Protection

Power protection in any computing, network, or peripheral equipment is
not optional. I have a standard reply when asked if such-and-such needs
to be protected with power-protection equipment: If you can afford to
throw away any or all of your equipment, replace it, and not be negative-
ly affected by downtime, then don't use power protection equipment.
With that in mind, let us examine the power protection equipment used
in this network.

Tripp Lite power-protection equipment is used in this network. A
wide variety of equipment is used, both rack mount and non-rack
mount. It is easiest to list the pieces of equipment used, then address
their functionality in the network:

- Isotel Premium surge suppressor with fax/modem protection
- Surge Alert plus tel
- Internet series uninterruptible power supplies (UPSs)
- BC Pro Series (UPS)
- Data Shield parallel dataline surge suppressors
- Data Shield 10BaseT surge suppressors
- Data Shield dial-up line surge suppressors
- Data Shield AUI 802.3 surge suppressors
- Data Shield DB9 surge suppressors
- Data Shield DB-25 serial surge suppressors
- SmartPro rack-mount 1050 UPS
- LCR 2400 rack-mount line conditioner (also called a voltage regulator)
- IBAR 12 rack-mount surge suppressors

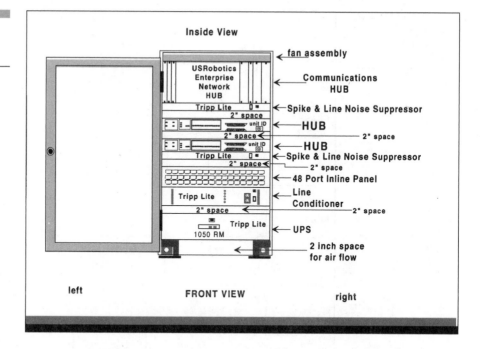

Figure 6-15
A populated rack enclosure.

All power protection in this list is integrated into the network at the appropriate places. The rack-mount pieces appear as in Fig. 6-15.

The rack enclosure houses the 1050 UPS, voltage regulator, and two spike and line-noise filters. Two spike and line-noise filters were needed because of the number of low-amperage devices requiring AC power. Another voltage regulator is used, but not shown. It is external to the rack enclosure. A voltage regulator is also used for the network printer.

The data shield protectors are used at all points in the network. These inline devices are used with serial, parallel, 10BaseT, AUI, modem cable, and DB-9 connections. Additionally, the electrical protection equipment also has telephone-line spike/surge protection to protect all incoming telephone lines.

Figure 6-16 shows the protection design used in this network. This figure shows AC wall outlets, spike and surge protection, and then either a voltage regulator, UPS, or another spike and surge protector. Next, other equipment connects to the second phase of power protection. This is *two-stage power protection*. I consider this the minimum level acceptable.

In addition to the power-protection equipment shown in the previous figures, some of the remaining equipment was used this way: Each IBM P70 monitor has a dedicated UPS with it, connected to a dedicated line and noise suppressor. The latter was connected into an isolated wall

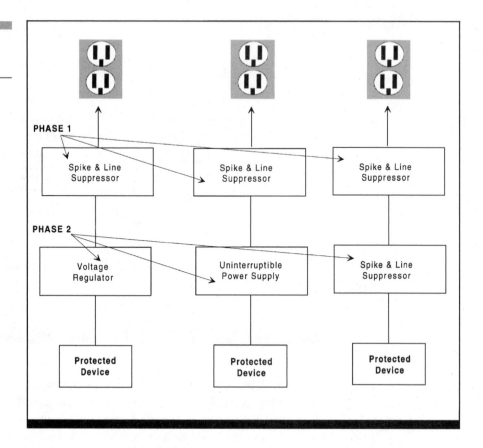

outlet. In this case, the isolated wall outlets were concentrated into a particular circuit breaker. Each notebook computer has a dedicated UPS connected into a spike and surge protector when it uses AC power.

Tripp Lite monitoring software provides administrators the ability to monitor and control power-protection equipment in a remote location. One example of the power behind the monitoring software included with Tripp Lite power protection is the ability to log information events. The following is an excerpt from a log of the network where this equipment is used:

DATE	TIME	INFORMATION
06/21/97	14:45	UPS monitoring started
06/21/97	14:45	Unable to communicate with UPS!
06/21/97	15:43	UPS monitoring started
06/21/97	15:43	Unable to communicate with UPS!
06/21/97	15:47	Unable to communicate with UPS!
06/21/97	15:50	Communications reestablished with UPS.

DATE	TIME	INFORMATION
06/21/97	15:50	The UPS is operating on batter power.
06/21/97	15:50	The utility power has been restored.
06/21/97	15:50	Self-Test Has Passed.
06/21/97	15:58	UPS monitoring terminated.
06/21/97	16:00	UPS monitoring started.
06/21/97	16:28	UPS monitoring started.
06/21/97	16:29	UPS monitoring terminated
06/21/97	16:35	UPS monitoring started
06/21/97	16:36	UPS monitoring terminated.
06/22/97	20:33	UPS monitoring started.
06/22/97	20:33	Unable to communicate with UPS!
06/22/97	20:34	UPS monitoring terminated.
06/22/97	21:04	UPS monitoring started.
06/22/97	21:05	Unable to communicate with UPS!
06/22/97	21:06	Communications reestablished with UPS.
06/22/97	21:07	Unable to communicate with UPS!
06/22/97	21:07	Communications reestablished with UPS.
06/23/97	10:30	UPS monitoring started.
06/23/97	10:31	Unable to communicate with UPS!
06/23/97	10:31	Communications reestablished with UPS.
06/23/97	20:10	Unable to communicate with UPS!
06/23/97	20:29	UPS monitoring terminated.
06/24/97	09:07	UPS monitoring started.
06/24/97	09:07	Unable to communicate with UPS!
06/24/97	09:13	UPS monitoring terminated.
06/24/97	16:10	UPS monitoring started.
06/24/97	16:10	Unable to communicate with UPS!
06/24/97	16:36	Communications reestablished withUPS.

This information was captured by the Tripp Lite software monitoring the Smart 1050 UPS. The software is capable of providing formatted data for printing and well as filing. Tripp Lite has an entire line of power-protection equipment that can be monitored by LAN equipment. For more information on Tripp Lite products, contact them at

Tripp Lite
500 North Orleans
Chicago, IL 60610
(312) 755-5400
www.tripplite.com

6.9 Communication Equipment

A U.S. Robotics enterprise network hub was selected to use in this network. Data-communication equipment is the single most critical link in any network. This is true because it is the central point of attachment between remote users and a backbone network, regardless of the size or location of the backbone. It is also true if all users are in the same physical location. Data-communication equipment is central to networks; as they go, so goes the network. At one time, remote computing meant having a device in one location and a terminal attached to it by a wire. U.S. Robotics revolutionized that definition by designing the enterprise network hub. This device, explained in greater detail forthwith, is powerful.

Consider Fig. 6-17, which illustrates a network designed in Dallas, with remote users and a remote network in Chicago. Notice that remote users are connecting directly into the Dallas network via the communications hub. In this case, the remote users use their modems and connect directly to the hub.

When remote users or remote networks are involved, multiple issues must be considered during the design phase. The following list is the minimum number of issues to be reviewed during your plans:

- Security

- Reliability

- Maintenance

- Ease of use

- Internal protocol compatibility

- Expandability

- Internal design architecture

- Interface standard compatibility

Security has become the single most important topic in networking, regardless of the type of network or its location. Networks can have a considerable degree of security built into the design if proper components are used that implement security. Where data-communication equipment is concerned, having a device that can provide a security firewall is best.

Consider Fig. 6-18. This figure shows a secure firewall implemented in the communications hub. Remote users are required to sign on to the hub. It is a point of isolation. Other devices on the network require sign-ons and passwords as well.

The U.S. Robotics communication hub used in this network has three possible configurations regarding its function in the network. U.S. Robotics refers to these configurations as *gateway application cards*. The three cards are referred to as follows:

- X.25

- NETServer card

- API card

Figure 6-17
A conceptual view of
the example
network.

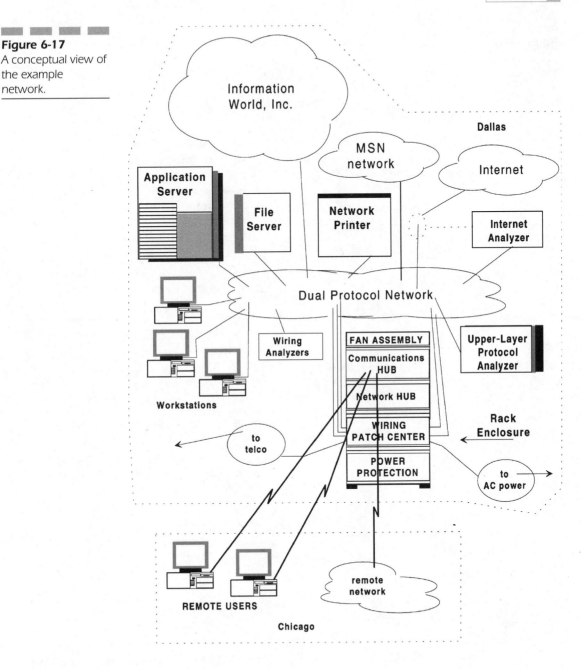

Figure 6-17
A conceptual view of the example network.

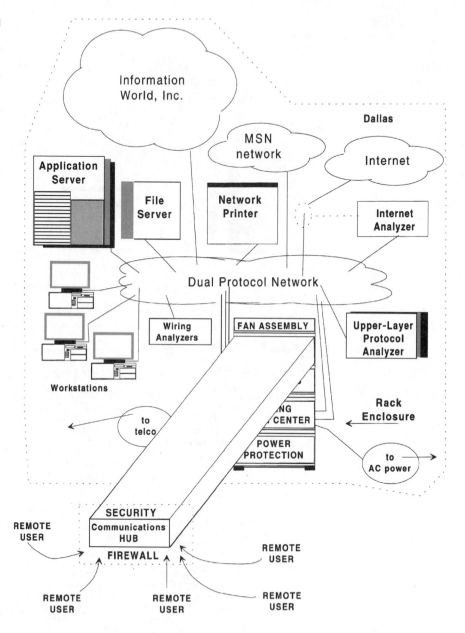

According to U.S. Robotics, the X.25 card provides access capability to packet-switched networks. This card uses an EIA-232/V.35 interface connection point.

The NETServer card functions as either a router, terminal server, or both. Ethernet and Token Ring NICs can be used with it. U.S. Robotics also refers to this card as the *EdgeServer* card. This card has Windows NT loaded onto it. The functionality of this card is explained in detail later in this chapter.

The API card is designed to let customers design their own applications by way of U.S. Robotics software-development kits.

Figure 6-19 illustrates the enterprise hub. Notice that it shows the hub with blank face panels. These panels can be removed and other cards inserted. A total of 17 slots exist. Slot 1 is the T1 card. Beginning with slot 2, there are analog or digital quad-modem cards, which have the equivalent of four modems on them. Slots 15 and 16 provide the Edge-Server location. Slot 17 is where the network management card is located. The remaining slots house two power supplies. Though not shown in this illustration, the undercradle portion of the hub houses approximately 16 fans to cool the components.

Reliability is another important factor for any communication equipment. The design of the U.S. Robotics hub has reliability built into it. For example, the hub has two power supplies, but only one is required to

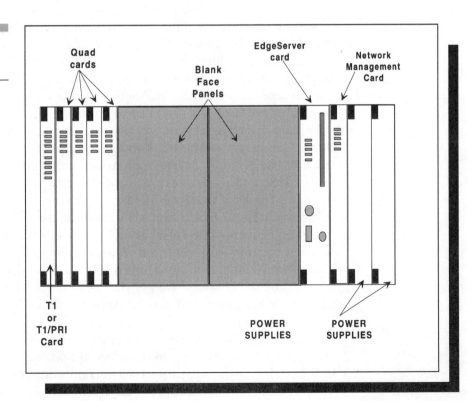

Figure 6-19
The 3Com/U. S.
Robotics hub.

operate the unit. It has built-in redundancy even to the level of the power supply unit.

Maintenance is another part of the equation for communication equipment. The hub used in this network has remote management capability, local management capability, and easy access for those components that might need removal.

Managing any communication device requires skill. Most require a fairly advanced level of skill to maximize use. The capability of any communication device has little to do with its ease of use. Ease of use is a design issue. With the hub used in this network, ease of use is designed into it. Ease of use can be measured in communication equipment by documentation provided, how thorough and detailed it is; by accessibility to configure ports; and the ability to use the equipment in a partially failed state (should that occur). My rule of thumb is that the more complex functions a device offers, the simpler the documentation should be. The simple fact is that data-communication equipment is complex enough without humans adding another layer of complexity to it.

Another factor to analyze with data-communication equipment is the protocol compatibility. This includes evaluation of upper- and lower-layer protocols. Because this hub has the EdgeServer card in it, NetBEUI, TCP/IP, and IPX upper-layer protocols are supported. Token Ring and Ethernet lower-layer protocols are supported as well. Use of Token Ring and Ethernet is more than sufficient; these two protocols are dominant lower-layer protocols used in networks today.

Expandability is very important with data-communication equipment. The design of the U.S. Robotics enterprise network hub is such that any size network can be built around this technology. It is possible to start a network with one or two enterprise hubs, then continue to add them until racks of them are filled.

Internal design architecture is also very important to data-communication equipment. The internal architecture of data-communication gear is the proverbial pivot upon which all communication transactions hinge. The internal-communications bus and the incoming port architecture provide the foundation of the device. These should be capable of handling a complete load on the device without causing hangups or system slowdowns.

Interface-standard compatibility is another matter to examine when you are evaluating data-communication equipment. In this network, the hub has flexibility regarding how certain connections are made. In some instances, options exist to make a connection. This alone makes for ease of use, installation, and maintenance. It also means some existing equipment at your site might be usable. That can save money.

The EdgeServer card installed in the 3Com/U.S. Robotics enterprise network hub has the following:

- 1.44 floppy drive
- Mouse port
- Keyboard port
- Display port
- SCSI port
- Minimum of 800 MB hard drive
- Minimum 100 MHz processor
- 10BaseT capability

The EdgeServer card also has Microsoft Windows NT Server 4.0 installed on it. Conceptually, the card and some of its functionality appears in Fig. 6-20.

The advantage of having NT Server on the EdgeServer card is manifold. First, when remote users access the communications hub for information purposes only, they can be stopped there and not access other systems that are part of the network. Second, remote users who require access to other systems can use the EdgeServer as a gateway, if you will, to access the network behind it. The EdgeServer card can function as an excellent firewall to protect the assets behind it, while permitting access to it.

Still another powerful feature of the EdgeServe card is the SCSI port it has located on the back. This feature makes it possible to connect a CD-ROM drive to it. Documentation is provided with enterprise network hub, but it is also provided via CD, which makes it convenient to access when manuals are not easily accessible.

A network management card is also part of the hub's component configuration. The card supports Ethernet and Token Ring as lower-layer protocols. It is a separate card, and it provides a console port that can be used for the following tasks:

1. Remote access, dialed into from a remote site
2. Local access via an RJ45 and DB-25 cable with a null-modem adapter provided with the hub
3. Software download to aid in the management aspect of the hub

The network management card supports 10BaseT, 10Base5, and 10Base2 connection points for Ethernet cable flexibility. Token Ring

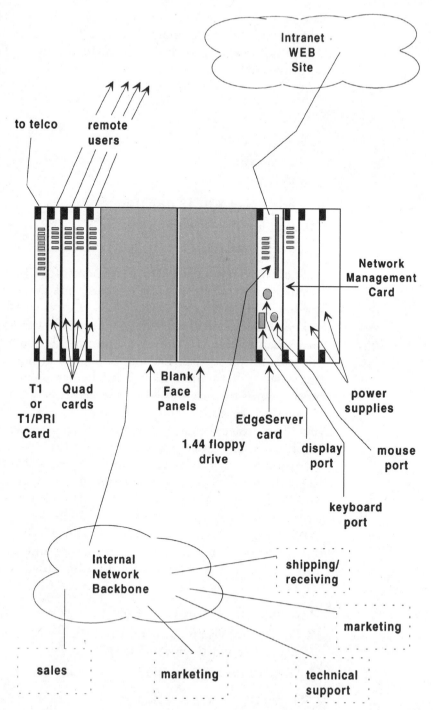

Figure 6-20
Data communica-
tion/hub integration.

cable support includes shielded twisted pair (also called IBM type 1) and unshielded twisted pair (also called IBM type 3).

The network management card does not run SNMP management agents directly on the card, but the support for SNMP is not compromised. The network management card technically functions as a proxy agent; the functionality and management ability with SNMP operation features the same support.

The enterprise hub used in this network has a T1 card. It operates as a Primary Rate Interface (PRI). The T1 card is managed by what is called Total Control Manager (TCM). It is SMP-based and works with Windows. The card itself is easily configurable. Either a dumb terminal, remote PC, LAN PC, or direct-connect PC can work with configuration-management parameters. Its operands function within the SNMP MIB standards; both GET and SET operations can be issued against the T1 card.

The T1 front panel includes LEDs to indicate the operational status of the card. Those LEDs include the following:

ALARM This LED is activated upon existence of any of the following states: alarm-indication signal, frame slip, out of frame, excessive CRC errors, change-of-frame alignment, line-format violation, or frame-alignment error.

CARRIER This LED indicates whether or not a carrier is present; an unframed signal LED indicates if an out-of-frame condition exists, when a loss-of-signal condition occurs, and if a signal is reported "not present."

LOOPBACK This LED indicates if a test is in operation, initiated from the local telephone company.

RUN/FAIL This LED indicates the operational mode of the T1 card; that is, if it is operating normally or in a critical mode due to a hardware and/or software fail condition.

The enterprise network hub modems are either analog or digital. Figure 6-21 illustrates this. This illustration shows the enterprise network hub with analog modems (quad cards) connected to analog telephone lines. The gateway interface card in this illustration shows a generic connectivity to a network. The gateway-connectivity portion of the enterprise network hub does not necessarily require Windows NT Server, but this implementation is popular.

Figure 6-22 shows a different hub implementation. This illustration shows a T1 link to the telephone company. It also shows digital modems used after a signal is in the hub. The gateway aspect of the hub indicates that users can have access outbound to a network, if such is configured. Technically, the network access a remote user has is configurable for either analog or digital modem connectivity.

The enterprise network hub is one of many products offered by 3Com/U.S. Robotics. For additional information contact them at:

3Com/U.S. Robotics
Corporate Systems Division
8100 North McCormick Blvd
Skokie, IL 60076
800-877-2677
www.usr.com

Figure 6-21
A highlighted view of analog modem connection.

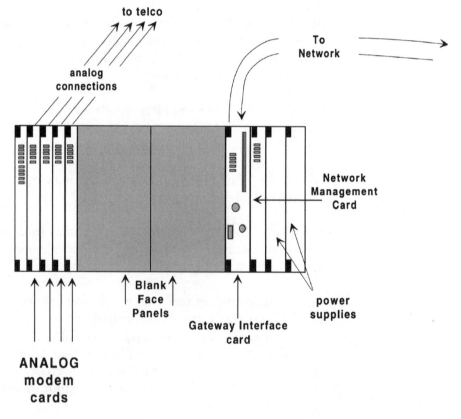

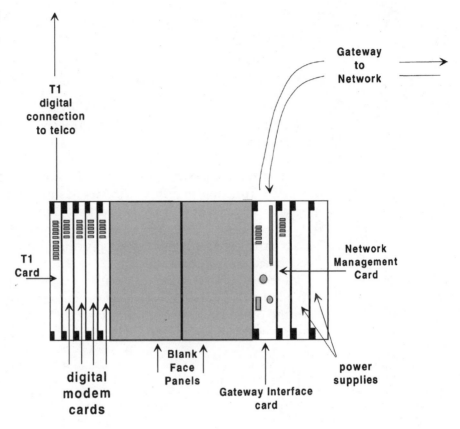

Figure 6-22
A highlighted view of digital modem connection.

6.10 **Operating System Software**

After considering the requirements made on the network, Microsoft Windows NT 4.0 Workstation and Server was chosen as the operating system software. Some workstations have Windows 95 as the operating system, but those are being upgraded to workstation software.

Windows NT 4.0 could become the standard by which other operating systems are judged. It has the look-and-feel of Windows 95, while having robust features for networking. This version of NT is separated into the NT Server and the NT Workstation. Some common characteristics between them are as follows:

- Advanced file-handling systems
- Back-up capabilities
- C2 security

- Graphical user interface tools
- Network capabilities
- Remote-access capabilities
- TCP/IP

The physical architecture of NT is divided into two components: Workstation and Server. Though they share commonalities, they operate independently of the other.

The NT Workstation is basically a stand-alone operating system. It does not require NT Server to operate. As shown in Fig. 6-23, NT Workstation can function in different modes to support a variety of software applications. It comes with the networking component integrated into it and can easily be configured to operate in a networked environment. NT Server is software capable of supporting a variety of workstations in a networked environment.

Figure 6-24 is a hypothetical example showing two segments of an intranet. In this example, each department can operate independently from the other. However, they are connected via a fiber connection. If required, workstations in segment 1 can access the NT Server in segment 2,

Figure 6-23
NT application support.

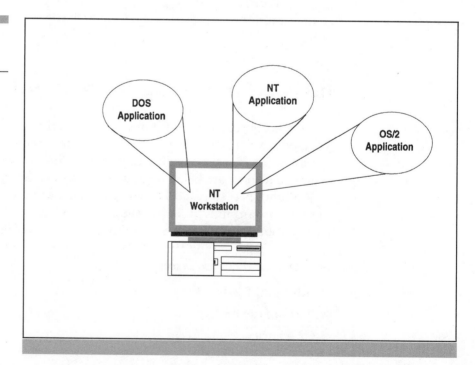

and vice-versa. In addition, each NT workstation can operate independent of the others.

Generally, NT is capable of supporting a significant amount of resources. For example, currently NT supports 4 GB of RAM per system. It also supports 2 GB of virtual memory and can address up to 402,000,000 (million) Terabytes of storage.

Windows NT architecture is considered modular because it contains multiple components. Figure 6-25 shows this idea. Figure 6-25 shows four distinct components of NT:

- Application Environment subsystems
- Hardware Abstraction Layer (HAL)
- Kernel
- NT Executive services

The Application Environment subsystem is that part above the WIN 32 subsystem. The WIN 32 subsystem is the main subsystem for NT. It includes WIN 32-bit application program interfaces (APIs). Beyond support for 32-bit programs, the WIN 32 subsystem supports application programs for other operating environments as well.

NT Executive is the highest order of control within the operating system. The NT Kernel is part of the Executive. The Kernel basically dispatches and schedules threads used in the operating system. The term *thread,* as used in this context, means an executable object that belongs to a process. An object can be something as concrete as a device port or application.

HAL is a dynamic load library (DLL) used between a system's hardware and the operating system software. The purpose of HAL is to keep NT from being concerned with I/O interrupts, thereby making NT easily portable. It is because of HAL that NT is considered portable between different types of operating systems. It also provides support for symmetrical multiprocessing (SMP). The result is NT can be used on Intel, MIPS, Alpha processors, and others.

The input/output part of NT handles input and output processing. The I/O manager coordinates all system I/O, drivers, installable file systems, network directors, and caching memory management.

The Local Procedure Call functions as an interface between all clients and servers on an NT system. Functionality of LPC is similar to Remote Program Call (RPC) facility. However, LPC and RPC are not equals because LPC permits exchange of information between two thread processes on a local machine.

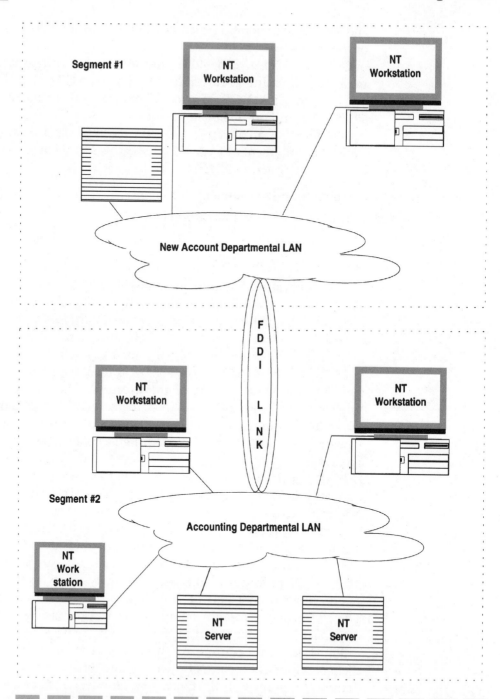

Figure 6-24 A single network link.

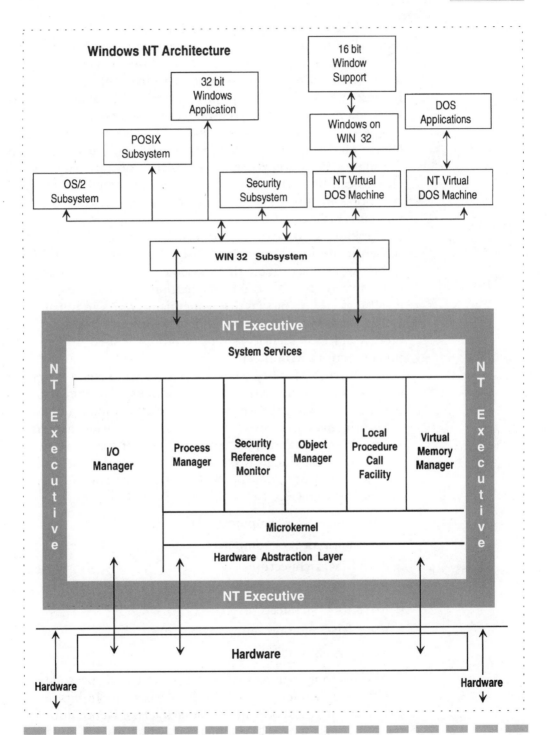

Windows NT Architecture

Figure 6-25 A conceptual view of NT architecture.

The microkernel is part of the NT Executive. It operates in kernel mode. The microkernel runs in kernel mode and communicates with the NT Executive via very low-level primitives. In a crude way, you could think of the microkernel as that part which controls the entire system. Because NT is based around preemptive multitasking, it controls time slices and manages to pass control to other processes.

The NT Executive uses the Object Manager to manage objects. The Object Manager creates, deletes, and modifies objects. Objects are nothing more than abstract data types used as operating system resources. Objects can be ports (physical) or threads (logical). The Object Manager also works in the system to clean up stray objects that could exist if a program crashes.

The Process Manager is involved in all processes and threads. General consensus defines a process as that which has an identifiable virtual address space, one or more threads, some system resources, and executable program code. In NT, when an application is started, the Object Manager is called to create a process that sequentially creates an initial thread.

The Security Reference Monitor is the core of NT security. A logon process and local security authority process is used in the implementation of security within NT.

The Virtual Memory Manager (VMM) translates a system's process memory address into actual memory addresses. In short, it manages virtual memory. Virtual memory in personal computers operates on the same principles as it does on large computers. Some aspects might be augmented or enhanced, but the foundation of the idea is the same.

NT's modular design makes it robust. It has a great degree of hardware independence. Some of the hardware platforms NT can operate on include the following:

- X86 uniprocessor computers
- X86 multiprocessor computers
- AXP RISC architecture
- AXP RISC multiprocessor computers
- MIPS RISC architecture
- MIPS RISC multiprocessor architecture
- Motorola PowerPC

Another powerful aspect of NT is its use of unicode rather than ASCII for a character set. Unicode is based on 16 bits, meaning this character set can represent 65,536 characters. Hence, unicode is more powerful than its ASCII counterpart, which uses 8 bits to yield only

256 characters. The inherent meaning of this is that multinational characters are supported, such as Japanese, Chinese, Russian, Swedish, and others.

Windows NT is clearly delineated into two distinct parts. Both Workstation and Server use what are considered advanced file systems. NT supports the NT file system (NTFS) and the file allocation table (FAT). NTFS supports long file names, file-level compression, file data-forking support (required for Macintosh systems), international file names, software-level sector support for fault tolerance, and file-level security permissions.

FAT support makes NT backwards-compatible with DOS. Under NT, floppy drives use FAT and RISC systems in the boot partition.

Both Workstation and Server support TCP/IP. NetBIOS is supported also, as defined in RFCs 1001 and 1002. This means logical naming at a session level is possible. The NetBIOS interface also supports dynamic data exchange (DDE). DDE enables sharing data embedded within documents. NT Workstation and Server also share the Dynamic Host Configuration Protocol (DHCP) client, so that an NT system can participate in an environment to obtain DNS addresses, IP addresses, gateway addresses, and netmasks.

NT provides support for SNMP. This makes workstations and servers capable of participating in environments that use management tools such as OpenView, SunNet Manager, and SystemView. Both support Remote Access Service (RAS), as well as point-to-point tunneling protocol. This function of NT enables a virtual network to be created over a wide distance. NT data encryption makes this part of NT popular.

Both Workstation and Server provide support for C2-level security. Protecting data on a system is achieved by access-control lists (ACLs). This enables directory- and file-level security maintenance.

NT Workstation and Server are administered the same way. Some of the tools by which this is done include the following:

- DHCP Manager
- Disk Manager
- Event Viewer
- Performance Monitor
- RAS Administrator
- Server Manager
- User Manager
- WINS Manager

The DHCP Manager is used to configure the station to obtain required information upon startup. In addition to the DHCP Manager, the DHCP Server Service permits remote control of DHCP servers.

The Disk Manager is the tool that handles disk partitions, mirrored disks, and volume set disk partitioning.

The Event Viewer enables a user to view events on local and remote systems. Information such as the system log, security log, and application log can be obtained.

The Performance Monitor provides a method for real-time monitoring. The Performance Monitor also makes sending administrative alerts to remote systems, monitoring performance counters, and maintaining performance logs possible.

The Remote Access Service Administrator is a configuration tool used to designate certain users as privileged to gain access to a network node. RAS Administrator can also be used to configure remote stations.

The Server Manager tool is used to create domain systems for NT workstations, stand-alone servers, and domain controllers. It is also used to determine the status of servers and users who are logged on.

The User Manager enables the creation and management of user accounts. When used on a domain controller, inter-domain trust relationships are set up by this tool. The User Manager is used to set up user rights, system-wide passwords, and auditing functions.

The Windows Internet Name Service (WINS) Manager is used to manage the WINS server service. This function of NT can be used to manage a local host or remote host. The basic purpose of WINS is dynamic name resolution.

NT security is probably the best of any PC-based operating system to date. More than one method can be used with NT security. One way to implement security in NT is with a logon ID and password. This is commonly referred to as *simple system access*. The next level, or way, security can be implemented is through resource access. This method of security is a double-edged sword; security can be tight, but it is difficult to maintain. The next level of security involves creating groups. For example, users in ABC group can access all files and programs in ABC group, but they cannot access files or programs in DEF group. NT security is powerful and flexible. I recommend that, once a site is set up with whatever security policy is in place, make records of the way it is set up and document changes as they are made to the system.

Domains are used with NT. Microsoft considers a domain as a secure workgroup with centralized maintenance. With a little thought, it is easy to understand how this can become complex, if a site is large.

Within the Microsoft domain, there are *domain controllers.* These are systems that maintain critical information to keep the domain operational, information such as resources, security access privileges, passwords, and logons. Many sites run the primary domain controller and a secondary or backup domain controller. Remember that the primary domain controller is actually made at the installation time and cannot be changed later, since the domain controller must be configured during installation. This means on any given network, only one primary domain controller can exist. Should two or more domain controllers think they are the primary controller, a contention scenario will ensue. The domain controller with the most "trust" relationships on the network will win the contention, and thus be considered the primary domain controller.

A trust relationship among domains simply means two or more domains trust the other. This does not mean that all users in one domain will have automatic access to resources in the other domain. In addition, the notion of a global domain is required for operation where multiple domains exist. One could think of this, loosely, as "the great global domain in the sky." The purpose behind it is to facilitate coordination of domains and users within various domains.

Another concept to remember about domains is that trust relationships among them are one-way, not two-way. For example, user A in domain A might trust resource B in domain B. This does not, however, mean that resource B in domain B trusts user A in domain A.

The simplest way to think of trust and domains is illustrated by three examples. First, a master domain has domains that have trust relationships with it. Second, two domains have bidirectional trust relationships, which could be called loosely peer-to-peer. Third, three or more domains have multiple trust relationships among more than one domain.

NT users have attributes associated with them, another powerful characteristic about this operating system. The following list reflects the attributes generally associated with a user:

- Account
- Password
- Application access
- Logon capabilities
- Home directory
- Group memberships

■ Profile

■ Rights

■ Remote access service capabilities

■ Policies

This information is used by administrators to customize a wide variety of user profiles. Some users might have much broader abilities than others due to their work requirements. This degree of information in user profiles enables great control over user access and provides system security.

These features, and functions of NT Workstation and Server contributed heavily to the decision to use it as the PC operating system in this network.

6.11   Network Printer

An IBM printer was selected to be the network printer. The Model 17 was determined to be the best fit. The following features/functions factored into that reasoning during the decision process. Before examining all features/functions of this printer, consider the first list of standard features/functions:

■ 17 pages per minute

■ 600×600 resolution

■ Up to five addressable input trays

■ 4 MB RAM (optional to 66 MB)

■ PCL5e standard language (PostScript, IPDS, and SCS optional)

■ Auto language-switching with options

■ Auto I/O switching

■ Standard parallel with two network interface slots

The following options were added to the printer to make it capable of meeting the needs of all users on the network.

■ 75-envelope feeder

■ Ethernet interface

■ Token Ring interface

■ 24 MB RAM

- PostScript language level 2
- 500-sheet second paper tray
- Duplex unit
- 10-bin secured mailbox unit

This printer arrived on a pallet weighing approximately 250 pounds (entire pallet weight). The printer itself is 40.9 pounds. With all options installed, the dimensions are: 31 inches high, 25 inches front-to-back, 17 inches wide, and 65 pounds in weight. (These measurements include space for rear cabling, etc., and are approximations.)

The printer was chosen because of its flexibility and power. It supports Intelligent Printer Data Stream (IPDS) and SCS character strings. This is valuable because, should the network need a system which uses either of these character strings for printing, the printer itself is already capable of handling it.

IPDS is used between an IBM host and a printer; generally, this refers to a SNA environment. This data stream is used with an all-points addressable printer. IPDS can intermix text and graphics, both vector- and raster-based. An SCS character string is a protocol used with printers and certain terminals in the SNA environment. LU1 and LU6.2 can use this data stream. One unique aspect of this data stream is its lack of data flow control functions. The significance of the Model 17 printer chosen for this network should not be overlooked. When the need arises for a host running MVS and VTAM, the current printer can be used with it. Here again is an example of designing success into the network.

Because the printer is on the network, all network users can take advantage of it. If a user is working on the network from a remote location, and wishes to print something to someone and have that document secure, it can be done.

A remote user connecting via a switched line can print a job to the network. The remote user works with a file on the NT Server. The NT Server then sends the file to the printer. The printer prints it and sends it to bin 3, where the owner of bin 3 must enter a code to receive the print.

Users onsite where the printer is installed have free access to it, with the exception of those who require secured access through the mailbox feature.

The IBM Model 17 printer arrived on a pallet, as described previously. From time of delivery until the printer was operational, one work day elapsed. It took about 2 hours for unpacking the printer and reading the material IBM recommends before beginning. Assembling the various components (accessories) for the printer was easy. IBM designed the printer so that minimal tools are needed to install it. More than likely, it will take longer to configure and integrate the network workstations and servers than set up the printer.

6.12 Network Security

Computer and network security is probably the single most important issue today, and in the future will become even more important. Viruses, bots, and all sorts of anti-data objects exist within the Internet. Most people have no idea how vulnerable parts of the Internet are. Even service providers are more vulnerable than they will admit. The sad fact is every company of any size has disinformation arsenals to make people feel secure. Management at the higher levels in most corporations operates in ignorance of these matters because this is legally safe for them, and it is prudent that they do so. Ask, and this statement will be denied, but remember where you read it. I point this out because I do not want you to be ignorant. There is no magic program or anything else that can make networks safe. Good programs exist; the ones chosen and implemented in this network are an example. However, no single program can make your network 100% immune.

Networks can have security designed into them from the outset. Security in your network needs to be factored into every area, from electricity provision through telephone access. The McAfee software suite was selected to meet the needs of this network. The first reason for this is the amount of anti-virus programs and information they have—and it works. The second reason for selecting McAfee was the frequency by which they update their anti-virus software. At present, McAfee has over 250 highly technical documents available on viruses; they claim to have information about the 1,000 most common viruses. When the security analysis for this network was complete, the following software packages were selected and are used in this network:

- VirusScan—This program may well be the most popular anti-virus software in the marketplace today. It operates with Windows 3.1, 95, NT 4.0, DOS, and OS/2. Once installed, it operates automatically upon power-up. It can be used at will once a system is operational. This program requires minimal space, but does a professional job. It is NCSAA certified.

- Desktop Security Suite—This suite of programs also operates with Windows 3.1, 95, and NT 4.0. It includes anti-virus software, backup abilities, and encryption technology. The virus program is VirusScan. QuickBackup operates with ZIP, JAZ, the Internet, and rewritable CD-ROMs. The backup program enables hourly or on-demand backup, whichever is best for you. The cryptographic part of the suite provides 160-bit encryption. This part of the suite enables users to encrypt files before they are sent over the Internet. The cryptographic part of the suite also permits network traffic to be encrypted between Windows-based computers and those running UNIX.

- Commuter—Commuter is more than just a communication software. It also includes virus protection, desktop storage management, electronic mail, a personal information organizer, a calendar, a to-do list, and a contact manager.

- QuickBackup—This backup program works with Windows 95 and NT 4.0. It enables transparent backup for files to SCSI, ZIP, and JAZ drives. Its icon-driven interface makes for ease of use. The program installs quickly and works well. It provides encryption protection and Internet support.

- McAfee Service Desk—This powerful product is actually multiple products in one box. It works with Windows 3.1, 95, and NT 4.0 to let customer-support personnel have access to information about the customer and make a remote connection to a system reported with a problem. The package comes with the ability to distribute software. The package also includes a system diagnostic part for support personnel to use with customers.

- NetShield—NetShield uses McAfee's proprietary codes, called Code Trace, Code Matrix and Code Poly. The product operates in an NT environment in native mode, taking full advantage of NT's server-client remote task-distribution capability. The product supports real-time scanning while other tasks run.

- WEBScan—This product is designed to detect viruses within a browser. It examines downloads and e-mail attachments, making it a powerful addition to any desktop or laptop system communicating in networks today. It also provides Cybersitter, which blocks out unwanted Web sites and chat groups. Coverage for the program includes examination of .doc, .zip, .exe, .zrc, .arj, and other file types.

- PCCrypto—PCCrypto is used to secure documents and other data files created by anyone using computers. It can encrypt graphics, spreadsheets, and text documents. It uses a 160-bit blowfish encryption mechanism. The package consumes a minimal amount of space and is one of the most powerful, if not *the* most powerful, tool of its kind on the market today.

These products have been implemented to varying degrees on each system.

I recommend you dedicate a system for testing software, initially. Then install McAfee VirusScan and scan each and every diskette you have. That's right. It does not matter if it takes someone two months. Even new diskettes out of the box should be scanned. In answer to the criticism that this is going a little too far, I'll say this: One year (I won't say which) I bought some software. It was new. It came from the original vendor in shrink-wrapped plastic. The shrink-wrapped diskettes were enclosed in a box with a seal on it. The box with a seal on it was shrink-wrapped. I didn't think a thing about it. Within 10 minutes after opening the software, it brought one of my systems to its knees. The diskette had a virus. How do I know? I checked it personally. How do I know the system it went into was clean? It was and is my benchmark system. Furthermore, those who know me know that nobody, not anyone, puts a disk in my systems except me. It took me two days to recover the system. Consider this the next time you stick a new program on diskette in your system. You can pay now or gamble, but remember the odds are against you.

McAfee has other products that might meet your needs. I recommend you contact them. My experience with them has always been pleasant and I have found their staff to be very informative. They can be reached through the following sources:

McAfee
2710 Walsh Avenue
Santa Clara, CA 95051
(408) 988-3832
www.mcafee.com

McAfee Canada
178 Main Street
Unionville, Ontario
Canada L3R 2G9

McAfee France S. A.
50 rue de Londres
75008 Paris
France

McAfee (UK) Ltd.
Hayley House, London Road
Bracknell, Berkshire
GR12 2TH United Kingdom

McAfee Europe B. V.
Orlypein 81 - Busitel 1
1043 DS Amsterdam
The Netherlands

McAfee Deutschland GmbH
Industriestrasse 1
D-82110 Germering
Germany

6.13 Multimedia Components

Creative Labs was chosen as the vendor for multimedia equipment. In the arena of multimedia, Creative Labs wrote the book. Some multimedia clone products exist. Many of these products attempt to copy what Creative Labs has designed. In the arena of multimedia, clones are the incorrect way to invest money. The nature of multimedia software is such that "clone" equipment might not be able to execute all of the necessary functions of multimedia. This might sound strange, but it is true.

Today, systems typically have CD-ROMs, speakers, microphones, line outputs for amplifiers, line inputs for peripheral integration, and software that enables a user to create, playback, and listen to or see various data streams.

All the desktop systems in this network are IBM 350 series. I selected these because they can be customized to deliver a robust workload. Another reason for choosing this series is the upgrade capability. The same is true with Creative Labs equipment. In each system, Creative Lab's equipment is the multimedia hardware and software. One system has a complete package of multimedia equipment from Creative Labs, including an interface board, speakers, cabling, microphone, CD-ROM, infrared remote control, software drivers, and various software titles for viewing and listening.

Creative Labs has designed the benchmark for multimedia systems. The significance of this should not be overlooked during the design phase of your network. Windows 95 and NT 4.0 acknowledges most, if not all, Creative Labs hardware and software. It is plug-n-play compatible. Another significant aspect of this equipment is its adaptability. Creative Labs is continually upgrading its equipment to stay in line with other vendors, but they also support equipment and systems that are not this year's product.

Multimedia is more than a CD-ROM and speakers. Today, multimedia typically encompasses a digital video disc (DVD) and enhanced display support. More than at any other time, displays need powerful drivers and memory to store the screen of information to be presented.

Creative Labs is based in California, but has offices around the world. I recommend contacting the one closest to you for additional information about multimedia products:

Creative Labs
1901 McCarthy Blvd
Milpitas, CA 95035
www.soundblaster.com

Creative Technology Ltd.
67 Ayer Rajah Crescent #03-18
Singapore 0513

Creative Labs Technical Support
1523 Cimarron Plaza
Stillwater, OK 74075

Creative Labs Ltd.
Blanchardstown Industrial Park
Blanardstown
Budlin 15
Ireland

6.14   Network Analyzer

To maintain a network requires a certain type of skill, a marriage between the abstract and practical, if you will. To keep a network operational at its peak is another thing altogether.

Obtaining information about an operational network is important. Being sure the information is accurate is even more important. It is important here to understand precisely what I mean. Figure 6-26 shows a conceptual view of network layers. At the lowest layer in the network is the physical layer. This layer is where the media connects to an interface board of some type. At this level, you need a tester to check cables and verify they are in good working order.

At the data link layer, and below, data link layer protocols operate. These include Ethernet, Token Ring, ATM, FDDI, Fast Ethernet, SDLC, SLIP, Frame Relay, ISDN, and others. At this layer, you need an analyzer to analyze how the network is talking among nodes. In order to understand this, you need to be fairly familiar with the lower-layer protocol that operates in the network.

From the network layer up, upper-layer protocols operate. A sophisticated network analyzer is needed here. Network protocols at the upper layers include TCP/IP, Novell, APPN, SNA, DECnet, AppleTalk, and NetBIOS. You might think that Windows NT should be on the list. The fact is that Windows NT uses one of these protocols to achieve its task.

For this network, the Hewlett Packard Internet Advisor was selected. It can provide a single screen of real-time information about the

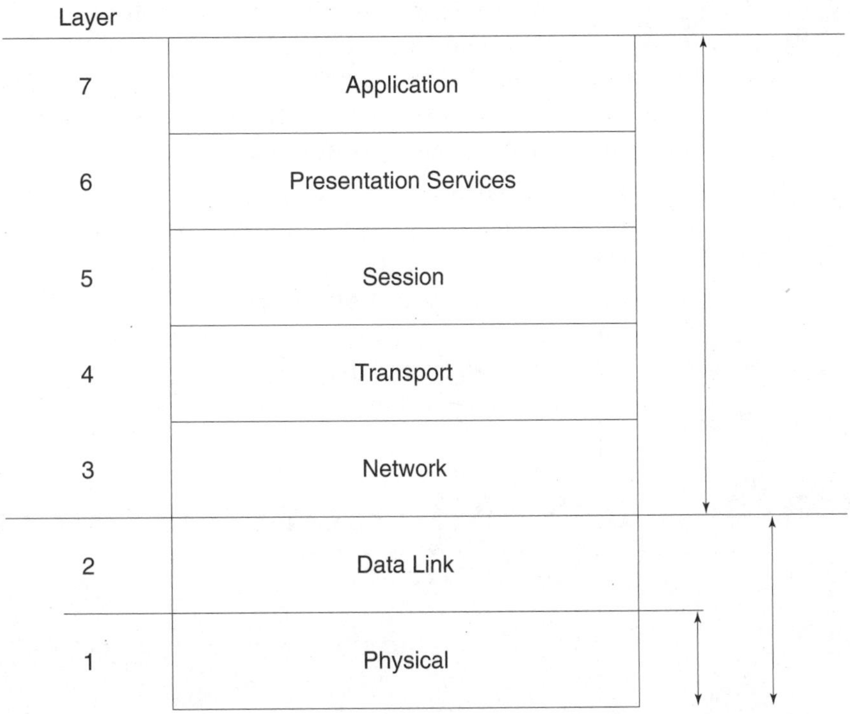

Figure 6-26

A conceptual view of network layers.

Layer	
7	Application
6	Presentation Services
5	Session
4	Transport
3	Network
2	Data Link
1	Physical

network. Frame errors, events, and other vital statistics can be viewed through it.

The Internet Advisor used in this network includes support for Ethernet, Fast Ethernet, and Token Ring. It provides a complete breakdown of the seven-layer protocol stack in the network. Results can be printed and/or saved to a file. A special feature of this device is its ability to select a single protocol and view it; even beyond this is its ability to select a single layer within a given protocol and view it.

One powerful aspect of the Internet Advisor is its ability to run in "promiscuous mode," as I refer to it. It can literally tap into the network and provide a window into network operations, and users never know they are being watched. Of course, technical concerns are the only ones worth watching. The Internet Advisor itself is small but powerful. It has the ability to accommodate all the major interfaces today. The Internet Advisor supports V.35, RS-232, 10BaseT, AUI, Token Ring, and other connectors.

The Internet Advisor comes configured from HP with the parts you need. The component that support Ethernet and Token Ring are combined, making for easy operation in dual-protocol networks. A quick switch of under-cradles makes operation in other network protocols easy. The device itself is very user friendly.

I recommend that you consider this particular device to maintain your network. The device is equal to the task of working with TCP/IP networks. The ease of use, documentation, and human support available make this device easy and fun to use. Contact Hewlett Packard at

Hewlett Packard Company
Colorado Communications Operation
5070 Centennial Blvd
Colorado Springs, CO 80919-2497
(800) 452-4844
www.hp.com/go/infoadvisor

6.15 Miscellaneous Devices and Tools

When you get into the design of your network, you might realize what I did: Some things you will need have to be looked for extensively or created.

External CD-ROM

First, a need in this network existed for an external CD-ROM that could connect via SCSI to the U.S. Robotics hub, laptops, and Hewlett Packard Internet Advisor. Sometimes finding things to make the network run is like the proverbial search for a needle in a haystack.

I selected a Sony CD-ROM, model PRD-650WN. I do not know whether this particular device will be available as you read this. However, should this specific device not be available, I suppose some other vendor will make a SCSI-attachable CD-ROM available. This device works well. The technology is genuine Sony and Adapted for the SCSI PCMCIA card. This device is a must-have for network installers and administrators.

Wire Testers

Another device or devices you will need is a cable tester. You need to be able to check AC cables, data cables of all sorts, RJ45 wire, RJ11 wire, RJ11 wall connectors, and so forth. Do not assume just because something is new that it is good. I have seen perfectly new equipment come right out of a vendor's box bad. No equipment or cable in this network was discovered bad, but that does not mean you should not test cables and connection points.

I selected Fluke meters to provide me with AC test ability. This equipment can also be used to obtain amperage readings, cable test, and so on. I ordered two devices to specifically check cable: one was for RJ45 and the other for RJ11 wire.

Break-out AC Test Cable

I make my own extension cords. Many ask why. If you saw them, you would know why. Any single one of them could be used to supply voltage and current to multiple dryers, ovens, or other high-amperage devices.

Who knows if my break-out AC test wire is UL listed or not, but it works. What I do is this: Buy some 10 gauge SO type wire—two conductor with ground. Connect a regular 110-volt three-prong connector on one end. On the other end, put one or two pairs of regular electrical outlets in a metal box with an enclosed top. Carefully, take about 6 to 8 inches of outer sheath off the conductors. Three conductors should be exposed with insulation. Tape these ends where the cuts were made. Presto! Now you can literally clamp an AMP probe on around the hot wire without going into any danger areas and get the amperage pull off the line.

Am I recommending you do this? Absolutely not. I'm telling you how I did it. This level of information assisted me in the ground-floor design phase of the network. I had to go back and draw four clean lines to the breaker box to accommodate the new equipment.

Again, make it your business to know the electrical part of your network. If you do not have the education and background for this, then get somebody who does. You cannot afford to not have this base covered.

6.16   Summary

Designing a TCP/IP network is challenging. TCP/IP is a powerful network protocol; it can be the single protocol upon which a network is built. The example network here uses TCP/IP as its network protocol.

Network design involves many different things. Electrical, technical, physical, human, and other factors go into network design. The purpose of this chapter was to present you a real network built from scratch that you can use as a baseline to gauge your network plans.

Internet Protocol (IP) Version 4

The Internet Protocol was originally designed for use in interconnected systems of packet-switched computer communication networks. This type of system has been called a *catenet*. Internet Protocol (IP) provides for transmitting blocks of data called *datagrams* from sources to destinations, where sources and destinations are hosts identified by fixed-length addresses. IP also provides for fragmentation and reassembly of long datagrams, if necessary, for transmission through small packet networks.

7.1 IP and its Functions

IP originally limited its purpose to providing the functions necessary to deliver a package of bits (internet datagrams) from a source to a destination through an interconnected group of networks. IP has no mechanisms to augment end-to-end data reliability, flow control, sequencing, or other services commonly found in host-to-host protocols. However, IP can capitalize on the services of its supporting networks to provide various types and qualities of service.

IP is invoked by a host-to-host protocol in the internet environment. This protocol calls on local network protocols to carry the internet datagram to the next gateway or destination host. In this case, a TCP module would call on the IP part to take a TCP segment (the TCP header and user data) as the *data* portion of an internet datagram.

IP implements two basic functions: addressing and fragmentation. Internet modules use the addresses carried in the internet header to transmit internet datagrams toward their destinations. The selection of a path for transmission is called *routing*. Fields in the IP are used to fragment and reassemble internet datagrams when necessary for transmission through small packet-oriented networks.

An internet module resides in each host engaged in internet communication and in each router that interconnects networks. These modules share common rules for interpreting address fields and for fragmenting and assembling internet datagrams. In addition, these modules have procedures for making routing decisions and other functions.

IP treats each internet datagram as an independent entity, unrelated to any other internet datagram. Hence, there are no connections or logical circuits. IP uses four key mechanisms in providing its service: type of service, time to live, options, and header checksum.

Type of Service   Type of service is used to indicate the quality of the service desired. This is an abstract or generalized set of parameters provided in the networks that make up the Internet. This type of service indication is to be used by routers to select the actual transmission parameters for a particular network, the network for the next hop, or the next router when routing an internet datagram.

Time-to-Live   Time-to-live indicates an upper boundary of time an IP datagram has to exist. It is set by the sender of the datagram and reduced at the points along the route where it is processed. If the time-to-live reaches zero before the internet datagram reaches its destination, the internet datagram is destroyed. The time-to-live is like a self-destruct mechanism.

Options   The options provide for control functions needed, useful in some situations but unnecessary for the most common communications. Options make timestamps, security, and special routing possible.

Header Checksum   The header checksum provides a verification that the information used in processing internet datagrams has been transmitted correctly. The data might contain errors. If the header checksum fails, the internet datagram is discarded at once by the entity that detects the error.

IP does not provide a reliable communication facility. There are no acknowledgments either end-to-end or hop-by-hop, neither is there any error control for data, only a header checksum. IP does not perform any retransmissions and has no flow control. Any errors detected are reported via Internet Control Message Protocol (ICMP). ICMP is a required component that accompanies IP.

7.2   IP Operation

Transmitting a datagram from one application program to another is best illustrated by an example. An intermediate router is a premise to this environment example. A sending application program prepares its data and calls on its IP module to send the data as a datagram. It passes the destination address and other parameters as arguments to this call.

The IP module prepares a datagram header and attaches the data to it. The IP module determines a local network address for this internet address. In this case, it is the address of a router.

It sends this datagram and the local network address to the network interface. The network interface, in turn, creates a local network header, and attaches the datagram to it. It then sends the result via the local network.

The datagram arrives at a router wrapped in the local network header. The local network interface strips off this header and turns the datagram over to the internet module. The internet module determines from the internet address that the datagram is to be forwarded to another host in a second network. The internet module determines a local net address for the destination host. It calls on the local network interface for that network to send the datagram.

This local network interface creates a local network header and attaches the datagram, sending the result to the destination host. Here, the destination host datagram is stripped of the local net header by the local network interface and handed to the internet module. The internet module determines that the datagram is for an application program in this host. It passes the data to the application program in response to a system call, passing the source address and other parameters as results of the call.

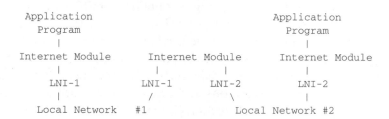

The purpose of IP is to move datagrams through a set of networks. Datagrams are passed from one internet module to another until the destination is reached. The internet modules reside in hosts and routers in the internet. Datagrams are routed from one internet module to another through individual networks based on the interpretation of an internet address. Thus, one important mechanism of the Internet Protocol is the internet address.

Because datagrams may have to traverse multiple intranetworks within the Internet, the routing of messages from one internet module to another may be achieved by *fragmentation*. Fragmentation is necessary to traverse a network whose maximum packet size is smaller than the size of the datagram.

Fragmentation

Fragmentation of an internet datagram is necessary when it originates in a local network that allows a large packet size and must traverse a local network that limits packets to a smaller size to reach its destination.

An internet datagram can be marked "don't fragment." Any internet datagram so marked is not to be fragmented under any circumstances. If an internet datagram marked "don't fragment" cannot be delivered to its destination without fragmenting, it is discarded instead. Fragmentation, transmission, and reassembly across a local network that is invisible to the IP module is called *intranet fragmentation*, and may be used.

The internet fragmentation and reassembly procedure needs to be able to break a datagram into an almost arbitrary number of pieces that can be later reassembled. The receiver of the fragments uses the identification field to ensure that fragments of different datagrams are not mixed. The fragment offset field tells the receiver the position of a fragment in the original datagram. The fragment offset and length determine the portion of the original datagram covered by this fragment. The more-fragments flag indicates (by being reset) the last fragment. These fields provide sufficient information to reassemble datagrams.

The identification field is used to distinguish the fragments of one datagram from those of another. The originating protocol module of an internet datagram sets the identification field to a value that must be unique for that source-destination pair and protocol for the time the datagram will be active in the internet system. The originating protocol module of a complete datagram sets the more-fragments flag to zero and the fragment offset to zero.

To fragment a long internet datagram, an IP module creates two new internet datagrams and copies the contents of the internet header fields from the long datagram into both new headers. The data of the long datagram is divided into two portions on an 8-octet (64-bit) boundary. The second portion might not be an integral multiple of eight octets, but the first must be. Call the number of 8-octet blocks in the first portion Number of Fragment Blocks (NFB). The first portion of the data is placed in the first new datagram, and the total length field is set to the length of the first datagram. The more-fragments flag is set to one. The second portion of the data is placed in the second new datagram, and the total length field is set to the length of the second datagram. The more-fragments flag carries the same value as the long datagram. The fragment offset field of the second new internet datagram is set to the value of that field in the long datagram, plus NFB.

This procedure can be generalized for an n-way split, rather than the two-way split described. To assemble the fragments of an internet datagram, an IP module (for example, at a destination host) combines datagrams that all have the same value for the following four fields:

- Identification
- Source
- Destination
- Protocol

The combination is done by placing the data portion of each fragment in the relative position indicated by the fragment offset in that fragment's internet header. The first fragment will have the fragment offset zero, and the last fragment will have the more-fragments flag reset to zero.

Addressing

A distinction is made between names, addresses, and route. A *name* indicates what we seek. An *address* indicates where it is. A *route* indicates how to get there. The Internet Protocol deals primarily with addresses. It is the task of higher-level protocols to make the mapping from names to addresses. The internet module maps internet addresses to local network addresses. It is the task of lower-level procedures (i.e., local networks or gateways) to make the mapping from local addresses to routes.

Addresses have a fixed length of four octets (32 bits). An address begins with a network number, followed by a local address (called the *rest* field). There are three formats or classes of internet addresses:

Class A The high-order bit is zero, the next seven bits are the network, and the last 24 bits are the local address.

Class B The two high-order bits are one-zero, the next 14 bits are the network, and the last 16 bits are the local address.

Class C The three high-order bits are one-one-zero, the next 21 bits are the network, and the last eight bits are the local address.

Care must be taken in mapping internet addresses to local addresses; a single physical host must be able to act as if it were several distinct hosts to the extent of using several distinct internet addresses. Some hosts will also have several physical interfaces; this is also called a *multi-homed host.* Provision must also be made for a host to have several physical interfaces to the network, with each having several logical internet addresses.

7.3 IP-Related Terminology

IP has terms related with it that have specific meanings. Sometimes these terms are misunderstood; for that reason, I am including this list for your reference here.

ARPANET Leader The control information on an ARPANET message at the host-IMP interface.

ARPANET Message The unit of transmission between a host and an IMP in the ARPANET. The maximum size is about 1012 octets (8096 bits).

ARPANET Packet A unit of transmission used internally in the ARPANET between IMPs. The maximum size is about 126 octets (1008 bits).

Destination The destination address, an internet header field.

DF The Don't Fragment bit carried in the flags field.

Flags An internet header field carrying various control flags.

Fragment Offset An internet header field indicating where in the datagram a fragment belongs.

GGP An abbreviation for *Gateway to Gateway Protocol,* the protocol used primarily between gateways to control routing and other gateway functions.

Header Control information at the beginning of a message, segment, datagram, packet, or block of data.

ICMP An abbreviation for *Internet Control Message Protocol.* Implemented in the internet module, the ICMP is used from gateways to hosts and between hosts to report errors and make routing suggestions.

Identification An internet header field carrying the identifying value assigned by the sender to aid in assembling the fragments of a datagram.

IHL An abbreviation for *Internet Header Length*, an internet header field that holds the length of the header, measured in 32-bit words.

IMP An abbreviation for *Interface Message Processor*, the packet switch of the ARPANET.

Internet Address A 4-octet (32-bit) source or destination address consisting of a network field and a local address field.

Internet Datagram The unit of data exchanged between a pair of internet modules (includes the header).

Internet Fragment A portion of the data of an internet datagram with an internet header.

Local Address The address of a host within a network. The actual mapping of an internet local address onto the host addresses in a network is quite general, allowing for many-to-one mappings.

MF The "more-fragments" flag, carried in the internet header flags field.

Module An implementation, usually in software, of a protocol or other procedure.

More-Fragments Flag A flag indicating whether or not this datagram contains the end of an internet datagram, carried in the internet header flags field.

NFB An abbreviation for *Number of Fragment Blocks*, the length of a portion of an internet fragment measured in 8-octet units.

Octet An 8-bit byte.

Options An internet header field that may contain several options. Each option may be several octets in length.

Padding The internet header field used to ensure that the data begins on a 32-bit word boundary. The padding is zero.

Protocol In this sense, the next-higher-level protocol identifier, an internet header field.

Rest The local address portion of an internet address.

Source The source address, an internet header field.

TCP An abbreviation for *Transmission Control Protocol*, a host-to-host protocol for reliable communication in internet environments.

TCP Segment The unit of data exchanged between TCP modules (including the TCP header).

TFTP An abbreviation for *Trivial File Transfer Protocol*, a simple file-transfer protocol built on UDP.

Time to Live An internet header field that indicates the upper bound on how long this datagram may exist.

TOS An abbreviation for *Type of Service*.

Total Length An internet header field for the length of the datagram in octets, including internet header and data.

TTL An abbreviation for *Time to Live*.

Type of Service An internet header field that indicates the type (or quality) of service for this datagram.

UDP An abbreviation for *User Datagram Protocol*, a user-level protocol for transaction-oriented applications.

User The user of the Internet Protocol. This may be a higher-level protocol module, an application program, or a gateway program.

Version A field that indicates the format of the internet header.

7.4 Routers and IP

Routers implement Internet Protocol to forward datagrams between networks. Routers also implement the Gateway-to-Gateway Protocol

(GGP) to coordinate routing and other internet control information. (Routers used to be called gateways.)

In a router, the higher-level protocols need not be implemented, and the GGP functions are added to the IP module.

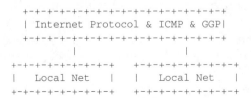

```
        +-+-+-+-+-+-+-+-+-+-+-+-+-+-+-+-+
        | Internet Protocol & ICMP & GGP|
        +-+-+-+-+-+-+-+-+-+-+-+-+-+-+-+-+
                 |                |
+-+-+-+-+-+-+-+-+    +-+-+-+-+-+-+-+-+
|  Local Net   |    |  Local Net   |
+-+-+-+-+-+-+-+-+    +-+-+-+-+-+-+-+-+
```

7.5   IP Header Format

The contents of the internet header, with explanations of its fields, follows:

```
0                   1                   2                   3
0 1 2 3 4 5 6 7 8 9 0 1 2 3 4 5 6 7 8 9 0 1 2 3 4 5 6 7 8 9 0 1
+-+-+-+-+-+-+-+-+-+-+-+-+-+-+-+-+-+-+-+-+-+-+-+-+-+-+-+-+-+-+-+-+
|Version|  IHL  |Type of Service|          Total Length         |
+-+-+-+-+-+-+-+-+-+-+-+-+-+-+-+-+-+-+-+-+-+-+-+-+-+-+-+-+-+-+-+-+
|         Identification        |Flags|      Fragment Offset    |
+-+-+-+-+-+-+-+-+-+-+-+-+-+-+-+-+-+-+-+-+-+-+-+-+-+-+-+-+-+-+-+-+
|  Time to Live |    Protocol   |         Header Checksum        |
+-+-+-+-+-+-+-+-+-+-+-+-+-+-+-+-+-+-+-+-+-+-+-+-+-+-+-+-+-+-+-+-+
|                         Source Address                         |
+-+-+-+-+-+-+-+-+-+-+-+-+-+-+-+-+-+-+-+-+-+-+-+-+-+-+-+-+-+-+-+-+
|                      Destination Address                       |
+-+-+-+-+-+-+-+-+-+-+-+-+-+-+-+-+-+-+-+-+-+-+-+-+-+-+-+-+-+-+-+-+
|                    Options                    |    Padding     |
+-+-+-+-+-+-+-+-+-+-+-+-+-+-+-+-+-+-+-+-+-+-+-+-+-+-+-+-+-+-+-+-+
```

Version: 4 bits   The Version field indicates the format of the internet header. This document describes version 4.

IHL: 4 bits   The Internet Header Length is the length of the header in 32-bit words, and thus points to the beginning of the data. Note that the minimum value for a correct header is 5.

Type of Service: 8 bits   Type of Service provides an indication of the abstract parameters of the quality of service desired. These parameters are to be used to guide the selection of the actual service parameters when transmitting a datagram through a particular network.

Several networks offer service precedence, which somehow treats high-precedence traffic as more important than other traffic (generally by accepting only traffic above a certain precedence at time of high load). The major choice is a three-way tradeoff between low delay, high reliability, and high throughput.

Bits 0-2:   Precedence

Bit 3:   0 = Normal delay, 1 = Low delay

Bit 4:   0 = Normal throughput, 1 = High throughput

Bit 5:   0 = Normal reliability, 1 = High reliability

Bits 6-7:   Reserved for future use

```
   0     1     2     3     4     5     6     7
+-+-+-+-+-+-+-+-+-+-+-+-+-+-+-+-+-+-+-+-+-+-+-+
|                 |     |     |     |     |     |
|   PRECEDENCE    |  D  |  T  |  R  |  0  |  0  |
|                 |     |     |     |     |     |
+-+-+-+-+-+-+-+-+-+-+-+-+-+-+-+-+-+-+-+-+-+-+-+
```

Precedence

111   Network Control

101   Internetwork Control

101   CRITIC/ECP

100   Flash Override

011   Flash

010   Immediate

001   Priority

000   Routine

Delay, throughput, and reliability indications might increase the cost (in some sense) of the service. In many networks, better performance for one of these parameters is coupled with worse performance on another. Except for very unusual cases, at most two of these three indications should be set. The type of service is used to specify the treatment of the datagram during its transmission through the internet system.

The Network Control precedence designation is intended to be used within a network only. The actual use and control of that designation is up to each network. The Internetwork Control designation is intended for use by gateway-control originators only. If the actual use of these precedence designations is of concern to a particular network, it is the responsibility of that network to control the access to, and use of, those precedence designations.

Total Length: 16 bits   Total Length is the length of the datagram, measured in octets, including internet header and data. This field allows the length of a datagram to be up to 65,535 octets. Such long datagrams are impractical for most hosts and networks. All hosts must be prepared to accept datagrams of up to 576 octets (whether they arrive whole or in fragments). It is recommended that hosts only send datagrams larger than 576 octets if they have assurance that the destination is prepared to accept the larger datagrams.

The number 576 is selected to allow a reasonably sized data block to be transmitted in addition to the required header information. For example, this size allows a data block of 512 octets plus 64 header octets to fit in a datagram. The maximal internet header is 60 octets, and a typical header is 20 octets, allowing a margin for headers of higher-level protocols.

Identification: 16 bits   An identifying value is assigned by the sender to aid in assembling the fragments of a datagram.

Flags: 3 bits   Here are three control flags:

Bit 0:   Reserved, must be zero

Bit 1:   (DF) 0 = May fragment, 1 = Don't fragment

Bit 2:   (MF) 0 = Last fragment, 1 = More fragments

Fragment Offset: 13 bits   The Fragment Offset field indicates where in the datagram this fragment belongs. The offset is measured in units of 8 octets (64 bits). The first fragment has an offset of zero.

Time to Live: 8 bits   The TTL field indicates the maximum time the datagram is allowed to remain in the internet system. If this field contains the value zero, then the datagram must be destroyed. This field is modified in header processing. The time is measured in units of seconds, but since every module that processes a datagram must decrease the TTL by at least one even if it processes the datagram in less than a second, the TTL must be thought of only as an upper bound on the time a datagram may exist. The intention is to cause undeliverable datagrams to be discarded, and to bound the maximum datagram lifetime.

Protocol: 8 bits   This field indicates the next level protocol used in the data portion of the internet datagram.

Header Checksum: 16 bits   A checksum is computed on the header only. Since some header fields change (e.g., Time to Live), this is recomputed and verified at each point that the internet header is processed.

The checksum field is the 16-bit one's complement of the one's complement sum of all 16-bit words in the header. For purposes of computing the checksum, the value of the checksum field is zero. This is an easy-to-compute checksum, and experimental evidence indicates it is adequate. However, it is provisional and may be replaced by a CRC procedure, depending on further experience.

Source and Destination Address: 32 bits each   The source and destination addresses are self-explanatory.

Options: Variable   Options may appear or not in datagrams. They must be implemented by all IP modules (host and routers). Their transmission in any particular datagram is optional, but not their implementation.

In some environments, the security option might be required in all datagrams. The option field is variable in length. There may be zero or more options. There are two ways to format an option: a single octet of option-type, or an option-type octet, an option-length octet, and the actual option-data octets.

The option-length octet counts the option-type octet and the option-length octet as well as the option-data octets. The option-type octet is viewed as having three fields:

- 1 bit for the copied flag
- 2 bits for the option class
- 5 bits for the option number

The copied flag indicates that this option is copied into all fragments on fragmentation. A zero indicates it is not copied, while a one indicates it is copied. Option classes are

- ■ 0 = Control
- ■ 1 = Reserved for future use
- ■ 2 = Debugging and measurement
- ■ 3 = Reserved for future use

Internet options are defined as follows:

Class	Number	Length	Description
0	0	--	End of Option list. This option occupies only one octet; it has no length octet.
0	1	--	No Operation. This option occupies only one octet; it has no length octet.
0	2	11	Security. Used to carry security, compartmentation, user group (TCC), and handling restriction codes compatible with DoD requirements.
0	3	Varies	Loose Source Routing. Used to route the internet datagram based on information supplied by the source.
0	9	Varies	Strict Source Routing. Used to route the internet datagram based on information supplied by the source.
0	7	Varies	Record Route. Used to trace the route an internet datagram takes.
0	8	4	Stream ID. Used to carry the stream identifier.
2	4	Varies	Internet Timestamp.

Specific Option Definitions

End of Option List This option indicates the end of the option list. This might not coincide with the end of the internet header according to the internet header length. This is used at the end of all options, not the end of each option, and needs to be used only if the end of the options would not otherwise coincide with the end of the header. It may be copied, introduced, or deleted on fragmentation, or for any other reason.

```
+-+-+-+-+-+-
|00000000|
+-+-+-+-+-+-
  Type=0
```

No Operation

```
+-+-+-+-+-+-
|00000001|
+-+-+-+-+-+-
  Type=1
```

This option may be used between options, for example, to align the beginning of a subsequent option on a 32-bit boundary. It may be copied, introduced, or deleted on fragmentation, or for any other reason.

Security This option provides a way for hosts to send security, compartmentation, handling restrictions, and TCC (closed user group) parameters. The format for this option is as follows:

```
+-+-+-+-+-+-+-+-+-+-+-+-//-+-+-+-//-+-+-+-//-+-+-+-//-+-+
|10000010|00001011|SSS  SSS|CCC  CCC|HHH  HHH|  TCC    |
+-+-+-+-+-+-+-+-+-+-+-+-//-+-+-+-//-+-+-+-//-+-+-+-//-+-+
Type=130 Length=11
```

Security (S field): 16 bits Specifies one of 16 levels of security (eight of which are reserved for future use):

Field	Security Level
00000000 00000000	Unclassified
11110001 00110101	Confidential
01111000 10011010	EFTO
10111100 01001101	MMMM
101011110 00100110	PROG
10101111 00010011	Restricted

Field	Security Level
11010111 10001000	Secret
01101011 11000101	Top Secret
00110101 11100010	Reserved for future use
10011010 11110001	Reserved for future use
01001101 01111000	Reserved for future use
00100100 10111101	Reserved for future use
00010011 01011110	Reserved for future use
10001001 10101111	Reserved for future use
11000100 11010110	Reserved for future use
11100010 01101011	Reserved for future use

Compartments (C field): 16 bits   The C field is an all-zero value used when the information transmitted is not compartmented. Other values for this field may be obtained from the Defense Intelligence Agency.

Handling Restrictions (H field): 16 bits   The values for the control and release markings are alphanumeric digraphs and are defined in the Defense Intelligence Agency Manual DIAM 65-19, "Standard Security Markings."

Transmission Control Code (TCC field): 24 bits   The TCC field provides a means to segregate traffic and define controlled communities of interest among subscribers. The TCC values are trigraphs, and are available from HQ DCA Code 530. They must be copied on fragmentation. This option appears at most once in a datagram.

Loose Source and Record Route

```
+-+-+-+-+-+-+-+-+-+-+-+-+-+-+-+-+-+-+//-+-+-+-+-
|10000011| length | pointer|    route data    |
+-+-+-+-+-+-+-+-+-+-+-+-+-+-+-+-+-+-+//-+-+-+-+-
Type=131
```

The loose source and record route (LSRR) option provides a means for the source of an internet datagram to supply routing information to be used by the gateways in forwarding the datagram to the destination, and to record the route information.

The option begins with the option type code. The second octet is the option length, which includes the option type code and the length octet, the pointer octet, and length-3 octets of route data. The third octet is the pointer into the route data, indicating the octet that begins the next source address to be processed. The pointer is relative to this option, and the smallest legal value for the pointer is four.

Route data is composed of a series of internet addresses. Each address is 32 bits or 4 octets. If the pointer is greater than the length, the source route is empty (and the recorded route full), and the routing is to be based on the destination address field. If the address in the destination address field has been reached and the pointer is not greater than the length, the next address in the source route replaces the address in the destination address field, the recorded route address replaces the source address just used, and the pointer is increased by four. The recorded route address is the internet module's own internet address as known in the environment into which this datagram is being forwarded.

This procedure of replacing the source route with the recorded route (although in the reverse of the order it must be in to be used as a source route) means the option, and the IP header as a whole, remains a constant length as the datagram progresses through the internet.

This option is a loose-source route because the gateway or host IP is allowed to use any route of any number of other intermediate gateways to reach the next address in the route. It must be copied on fragmentation and appears at most once in a datagram.

Strict Source and Record Route

```
+-+-+-+-+-+-+-+-+-+-+-+-+-+-+-+-+-+//-+-+-+-+-
|10001001| length | pointer|     route data    |
+-+-+-+-+-+-+-+-+-+-+-+-+-+-+-+-+-+//-+-+-+-+-
Type=137
```

The strict source and record route (SSRR) option provides a means for the source of an internet datagram to supply routing information to be used by the gateways in forwarding the datagram to the destination, and to record the route information.

The option begins with the option type code. The second octet is the option length, which includes the option type code and the length octet, the pointer octet, and length-3 octets of route data. The third octet is the pointer into the route data, indicating the octet that begins the next source address to be processed. The pointer is relative to this option, and the smallest legal value for the pointer is four.

A route data is composed of a series of internet addresses. Each address is 32 bits or 4 octets. If the pointer is greater than the length, the source route is empty (and the recorded route full), and the routing is to be based on the destination address field.

If the address in the destination address field has been reached and the pointer is not greater than the length, the next address in the source route replaces the address in the destination address field, the recorded route address replaces the source address just used, and pointer is increased by four. The recorded route address is the internet module's own internet address, as known in the environment into which this datagram is being forwarded.

This procedure of replacing the source route with the recorded route (although in the reverse of the order it must be in to be used as a source route) means the option, and the IP header as a whole, remains a constant length as the datagram progresses through the internet.

This option is a strict-source route because the gateway or host IP must send the datagram directly to the next address in the source route through only the directly connected network indicated in the next address to reach the next gateway or host specified in the route. It must be copied on fragmentation and appears at most once in a datagram.

Record Route

```
+-+-+-+-+-+-+-+-+-+-+-+-+-+-+-+-+-+-+-+-+//-+-+-+-+-
|00000111| length | pointer|      route data      |
+-+-+-+-+-+-+-+-+-+-+-+-+-+-+-+-+-+-+-+-+//-+-+-+-+-
Type=7
```

The record route option provides a means to record the route of an internet datagram. The option begins with the option type code. The second octet is the option length, which includes the option type code and the length octet, the pointer octet, and length-3 octets of route data. The third octet is the pointer into the route data, indicating the octet that begins the next area to store a route address. The pointer is relative to this option, and the smallest legal value for the pointer is four.

A recorded route is composed of a series of internet addresses. Each address is 32 bits or 4 octets. The originating host must compose this option with a large enough route data area to hold all the address expected. The size of the option does not change due to adding addresses. The initial contents of the route data area must be zero.

When an internet module routes a datagram, it checks to see if the record route option is present. If it is, it inserts its own internet address (as known in the environment into which this datagram is being forwarded) into the recorded route, beginning at the octet indicated by the pointer, and increments the pointer by four. If the route data area is already full (the pointer exceeds the length), the datagram is forwarded without inserting the address into the recorded route. If there is some room, but not enough room for a full address to be inserted, the original datagram is considered to be in error and is discarded. In either case, an ICMP parameter problem message may be sent to the source host.

It is not copied on fragmentation, goes in the first fragment only, and appears at most once in a datagram.

Stream Identifier

```
+-+-+-+-+-+-+-+-+-+-+-+-+-+-+-+-+
|10001000|00000010|    Stream ID    |
+-+-+-+-+-+-+-+-+-+-+-+-+-+-+-+-+
Type=136 Length=4
```

This option provides a way for the 16-bit SATNET stream identifier to be carried through networks that do not support the stream concept. It must be copied on fragmentation and appears at most once in a datagram.

7.6   Internet Timestamp

The internet timestamp is an integral tool in the implementation and use of IP. The internet timestamp appears as follows:
The Option Length is the number of octets in the option, counting the type, length, pointer, and overflow/flag octets (maximum length 40). The Pointer is the number of octets from the beginning of this option to the end of timestamp, plus one (i.e., it points to the octet beginning the space for the next timestamp). The smallest legal value is 5. The timestamp area is full when the pointer is greater than the length.

```
+-+-+-+-+-+-+-+-+-+-+-+-+-+-+-+-+-+-+
|01000100| length | pointer|oflw|flg|
+-+-+-+-+-+-+-+-+-+-+-+-+-+-+-+-+-+-+
|           internet address        |
+-+-+-+-+-+-+-+-+-+-+-+-+-+-+-+-+-+-+
|              timestamp            |
+-+-+-+-+-+-+-+-+-+-+-+-+-+-+-+-+-+-+
|                    .              |
                     .
                     .
```

Type = 68

The 4-bit Overflow (oflw) is the number of IP modules that cannot register timestamps due to lack of space.

The 4-bit Flag (flg) values are as follows:

■ 0 = Timestamps only are stored in consecutive 32-bit words.

■ 1 = Each timestamp is preceded by an internet address of the registering entity.

■ 3 = The internet address fields are pre-specified.

An IP module only registers its timestamp if it matches its own address with the next specified internet address.

The timestamp is a right-justified, 32-bit number in milliseconds since midnight UT. If the time is not available in milliseconds or cannot be provided with respect to midnight UT, then any time may be inserted as a timestamp, provided the high-order bit of the timestamp field is set to one to indicate the use of a nonstandard value.

The originating host must compose this option with a large enough data area to hold all the timestamp information expected. The size of the option does not change due to adding timestamps. The initial contents of the timestamp data area must be zero or internet address/zero pairs.

If the timestamp data area is already full (the pointer exceeds the length), the datagram is forwarded without inserting the timestamp, but the overflow count is incremented by one. If there is some room, but not enough room for a full timestamp to be inserted, or the overflow count itself overflows, the original datagram is considered to be in error and is discarded. In either case, an ICMP parameter problem message may be sent to the source host.

The timestamp option is not copied upon fragmentation. It is carried in the first fragment and appears at most once in a datagram.

Padding: variable The internet header padding is used to ensure that the header ends on a 32-bit boundary. The padding is zero.

Fragmentation and Reassembly

The internet Identification field (ID) is used together with the source and destination address, and the protocol fields, to identify datagram fragments for reassembly.

The More Fragments flag bit (MF) is set if the datagram is not the last fragment. The Fragment Offset field identifies the fragment location, relative to the beginning of the original unfragmented datagram. Fragments are counted in units of 8 octets. The fragmentation strategy is designed so that an unfragmented datagram has all zero fragmentation information (MF = 0, fragment offset = 0). If an internet datagram is fragmented, its data portion must be broken on 8-octet boundaries.

This format allows 2^{13}, or 8192, fragments of 8 octets each, for a total of 65,536 octets. Note that this is consistent with the datagram total length field (of course, the header is counted in the total length and not in the fragments). When fragmentation occurs, some options are copied, but others remain with the first fragment only. Every internet module must be able to forward a datagram of 68 octets without further fragmentation. This is because an internet header may be up to 60 octets, and the minimum fragment is 8 octets. Every internet destination must be able to receive a datagram of 576 octets, either in one piece or in fragments to be reassembled.

The fields which may be affected by fragmentation include the following:

- Options field
- More-Fragments flag
- Fragment offset
- Internet Header Length field
- Total Length field
- Header checksum

If the DF (Don't Fragment) flag bit is set, then internet fragmentation of this datagram is not permitted, although it may be discarded. This can be used to prohibit fragmentation in cases where the receiving host does not have sufficient resources to reassemble internet fragments.

One example of use of the DF flag is to down-line load a small host. A small host could have a bootstrap program that accepts a datagram, stores it in memory, and then executes it.

The fragmentation and reassembly procedures are most easily described by examples, given as "psuedo-programs" in the following sections. General notation in the pseudo-programs:

Symbol	Meaning
=<	Less than or equal
#	Not equal
=	Equal
<-	Is set to
x to *y*	Includes x and excludes y, for example, "4 to 7" would include 4, 5, and 6, but not 7

Fragmentation Procedure Example

The maximum-size datagram that can be transmitted through the next network is called the *maximum transmission unit* (MTU). If the total length is less than or equal to the MTU, then submit this datagram to the next step in datagram processing; otherwise, cut the datagram into two fragments, the first fragment being the maximum size, and the second fragment being the rest of the datagram. The first fragment is submitted to the next step in datagram processing, while the second fragment is submitted to this procedure in case it is still too large.

Notation reference:

- FO—Fragment Offset
- IHL—Internet Header Length
- DF—Don't Fragment flag
- MF—More Fragments flag
- TL—Total Length
- OFO—Old Fragment Offset
- OIHL—Old Internet Header Length
- OMF—Old More Fragments flag
- OTL—Old Total Length
- NFB—Number of Fragment Blocks
- MTU—Maximum Transmission Unit

Procedure:
  IF TL =< MTU
    THEN submit this datagram to the next step in datagram processing
  ELSE IF DF = 1
    THEN discard the datagram ELSE

To produce the first fragment:

1. Copy the original internet header.
2. OIHL <- IHL; OTL <- TL; OFO <- FO; OMF <- MF
3. NFB <- (MTU-IHL*4)/8
4. Attach the first NFB*8 data octets.
5. Correct the header:

    MF <- 1; TL <- (IHL*4)+(NFB*8); recompute checksum

6. Submit this fragment to the next step in datagram processing.

    To produce the second fragment:

7. Selectively copy the internet header. (Some options are not copied, see option definitions.)
8. Append the remaining data.
9. Correct the header:

    IHL <- (((OIHL*4)-(length of options not copied))+3)/4; TL <- OTL - NFB*8 - (OIHL-IHL)*4); FO <- OFO + NFB; MF <- OMF; recompute checksum

10. Submit this fragment to the fragmentation test; DONE.

In this procedure, each fragment (except the last) was made the maximum allowable size. An alternative might produce less than the maximum size datagrams. For example, you could implement a fragmentation procedure that repeatedly divided large datagrams in half until the resulting fragments were less than the maximum transmission-unit size.

Reassembly Procedure Example

For each datagram, the buffer identifier is computed as the concatenation of the source, destination, protocol, and identification fields. If this is a whole datagram (that is, both the fragment offset and the more fragments fields are zero), then any reassembly resources associated with this buffer identifier are released and the datagram is forwarded to the next step in datagram processing.

If no other fragment with this buffer identifier is on hand, then reassembly resources are allocated. The reassembly resources consist of a data buffer, a header buffer, a fragment-block bit table, a total data length field, and a timer. The data from the fragment is placed in the data buffer according to its fragment offset and length, and bits are set in the fragment-block bit table corresponding to the fragment blocks received.

If this is the first fragment (that is, the fragment offset is zero), this header is placed in the header buffer. If this is the last fragment (the More Fragments field is zero), the total data length is computed. If this fragment completes the datagram (tested by checking the bits set in the fragment block table), then the datagram is sent to the next step in datagram processing; otherwise the timer is set to the maximum of the current timer value and the value of the Time to Live field from this fragment, and the reassembly routine gives up control.

If the timer runs out, all reassembly resources for this buffer identifier are released. The initial setting of the timer is a lower bound on the reassembly waiting time. This is because the waiting time will be increased if the Time to Live in the arriving fragment is greater than the current timer value but will not be decreased if it is less. The maximum this timer value could reach is the maximum Time to Live (approximately 4.25 minutes).

The current recommendation for the initial timer setting is 15 seconds. This may be changed as experience with this protocol accumulates. Note that the choice of this parameter value is related to the buffer capacity available and the data rate of the transmission medium; that is, data rate times timer value equals buffer size ($10\text{Kb/s} \times 15\text{s} = 150\text{Kb}$).

Notation Reference:

- FO—Fragment Offset
- IHL—Internet Header Length
- MF—More Fragments flag
- TTL—Time To Live
- NFB—Number of Fragment Blocks
- TL—Total Length
- TDL—Total Data Length
- BUFID—Buffer Identifier
- RCVBT—Fragment Received Bit Table
- TLB—Timer Lower Bound

Procedure:

1. BUFID <- source ¦ destination ¦ protocol ¦ identification

2. IF FO = 0 AND MF = 0

3. THEN IF buffer with BUFID is allocated

4. THEN flush all reassembly for this BUFID

5. Submit datagram to next step; DONE.

6. ELSE IF no buffer with BUFID is allocated

7. THEN allocate reassembly resources with BUFID; TIMER <- TLB; TDL <- 0

8. Put data from fragment into data buffer with BUFID from octet FO*8 to octet (TL-(IHL*4))+FO*8

9. Set RCVBT bits from FO to FO+((TL-(IHL*4)+7)/8)

10. IF MF = 0 THEN TDL <- TL-(IHL*4)+(FO*8)

11. IF FO = 0 THEN put header in header buffer

12. IF TDL # 0

13. AND all RCVBT bits from 0 to (TDL+7)/8 are set

14. THEN TL <- TDL+(IHL*4)

15. Submit datagram to next step.

16. Free all reassembly resources for this BUFID; DONE.

17. TIMER <- MAX(TIMER,TTL)

18. Give up until next fragment or timer expires.

19. Timer expires: flush all reassembly with this BUFID; DONE.

In the case that two or more fragments contain the same data, either identically or through a partial overlap, this procedure will use the more recently arrived copy in the data buffer and datagram delivered.

Identification

The choice of the identifier for a datagram is based on the need to provide a way to uniquely identify the fragments of a particular datagram. The protocol module for assembling fragments judges fragments to belong to the same datagram if they have the same source, destination, protocol, and identifier. Thus, the sender must choose the identifier to be unique for this source, destination pair, and protocol for the time the datagram (or any fragment of it) could be alive in the internet.

It seems, then, that a sending protocol module needs to keep a table of identifiers, one entry for each destination it has communicated within the last maximum packet lifetime for the internet. However, since the identifier field allows 65,536 different values, some hosts might be able to simply use unique identifiers independent of destination.

It is appropriate for some higher-level protocols to choose the identifier. For example, TCP protocol modules may retransmit an identical TCP segment, and the probability for correct reception would be enhanced if the retransmission carried the same identifier as the original transmission, since fragments of either datagram could be used to construct a correct TCP segment.

Type of Service

The Type of Service field (TOS) is for internet service quality selection. TOS is specified along the abstract parameters precedence, delay, throughput, and reliability:

- Precedence—An independent measure of the importance of this datagram.

- Delay—Prompt delivery is important for datagrams with this indication.

- Throughput—High data rate is important for datagrams with this indication.

- Reliability—A higher level of effort to ensure delivery is important for datagrams with this indication.

These abstract parameters are to be mapped into the actual service parameters of the particular networks that the datagram traverses.

Time to Live

The Time to Live (TTL) is set by the sender to the maximum time the datagram is allowed to be in the internet system. If the datagram is in the internet system longer than the TTL, then the datagram must be destroyed.

This field must be decreased at each point that the internet header is processed to reflect the time spent processing the datagram. Even if no local information is available on the time actually spent, the field must be decremented by one. The time is measured in units of seconds (i.e. the value 1 means one second). Thus, the maximum TTL is 255 seconds, or

4.25 minutes. Since every module that processes a datagram must decrease the TTL by at least one even if it process the datagram in less than a second, the TTL must be thought of only as an upper bound on the time a datagram may exist. The intention is to cause undeliverable datagrams to be discarded, and to bound the maximum datagram lifetime.

Some higher-level reliable connection protocols are based on assumptions that old duplicate datagrams will not arrive after a certain time elapses. The TTL is a way for such protocols to have an assurance that their assumption is met.

Options

The options are optional in each datagram, but required in implementations. That is, the presence or absence of an option is the choice of the sender, but each internet module must be able to parse every option. There can be several options present in the option field.

The options might not end on a 32-bit boundary. The internet header must be filled out with octets of zeros. The first of these would be interpreted as the end-of-options option, and the remainder as internet header padding. Every internet module must be able to act on every option. The Security Option is required if classified, restricted, or compartmented traffic is to be passed.

Checksum

The internet header checksum is recomputed if the header is changed, for example, by a reduction of the Time to Live, additions or changes to internet options, or due to fragmentation. This checksum at the internet level is intended to protect the internet header fields from transmission errors.

There are some applications where a few data-bit errors are acceptable while retransmission delays are not. If the Internet Protocol enforced data correctness, such applications could not be supported.

Errors

Internet Protocol errors may be reported via the ICMP messages.

7.7 Interfaces and IPv4

The functional description of user interfaces to the IP is, at best, fictional, since every operating system will have different facilities. Consequently, I must warn you that different IP implementations may have different user interfaces. However, all IPs must provide a certain minimum set of services to guarantee that all IP implementations can support the same protocol hierarchy. This section specifies the functional interfaces required of all IP implementations.

The Internet Protocol provides interfaces, on one side, to the local network, and on the other side, to either a higher-level protocol or an application program. In the following, the higher-level protocol or application program (or even a gateway program) will be called the "user" since it is using the internet module. Since IP is a datagram protocol, there is minimal memory or state maintained between datagram transmissions. Each call on the IP module by the user supplies all information necessary for the IP to perform the service requested.

Upper-Level Interface Example

The following two example calls satisfy the requirements for the user-to-IP-module communication (where => means returns):

```
SEND (src, dst, prot, TOS, TTL, BufPTR, len, Id,
DF, opt => result)
```

where

- src = Source address
- dst = Destination address
- prot = Protocol
- TOS = Type of service
- TTL = Time to live
- BufPTR = Buffer pointer
- len = Length of buffer
- Id = Identifier
- DF = Don't fragment
- opt = Option data
- result = Response
- OK = Datagram sent ok

- Error = Error in arguments or local network error
- Note that the precedence is included in the TOS, and the security/compartment is passed as an option.

```
RECV (BufPTR, prot, => result, src, dst, TOS,
len, opt)
```

where

- BufPTR = Buffer pointer
- prot = Protocolresult = Response
- OK = Datagram received ok
- Error = Error in arguments
- len = Length of buffer
- src = Source address
- dst = Destination address
- TOS = Type of service
- opt = Option data

When the user sends a datagram, it executes the SEND call, supplying all the arguments. The Internet Protocol module, on receiving this call, checks the arguments and prepares and sends the message. If the arguments are good and the datagram is accepted by the local network, the call returns successfully. If either the arguments are bad or the datagram is not accepted by the local network, the call returns unsuccessfully. On unsuccessful returns, a reasonable report must be made as to the cause of the problem, but the details of such reports are up to individual implementations.

When a datagram arrives at the IP module from the local network, either there is a pending RECV call from the user addressed or there is not. In the first case, the pending call is satisfied by passing the information from the datagram to the user. In the second case, the user addressed is notified of a pending datagram. If the user addressed does not exist, an ICMP error message is returned to the sender, and the data is discarded. The notification of a user may be via a pseudo-interrupt or similar mechanism, as appropriate in the particular operating system environment of the implementation.

A user's RECV call may then either be immediately satisfied by a pending datagram, or the call may be pending until a datagram arrives. The source address is included in the send call in case the sending host has several addresses (multiple physical connections or logical addresses). The IP module must check to see that the source address is one of the legal addresses for this host. An implementation may also allow or require a call to the module to indicate interest in or reserve exclusive use of a class of datagrams (e.g., all those with a certain value in the protocol field).

This section functionally characterizes a USER/IP interface. The notation used is similar to most procedure or function calls in high-level languages, but this usage is not meant to rule out trap-type service calls (e.g., SVCs, UUOs, EMTs), or any other form of interprocess communication.

IPv4 Datagram

This is an example of the minimal data-carrying internet datagram:

```
0                   1                   2                   31
0 1 2 3 4 5 6 7 8 9 0 1 2 3 4 5 6 7 8 9 0 1 2 3 4 5 6 7 8 9 0 1
+-+-+-+-+-+-+-+-+-+-+-+-+-+-+-+-+-+-+-+-+-+-+-+-+-+-+-+-+-+-+-+-+
|Ver= 4 |IHL= 5 |Type of Service|      Total Length = 21       |
+-+-+-+-+-+-+-+-+-+-+-+-+-+-+-+-+-+-+-+-+-+-+-+-+-+-+-+-+-+-+-+-+
|      Identification = 111     |Flg=0|  Fragment Offset = 0   |
+-+-+-+-+-+-+-+-+-+-+-+-+-+-+-+-+-+-+-+-+-+-+-+-+-+-+-+-+-+-+-+-+
|  Time = 123   | Protocol = 1  |      header checksum          |
+-+-+-+-+-+-+-+-+-+-+-+-+-+-+-+-+-+-+-+-+-+-+-+-+-+-+-+-+-+-+-+-+
|                     source address                           |
+-+-+-+-+-+-+-+-+-+-+-+-+-+-+-+-+-+-+-+-+-+-+-+-+-+-+-+-+-+-+-+-+
|                  destination address                         |
+-+-+-+-+-+-+-+-+-+-+-+-+-+-+-+-+-+-+-+-+-+-+-+-+-+-+-+-+-+-+-+-+
|    data       |
+-+-+-+-+-+-+-+-+-+
```

Each tick mark represents one bit position. This internet datagram reflects version 4 of IP. The IP header consists of five 32-bit words, and the total length of the datagram is 21 octets. This datagram is a complete datagram, not a fragment.

IPv4 Datagram Fragment

The following example shows a moderately sized IP datagram (452 data octets), then two internet fragments that might result from the fragmentation of this datagram if the maximum transmission size allowed were 280 octets.

```
 0                   1                   2                   31
 0 1 2 3 4 5 6 7 8 9 0 1 2 3 4 5 6 7 8 9 0 1 2 3 4 5 6 7 8 9 0 1
+-+-+-+-+-+-+-+-+-+-+-+-+-+-+-+-+-+-+-+-+-+-+-+-+-+-+-+-+-+-+-+-+
|Ver= 4 |IHL= 5 |Type of Service|       Total Length = 472      |
+-+-+-+-+-+-+-+-+-+-+-+-+-+-+-+-+-+-+-+-+-+-+-+-+-+-+-+-+-+-+-+-+
|      Identification = 111     |Flg=0|   Fragment Offset = 0   |
+-+-+-+-+-+-+-+-+-+-+-+-+-+-+-+-+-+-+-+-+-+-+-+-+-+-+-+-+-+-+-+-+
|  Time = 123   | Protocol = 6  |         header checksum        |
+-+-+-+-+-+-+-+-+-+-+-+-+-+-+-+-+-+-+-+-+-+-+-+-+-+-+-+-+-+-+-+-+
|                         source address                        |
+-+-+-+-+-+-+-+-+-+-+-+-+-+-+-+-+-+-+-+-+-+-+-+-+-+-+-+-+-+-+-+-+
|                       destination address                     |
+-+-+-+-+-+-+-+-+-+-+-+-+-+-+-+-+-+-+-+-+-+-+-+-+-+-+-+-+-+-+-+-+
|                             data                              |
+-+-+-+-+-+-+-+-+-+-+-+-+-+-+-+-+-+-+-+-+-+-+-+-+-+-+-+-+-+-+-+-+
|                             data                              |
\                                                               \
\                                                               \
|                             data                              |
+-+-+-+-+-+-+-+-+-+-+-+-+-+-+-+-+-+-+-+-+-+-+-+-+-+-+-+-+-+-+-+-+
|             data              |
+-+-+-+-+-+-+-+-+-+-+-+-+-+-+-+-+
```

IPv4 First Datagram Fragment

The following is an example of the first fragment that results from splitting the datagram after 256 data octets:

```
 0                   1                   2                   31
 0 1 2 3 4 5 6 7 8 9 0 1 2 3 4 5 6 7 8 9 0 1 2 3 4 5 6 7 8 9 0 1
+-+-+-+-+-+-+-+-+-+-+-+-+-+-+-+-+-+-+-+-+-+-+-+-+-+-+-+-+-+-+-+-+
|Ver= 4 |IHL= 5 |Type of Service|       Total Length = 276      |
+-+-+-+-+-+-+-+-+-+-+-+-+-+-+-+-+-+-+-+-+-+-+-+-+-+-+-+-+-+-+-+-+
|      Identification = 111     |Flg=1|   Fragment Offset = 0   |
+-+-+-+-+-+-+-+-+-+-+-+-+-+-+-+-+-+-+-+-+-+-+-+-+-+-+-+-+-+-+-+-+
|  Time = 119   | Protocol = 6  |        Header Checksum         |
+-+-+-+-+-+-+-+-+-+-+-+-+-+-+-+-+-+-+-+-+-+-+-+-+-+-+-+-+-+-+-+-+
|                         source address                        |
+-+-+-+-+-+-+-+-+-+-+-+-+-+-+-+-+-+-+-+-+-+-+-+-+-+-+-+-+-+-+-+-+
|                       destination address                     |
+-+-+-+-+-+-+-+-+-+-+-+-+-+-+-+-+-+-+-+-+-+-+-+-+-+-+-+-+-+-+-+-+
|                             data                              |
+-+-+-+-+-+-+-+-+-+-+-+-+-+-+-+-+-+-+-+-+-+-+-+-+-+-+-+-+-+-+-+-+
|                             data                              |
\                                                               \
\                                                               \
|                             data                              |
+-+-+-+-+-+-+-+-+-+-+-+-+-+-+-+-+-+-+-+-+-+-+-+-+-+-+-+-+-+-+-+-+
|                             data                              |
+-+-+-+-+-+-+-+-+-+-+-+-+-+-+-+-+-+-+-+-+-+-+-+-+-+-+-+-+-+-+-+-+
```

IPv4 Second Datagram Fragment

An example of the second fragment is as follows:

```
0                   1                   2                   30
01 2 3 4 5 6 7 8 9 0 1 2 3 4 5 6 7 8 9 0 1 2 3 4 5 6 7 8 9 0 1
+-+-+-+-+-+-+-+-+-+-+-+-+-+-+-+-+-+-+-+-+-+-+-+-+-+-+-+-+-+-+-+
|Ver= 4 |IHL= 5 |Type of Service|      Total Length = 216     |
+-+-+-+-+-+-+-+-+-+-+-+-+-+-+-+-+-+-+-+-+-+-+-+-+-+-+-+-+-+-+-+
|    Identification = 111        |Flg=0| Fragment Offset  = 32 |
+-+-+-+-+-+-+-+-+-+-+-+-+-+-+-+-+-+-+-+-+-+-+-+-+-+-+-+-+-+-+-+
|  Time = 119  | Protocol = 6   |        Header Checksum        |
+-+-+-+-+-+-+-+-+-+-+-+-+-+-+-+-+-+-+-+-+-+-+-+-+-+-+-+-+-+-+-+
|                        source address                        |
+-+-+-+-+-+-+-+-+-+-+-+-+-+-+-+-+-+-+-+-+-+-+-+-+-+-+-+-+-+-+-+
|                      destination address                     |
+-+-+-+-+-+-+-+-+-+-+-+-+-+-+-+-+-+-+-+-+-+-+-+-+-+-+-+-+-+-+-+
|                             data                             |
+-+-+-+-+-+-+-+-+-+-+-+-+-+-+-+-+-+-+-+-+-+-+-+-+-+-+-+-+-+-+-+
|                             data                             |
\                                                              \
\                                                              \
|                             data                             |
+-+-+-+-+-+-+-+-+-+-+-+-+-+-+-+-+-+-+-+-+-+-+-+-+-+-+-+-+-+-+-+
|          data            |
+-+-+-+-+-+-+-+-+-+-+-+-+-+-+
```

IPv4 Datagram with Options

```
0                   1                   2                   30
01 2 3 4 5 6 7 8 9 0 1 2 3 4 5 6 7 8 9 0 1 2 3 4 5 6 7 8 9 0 1
+-+-+-+-+-+-+-+-+-+-+-+-+-+-+-+-+-+-+-+-+-+-+-+-+-+-+-+-+-+-+-+
|Ver= 4 |IHL= 8 |Type of Service|      Total Length = 576     |
+-+-+-+-+-+-+-+-+-+-+-+-+-+-+-+-+-+-+-+-+-+-+-+-+-+-+-+-+-+-+-+
|      Identification = 111      |Flg=0|   Fragment Offset = 0 |
+-+-+-+-+-+-+-+-+-+-+-+-+-+-+-+-+-+-+-+-+-+-+-+-+-+-+-+-+-+-+-+
|  Time = 123  | Protocol = 6   |        Header Checksum        |
+-+-+-+-+-+-+-+-+-+-+-+-+-+-+-+-+-+-+-+-+-+-+-+-+-+-+-+-+-+-+-+
|                        source address                        |
+-+-+-+-+-+-+-+-+-+-+-+-+-+-+-+-+-+-+-+-+-+-+-+-+-+-+-+-+-+-+-+
|                      destination address                     |
+-+-+-+-+-+-+-+-+-+-+-+-+-+-+-+-+-+-+-+-+-+-+-+-+-+-+-+-+-+-+-+
| Opt. Code = x | Opt.  Len.= 3 | option value | Opt. Code = x |
+-+-+-+-+-+-+-+-+-+-+-+-+-+-+-+-+-+-+-+-+-+-+-+-+-+-+-+-+-+-+-+
| Opt. Len. = 4 |         option value         | Opt. Code = 1 |
+-+-+-+-+-+-+-+-+-+-+-+-+-+-+-+-+-+-+-+-+-+-+-+-+-+-+-+-+-+-+-+
| Opt. Code = y | Opt. Len. = 3 | option value | Opt. Code = 0 |
+-+-+-+-+-+-+-+-+-+-+-+-+-+-+-+-+-+-+-+-+-+-+-+-+-+-+-+-+-+-+-+
|                             data                             |
\                                                              \
\                                                              \
|                             data                             |
+-+-+-+-+-+-+-+-+-+-+-+-+-+-+-+-+-+-+-+-+-+-+-+-+-+-+-+-+-+-+-+
|                             data                             |
+-+-+-+-+-+-+-+-+-+-+-+-+-+-+-+-+-+-+-+-+-+-+-+-+-+-+-+-+-+-+-+
```

Order of IP Data Transmission

The order of transmission of the header and data described here is resolved to the octet level. Whenever a diagram shows a group of octets, the order of transmission of those octets is the normal order in which they are read in English. For example, in the following diagram, the octets are transmitted in the order they are numbered:

```
0                   1                   2                   3
0 1 2 3 4 5 6 7 8 9 0 1 2 3 4 5 6 7 8 9 0 1 2 3 4 5 6 7 8 9 0 1
+-+-+-+-+-+-+-+-+-+-+-+-+-+-+-+-+-+-+-+-+-+-+-+-+-+-+-+-+-+-+-+-+
|       1       |       2       |       3       |       4       |
+-+-+-+-+-+-+-+-+-+-+-+-+-+-+-+-+-+-+-+-+-+-+-+-+-+-+-+-+-+-+-+-+
|       5       |       6       |       7       |       8       |
+-+-+-+-+-+-+-+-+-+-+-+-+-+-+-+-+-+-+-+-+-+-+-+-+-+-+-+-+-+-+-+-+
|       9       |      10       |      11       |      12       |
+-+-+-+-+-+-+-+-+-+-+-+-+-+-+-+-+-+-+-+-+-+-+-+-+-+-+-+-+-+-+-+-+
```

Whenever an octet represents a numeric quantity, the leftmost bit in the diagram is the high order, or most significant, bit. That is, the bit labeled 0 is the most significant bit. For example, the following diagram represents the value 170 (decimal):

```
 0 1 2 3 4 5 6 7
+-+-+-+-+-+-+-+-+
|1 0 1 0 1 0 1 0|
+-+-+-+-+-+-+-+-+
```

Whenever a multi-octet field represents a numeric quantity, the left-most bit of the whole field is the most significant bit. When a multi-octet quantity is transmitted, the most significant octet is transmitted first.

Internet Protocol
Version 6 (IPv6)

IP version 6 (IPv6) is a new version of the Internet Protocol, designed as a successor to IP version 4 (IPv4) [RFC-791]. The changes from IPv4 to IPv6 fall primarily into the following categories:

Expanded Addressing Capabilities   IPv6 increases the IP address size from 32 bits to 128 bits, to support more levels of addressing hierarchy, a much greater number of addressable nodes, and simpler auto-configuration of addresses. The scalability of multicast routing is improved by adding a "scope" field to multicast addresses. And a new type of address called an "anycast address" is defined, used to send a packet to any one of a group of nodes.

Header Format Simplification   Some IPv4 header fields have been dropped or made optional, to reduce the common-case processing cost of packet handling and to limit the bandwidth cost of the IPv6 header.

Improved Support for Extensions and Options   Changes in the way IP header options are encoded allows for more efficient forwarding, less stringent limits on the length of options, and greater flexibility for introducing new options in the future.

Flow Labeling Capability   A new capability is added to enable the labeling of packets belonging to particular traffic "flows" for which the sender requests special handling, such as non-default quality of service or "real-time" service.

Authentication and Privacy Capabilities   Extensions to support authentication, data integrity, and data confidentiality are specified for IPv6. This document specifies the basic IPv6 header and the initially-defined IPv6 extension headers and options. It also discusses packet size issues, the semantics of flow labels and priority, and the effects of IPv6 on upper-layer protocols. The format and semantics of IPv6 addresses are specified separately in [RFC-1884]. The IPv6 version of ICMP, which all IPv6 implementations are required to include, is specified in [RFC-1885].

8.1   IPv6 Terminology

The following terms are new in IPv6:

- *node*—A device that implements IPv6.
- *router*—A node that forwards IPv6 packets not explicitly addressed to itself.
- *host*—Any node that is not a router.
- *upper layer*—A protocol layer immediately above IPv6. Examples are transport protocols such as TCP and UDP, control protocols such as ICMP, routing protocols such as OSPF, and Internet or lower-layer protocols being "tunneled" over (i.e., encapsulated in) IPv6, such as IPX, AppleTalk, or IPv6 itself.
- *link*—A communication facility or medium over which nodes can communicate at the link layer, i.e., the layer immediately below IPv6. Examples are Ethernets (simple or bridged); PPP links; X.25, Frame Relay, or ATM networks; and Internet (or higher) layer tunnels, such as tunnels over IPv4 or IPv6 itself.
- *neighbors*—Nodes attached to the same link.
- *interface*—A node's attachment to a link.
- *address*—An IPv6-layer identifier for an interface or a set of interfaces.
- *packet*—An IPv6 header plus payload.
- *link MTU*—The maximum transmission unit, i.e., the maximum packet size, in octets, that can be conveyed in one piece over a link.
- *path MTU*—The minimum link MTU of all the links in a path between a source node and a destination node.

It is possible for a device with multiple interfaces to be configured to forward non-self-destined packets arriving from some set (fewer than all) of its interfaces, and to discard non-self-destined packets arriving from its other interfaces. Such a device must obey the protocol requirements for routers when receiving packets from, and interacting with neighbors over, the former (forwarding) interfaces. It must obey the protocol requirements for hosts when receiving packets from, and interacting with neighbors over, the latter (non-forwarding) interfaces.

8.2 IPv6 Header Format

The following is a representation of the IPv6 header:

```
+-+-+-+-+-+-+-+-+-+-+-+-+-+-+-+-+-+-+-+-+-+-+-+-+-+-+-+-+-+-+-+-+
|Version| Prio. |                Flow Label                     |
+-+-+-+-+-+-+-+-+-+-+-+-+-+-+-+-+-+-+-+-+-+-+-+-+-+-+-+-+-+-+-+-+
|         Payload Length        |  Next Header  |   Hop Limit   |
+-+-+-+-+-+-+-+-+-+-+-+-+-+-+-+-+-+-+-+-+-+-+-+-+-+-+-+-+-+-+-+-+
|                                                               |
+                                                               +
|                                                               |
+                       Source Address                          +
|                                                               |
+                                                               +
|                                                               |
+-+-+-+-+-+-+-+-+-+-+-+-+-+-+-+-+-+-+-+-+-+-+-+-+-+-+-+-+-+-+-+-+
|                                                               |
+                                                               +
|                                                               |
+                     Destination Address                       +
|                                                               |
+                                                               +
|                                                               |
+-+-+-+-+-+-+-+-+-+-+-+-+-+-+-+-+-+-+-+-+-+-+-+-+-+-+-+-+-+-+-+-+
```

- *Version* is the 4-bit Internet Protocol version number, 6.
- *Priority* is a 4-bit priority value.
- *Flow Label* is a 24-bit flow label.
- *Payload Length* is a 16-bit unsigned integer containing the length of payload, i.e., the rest of the packet following the IPv6 header, in octets. A value of zero indicates that the payload length is carried in a Jumbo Payload hop-by-hop option.
- *Next Header* is an 8-bit selector that identifies the type of header immediately following the IPv6 header. It uses the same values as the IPv4 Protocol field.
- *Hop Limit* is an 8-bit unsigned integer that is decremented by one by each node that forwards the packet. The packet is discarded if the Hop Limit is decremented to zero.

■ *Source Address* is a 128-bit address of the originator of the packet.

■ *Destination Address* is a 128-bit address of the intended recipient of the packet, possibly not the ultimate recipient, if a routing header is present.

8.3   IPv6 Extension Headers

In IPv6, optional Internet-layer information is encoded in separate headers that may be placed between the IPv6 header and the upper-layer header in a packet. There are a small number of such extension headers, each identified by a distinct Next Header value. As illustrated in these examples, an IPv6 packet may carry zero, one, or more extension headers, each identified by the Next Header field of the preceding header:

```
+-+-+-+-+-+-+-+-+-+-+-+-+-+-+-+-+-+-+
|  IPv6 header  | TCP header + data
|               |
| Next Header = |
|      TCP      |
+-+-+-+-+-+-+-+-+-+-+-+-+-+-+-+-+-+-+
+-+-+-+-+-+-+-+-+-+-+-+-+-+-+-+-+-++-+-+-+-+-+-+-+-+-+-+-+
|  IPv6 header  | Routing header | TCP header + data
|               |                |
| Next Header = |  Next Header = |
|    Routing    |       TCP      |
+-+-+-+-+-+-+-+-+-+-+-+-+-+-+-+-++-+-+-+-+-+-+-+-+-+-+-+-+
+-+-+-+-++-+-+-+-+-+-+-+-++-+-+-+-+-+-+-+-+-+-+-+-+-+-+-+-
|IPv6 header    | Routing header | Fragment header | fragment of TCP
|               |                |                 | header + data
| Next Header = |  Next Header = |  Next Header =  |
|    Routing    |    Fragment    |       TCP       |
+-+-+-+-+-+-+-+-+-+-+-+-+-+-+-+-++-+-+-+-+-+-+-+-+-+-+-+-+
```

With one exception, extension headers are not examined or processed by any node along a packet's delivery path until the packet reaches the node (or each of the set of nodes, in the case of multicast) identified in the Destination Address field of the IPv6 header. There, normal demultiplexing on the Next Header field of the IPv6 header invokes the module to process the first extension header, or the upper-layer header if no extension header is present. The contents and semantics of each extension header determine whether or not to proceed to the next header. Therefore, extension headers must be processed strictly in the order they appear in the packet; a receiver must not, for example, scan through a packet looking for a particular kind of extension header and process that header prior to processing all preceding ones.

The exception referred to in the preceding paragraph is the Hop-by-Hop Options header, which carries information that must be examined and processed by every node along a packet's delivery path, including the source and destination nodes. The Hop-by-Hop Options header, when present, must immediately follow the IPv6 header. Its presence is indicated by the value zero in the Next Header field of the IPv6 header.

If, as a result of processing a header, a node is required to proceed to the next header but the Next Header value in the current header is unrecognized by the node, it should discard the packet and send an ICMP Parameter Problem message to the source of the packet, with an ICMP Code value of 2 ("unrecognized Next Header type encountered") and the ICMP Pointer field containing the offset of the unrecognized value within the original packet. The same action should be taken if a node encounters a Next Header value of zero in any header other than an IPv6 header.

Each extension header is an integer multiple of 8 octets, in order to retain 8-octet alignment for subsequent headers. Multi-octet fields within each extension header are aligned on their natural boundaries, i.e., fields of width n octets are placed at an integer multiple of n octets from the start of the header, for $n = 1, 2, 4,$ or 8.

A full implementation of IPv6 includes implementation of the following extension headers:

- Hop-by-Hop Options
- Routing (Type 0)
- Fragment
- Destination Options
- Authentication
- Encapsulating Security Payload

8.4   Extension Header Order

When more than one extension header is used in the same packet, it is recommended that those headers appear in the following order:

1. IPv6 header
2. Hop-by-Hop Options header

3. Destination Options header

4. Routing header

5. Fragment header

6. Authentication header

7. Encapsulating Security Payload header

8. Destination Options header

9. Upper-Layer header

Each extension header should occur at most once, except for the Destination Options header, which should occur at most twice, once before a Routing header and once before the upper-layer header. If the upper-layer header is another IPv6 header (in the case of IPv6 being tunneled over or encapsulated in IPv6), it may be followed by its own extensions headers, which are separately subject to the same ordering recommendations. If and when other extension headers are defined, their ordering constraints relative to the listed headers must be specified.

IPv6 nodes must accept and attempt to process extension headers in any order and occurring any number of times in the same packet, except for the Hop-by-Hop Options header, which is restricted to appear immediately after an IPv6 header only. Nonetheless, it is strongly advised that sources of IPv6 packets adhere to the recommended order until and unless subsequent specifications revise that recommendation.

Options

Two of the currently-defined extension headers, the Hop-by-Hop Options header and the Destination Options header, carry a variable number of type-length-value (TLV) encoded options, of the following format:

```
+-+-+-+-+-+-+-+-+-+-+-+-+-+-+-+-+- - - - - - - -
|  Option Type  |  Opt Data Len |  Option Data
+-+-+-+-+-+-+-+-+-+-+-+-+-+-+-+-+- - - - - - - -
```

- *Option Type* is an 8-bit identifier of the type of option.
- *Optional Data Length* is an 8-bit unsigned integer specifying the data field of this option, in octets.
- *Option Data* is a variable-length field of option-types-specific data.

The sequence of options within a header must be processed strictly in the order they appear in the header; a receiver must not, for example, scan through the header looking for a particular kind of option and process that option prior to processing all preceding ones.

The Option Type identifiers are internally encoded such that their highest-order two bits specify the action that must be taken if the processing IPv6 node does not recognize the Option Type:

- 00: Skip over this option and continue processing the header.

- 01: Discard the packet.

- 10: Discard the packet and, regardless of whether or not the packets's destination address was a multicast address, send an ICMP Parameter Problem, Code 2, message to the packet's source address, pointing to the unrecognized Option Type.

- 11: Discard the packet and, only if the packet's destination address was not a multicast address, send an ICMP Parameter Problem, Code 2, message to the packet's source address, pointing to the unrecognized Option Type.

The third-highest-order bit of the Option Type specifies whether or not the Option Data of that option can change en route to the packet's final destination. When an Authentication header is present in the packet, for any option whose data may change en route, its entire Option Data field must be treated as zero-valued octets when computing or verifying the packet's authenticating value:

- 0: Option Data does not change en route.

- 1: Option Data may change en route.

Individual options may have specific alignment requirements, to ensure that multi-octet values within Option Data fields fall on natural boundaries. The alignment requirement of an option is specified using the notation *xn+y*, meaning the Option Type must appear at an integer multiple of *x* octets from the start of the header, plus *y* octets. For example, *2n* means any 2-octet offset from the start of the header and *8n+2* means any 8-octet offset from the start of the header, plus two octets.

There are two padding options that are used when necessary to align subsequent options and to pad out the containing header to a multiple of eight octets in length. These padding options must be recognized by all IPv6 implementations:

```
Pad1 option (alignment requirement: none)
+-+-+-+-+-+-+-+-+
|      0        |
+-+-+-+-+-+-+-+-+
```

The format of the Pad1 option is a special case—it does not have length and value fields. The Pad1 option is used to insert one octet of padding into the Options area of a header. If more than one octet of padding is required, the PadN option, described next, should be used, rather than multiple Pad1 options.

```
PadN option  (alignment requirement: none)
+-+-+-+-+-+-+-+-+-+-+-+-+-+-+-+-+- - - - - - - -
|      1       | Opt Data Len |  Option Data
+-+-+-+-+-+-+-+-+-+-+-+-+-+-+-+-+- - - - - - - -
```

The PadN option is used to insert two or more octets of padding into the Options area of a header. For N octets of padding, the Opt Data Len field contains the value N-2, and the Option Data consists of N-2 zero-valued octets.

8.5   IPv6 Options Header (Hop-by-Hop)

The Hop-by-Hop Options header is used to carry optional information that must be examined by every node along a packet's delivery path. The Hop-by-Hop Options header is identified by a Next Header value of zero in the IPv6 header, and has the following format:

```
+-+-+-+-+-+-+-+-+-+-+-+-+-+-+-+-+-+-+-+-+-+-+-+-+-+-+-+-+-+-+-+-+
| Next Header  |  Hdr Ext Len  |                               |
+-+-+-+-+-+-+-+-+-+-+-+-+-+-+-+-+                               +
|                                                              |
.                                                              .
.                           Options                            .
.                                                              .
|                                                              |
+-+-+-+-+-+-+-+-+-+-+-+-+-+-+-+-+-+-+-+-+-+-+-+-+-+-+-+-+-+-+-+-+
```

- *Next Header* is an 8-bit selector that identifies the type of header immediately following the Hop-by-Hop Options header. It uses the same values as the IPv4 protocol field.

- *Hdr Ext LenG* is an 8-bit unsigned integer that specifies the length of the Hop-by-Hop Options header in 8-octet units, not including the first eight octets.

■ *Options* is a variable-length field, of length such that the complete Hop-by-Hop Options header is an integer multiple of eight octets. It contains one or more TLV-encoded options.

In addition to the Pad1 and PadN options, the Jumbo Payload hop-by-hop option is defined with an alignment requirement of $4n+2$.

```
+-+-+-+-+-+-+-+-+-+-+-+-+-+-+-+-+
|      194      |Opt Data Len=4 |
+-+-+-+-+-+-+-+-+-+-+-+-+-+-+-+-+-+-+-+-+-+-+-+-+-+-+-+-+-+-+-+-+
|                       Jumbo Payload Length                    |
+-+-+-+-+-+-+-+-+-+-+-+-+-+-+-+-+-+-+-+-+-+-+-+-+-+-+-+-+-+-+-+-+
```

The Jumbo Payload option is used to send IPv6 packets with payloads longer than 65,535 octets. The Jumbo Payload Length is the length of the packet in octets, excluding the IPv6 header but including the Hop-by-Hop Options header; it must be greater than 65,535. If a packet is received with a Jumbo Payload option containing a Jumbo Payload Length less than or equal to 65,535, an ICMP Parameter Problem message, Code 0, should be sent to the packet's source, pointing to the high-order octet of the invalid Jumbo Payload Length field.

The Payload Length field in the IPv6 header must be set to zero in every packet that carries the Jumbo Payload option. If a packet is received with a valid Jumbo Payload option present and a non-zero IPv6 Payload Length field, an ICMP Parameter Problem message, Code 0, should be sent to the packet's source, pointing to the Option Type field of the Jumbo Payload option. The Jumbo Payload option must not be used in a packet that carries a Fragment header. If a Fragment header is encountered in a packet that contains a valid Jumbo Payload option, an ICMP Parameter Problem message, Code 0, should be sent to the packet's source, pointing to the first octet of the Fragment header.

An implementation that does not support the Jumbo Payload option cannot have interfaces to links whose link MTU is greater than 65,575 (40 octets of IPv6 header plus 65,535 octets of payload).

8.6   IPv6 Routing Header

The Routing header is used by an IPv6 source to list one or more intermediate nodes to be "visited" on the way to a packet's destination. This function is very similar to IPv4's Source Route options. The Routing header is identified by a Next Header value of 43 in the immediately preceding header, and has the following format:

```
+-+-+-+-+-+-+-+-+-+-+-+-+-+-+-+-+-+-+-+-+-+-+-+-+-+-+-+-+-+-+-+-+
|  Next Header  |  Hdr Ext Len  |  Routing Type | Segments Left |
+-+-+-+-+-+-+-+-+-+-+-+-+-+-+-+-+-+-+-+-+-+-+-+-+-+-+-+-+-+-+-+-+
|                                                               |
.                                                               .
.                      Type-Specific Data                       .
.                                                               .
| .                                                             |
+-+-+-+-+-+-+-+-+-+-+-+-+-+-+-+-+-+-+-+-+-+-+-+-+-+-+-+-+-+-+-+-+
```

- *Next Header* is an 8-bit selector that identifies the type of header immediately following the Routing header. It uses the same values as the IPv4 Protocol field.

- *Hdr Ext Len* is an 8-bit unsigned integer with the length of the Routing header in 8-octet units, not including the first 8 octets.

- *Routing Type* is an 8-bit identifier of a particular Routing header variant.

- *Segments Left* is an 8-bit unsigned integer with the number of route segments remaining, i.e., the number of explicitly listed intermediate nodes still to be visited before reaching the final destination.

- *Type-Specific Data* is a variable-length field, of a format determined by the Routing Type, and of a length such that the complete Routing header is an integer multiple of eight octets.

If, while processing a received packet, a node encounters a Routing header with an unrecognized Routing Type value, the required behavior of the node depends on the value of the Segments Left field, as follows:

- If Segments Left is zero, the node must ignore the Routing header and proceed to process the next header in the packet, whose type is identified by the Next Header field in the Routing header.

- If Segments Left is non-zero, the node must discard the packet and send an ICMP Parameter Problem, Code 0, message to the packet's Source Address, pointing to the unrecognized Routing Type.

The Type-0 Routing header has the following format:

```
+-+-+-+-+-+-+-+-+-+-+-+-+-+-+-+-+-+-+-+-+-+-+-+-+-+-+-+-+-+-+-+-+
|  Next Header  |  Hdr Ext Len  | Routing Type=0| Segments Left |
+-+-+-+-+-+-+-+-+-+-+-+-+-+-+-+-+-+-+-+-+-+-+-+-+-+-+-+-+-+-+-+-+
|   Reserved    |               Strict/Loose Bit Map            |
+-+-+-+-+-+-+-+-+-+-+-+-+-+-+-+-+-+-+-+-+-+-+-+-+-+-+-+-+-+-+-+-+
|                                                               |
+                                                               +
|                                                               |
+                           Address 1                           +
|                                                               |
+                                                               +
|                                                               |
+-+-+-+-+-+-+-+-+-+-+-+-+-+-+-+-+-+-+-+-+-+-+-+-+-+-+-+-+-+-+-+-+
|                                                               |
+                                                               +
|                                                               |
+                           Address 2                           +
|                                                               |
+                                                               +
|                                                               |
+-+-+-+-+-+-+-+-+-+-+-+-+-+-+-+-+-+-+-+-+-+-+-+-+-+-+-+-+-+-+-+-+
.                               .                               .
.                               .                               .
.                               .                               .
+-+-+-+-+-+-+-+-+-+-+-+-+-+-+-+-+-+-+-+-+-+-+-+-+-+-+-+-+-+-+-+-+
|                                                               |
+                                                               +
|                                                               |
+                          Address[n]                           +
|                                                               |
+                                                               +
|                                                               |
+-+-+-+-+-+-+-+-+-+-+-+-+-+-+-+-+-+-+-+-+-+-+-+-+-+-+-+-+-+-+-+-+
```

- *Next Header* is an 8-bit selector that identifies the type of header immediately following the Routing header. Uses the same values as the IPv4 Protocol field.

- *Hdr Ext Len* is an 8-bit unsigned integer with the length of the Routing header in 8-octet units, not including the first 8 octets. For the Type 0 Routing header, Hdr Ext Len is equal to two times the number of addresses in the header, and must be an even number less than or equal to 46.

- *Routing Type* is zero.

- *Segments Left* is an 8-bit unsigned integer with the number of route segments remaining, i.e., the number of explicitly listed intermediate nodes still to be visited before reaching the final destination. Its maximum legal value is 23.

- *Reserved* is an 8-bit reserved field that is initialized to zero for transmission and ignored on reception.
- *Strict/Loose Bit Map* is a 24-bit bitmap, numbered 0 to 23, left-to-right, that indicates, for each segment of the route, whether or not the next destination address must be a neighbor of the preceding address. A 1 means strict (must be a neighbor), 0 means loose (need not be a neighbor).
- *Address[1..n]* is a vector of 128-bit addresses, numbered 1 to n.

Multicast addresses must not appear in a Routing header of Type 0, or in the IPv6 Destination Address field of a packet carrying a Routing header of Type 0. If bit 0 of the Strict/Loose Bit Map has a value of one, the Destination Address field of the IPv6 header in the original packet must identify a neighbor of the originating node. If bit 0 has a value of zero, the originator may use any legal, non-multicast address as the initial Destination Address. Bits numbered greater than n, where n is the number of addresses in the Routing header, must be set to zero by the originator and ignored by receivers.

A Routing header is not examined or processed until it reaches the node identified as the destination address. In that node, dispatching on the Next Header field of the immediately preceding header causes the Routing header module to be invoked, which, in the case of Routing Type 0, performs the following algorithm:

```
IF Segments Left = 0
{
Proceed to process the next header in the packet, whose type is
identified by the Next Header field in the Routing header.
}ELSE IF Hdr Ext Len is odd or greater than 46
{
Send an ICMP Parameter Problem, Code 0, message to the Source
Address, pointing to the Hdr Ext Len field, and discard the packet.
}
ELSE
{
Compute n, the number of addresses in the Routing header, by divid-
ing Hdr Ext Len by two. IF Segments Left is greater than n
{
Send an ICMP Parameter Problem, Code 0, message to the Source
Address, pointing to the Segments Left field, and discard the pack-
et.
}
ELSE
{
Decrement Segments Left by one; compute i, the index of the next
address to be visited in the address vector, by subtracting Seg-
```

```
ments Left from n. IF Address [i] or the IPv6 Destination Address
is multicast
{
Discard the packet.
}
ELSE
{
Swap the IPv6 Destination Address and Address[i]. IF bit i of the
Strict/Loose Bit Map has a value of one and the new Destination
Address is not the address of a neighbor of this node
{
Send an ICMP Destination Unreachable—Not a Neighbor message to the
Source Address and discard the packet.
}
ELSE IF the IPv6 Hop Limit is less than or equal to one
{
Send an ICMP Time Exceeded—Hop Limit Exceeded in Transit message to
the Source Address and discard the packet
}
ELSE
{
Decrement the Hop Limit by one. Resubmit the packet to the IPv6
module for transmission to the new destination.
}
}
}
}
```

Consider the case of a source node S sending a packet to destination node
D, using a Routing header to cause the packet to be routed via intermediate
nodes I1, I2, and I3. The values of the relevant IPv6 header and Routing
header fields on each segment of the delivery path would be as follows:

As the packet travels from S to I1:

- Source Address = S
- Hdr Ext Len = 6
- Destination Address = I1
- Segments Left = 3
- Address[1] = I2
- Address[2] = I3
- Address[3] = D

If bit 0 of the Bit Map is one, S and I1 must be neighbors; this is
checked by S. As the packet travels from I1 to I2:

- Source Address = S
- Hdr Ext Len = 6
- Destination Address = I2
- Segments Left = 2
- Address[1] = I1Address[2] = I3

- Address[3] = D

If bit 1 of the Bit Map is one, I1 and I2 must be neighbors; this is checked by I1. As the packet travels from I2 to I3:

- Source Address = S
- Hdr Ext Len = 6
- Destination Address = I3
- Segments Left = 1
- Address[1] = I1
- Address[2] = I2
- Address[3] = D

If bit 2 of the Bit Map is one, I2 and I3 must be neighbors; this is checked by I2. As the packet travels from I3 to D:

- Source Address = S
- Hdr Ext Len = 6
- Destination Address = D
- Segments Left = 0
- Address[1] = I1
- Address[2] = I2
- Address[3] = I3

If bit 3 of the Bit Map is one, I3 and D must be neighbors; this is checked by I3.

8.7 IPv6 Fragment Header

The Fragment header is used by an IPv6 source to send packets larger than would fit in the path MTU to their destinations. Unlike IPv4, fragmentation in IPv6 is performed only by source nodes, not by routers along a packet's delivery path. The Fragment header is identified by a Next Header value of 44 in the immediately preceding header, and has the following format:

```
+-+-+-+-+-+-+-+-+-+-+-+-+-+-+-+-+-+-+-+-+-+-+-+-+-+-+-+-+-+-+-+-+
|  Next Header  |   Reserved    |      Fragment Offset    |Res|M|
+-+-+-+-+-+-+-+-+-+-+-+-+-+-+-+-+-+-+-+-+-+-+-+-+-+-+-+-+-+-+-+-+
|                         Identification                        |
+-+-+-+-+-+-+-+-+-+-+-+-+-+-+-+-+-+-+-+-+-+-+-+-+-+-+-+-+-+-+-+-+
```

- *Next Header* is an 8-bit selector that identifies the initial header type of the fragmentable part of the original packet. It uses the same values as the IPv4 Protocol field.

- *Reserved* is an 8-bit reserved field that is initialized to zero for transmission and ignored on reception.

- *Fragment Offset* is a 13-bit unsigned integer. The offset, in 8-octet units, of the data following this header, relative to the start of the Fragmentable Part of the original packet.

- *Res* is a 2-bit reserved field that is initialized to zero for transmission and ignored on reception.

- *M flag* is set to one to indicate more fragments or zero to indicate the last fragment.

- *Identification* is a 32-bit field described later in this chapter.

In order to send a packet that is too large to fit in the MTU of the path to its destination, a source node may divide the packet into fragments and send each fragment as a separate packet, to be reassembled at the receiver.

For every packet that is to be fragmented, the source node generates an identification value. The identification must be different than that of any other fragmented packet sent recently with the same source address and destination address. If a Routing header is present, the destination address of concern is that of the final destination.

The initial, unfragmented packet is referred to as the original packet, and it is considered to consist of two parts.

```
+-+-+-+-+-+-+-+-+-+-+-+-+-+-+-+-+-+-+-+-+-+-//-+-+-+-+-+-+-+-+-+-
|   Unfragmentable  |                Fragmentable               |
|       Part        |                    Part                   |
+-+-+-+-+-+-+-+-+-+-+-+-+-+-+-+-+-+-+-+-+-+-//-+-+-+-+-+-+-+-+-+-
```

The unfragmentable part consists of the IPv6 header plus any extension headers that must be processed by nodes en route to the destina-

---

"Recently" means within the maximum likely lifetime of a packet, including transit time from source to destination and time spent awaiting reassembly with other fragments of the same packet. However, it is not required that a source node know the maximum packet lifetime. Rather, it is assumed that the requirement can be met by maintaining the Identification value as a simple, 32-bit, "wrap-around" counter, incremented each time a packet must be fragmented. It is an implementation choice whether to maintain a single counter or multiple counters for the node, e.g., one for each of the node's possible source addresses, or one for each active (source address, destination address) combination.

tion, that is, all headers up to and including the Routing header if present, the Hop-by-Hop Options header if present, or no extension headers. The fragmentable part consists of the rest of the packet, that is, any extension headers that need to be processed only by the final destination node(s), plus the upper-layer header and data.

The fragmentable part of the original packet is divided into fragments. Each fragment, except possibly the last (rightmost) one, is an integer multiple of eight octets. The fragments are transmitted in separate "fragment packets," as shown:

Original Packet
```
+-+-+-+-+-+-+-+-+-+-+-+-+-+-+-+-+-+-+-+-+-+-+-+-+-+-//-+-+-+-+-+-+-
|  Unfragmentable  |    First     |    Second    |      |  Last   |
|      Part        |   Fragment   |   Fragment   | .... | Fragment|
+-+-+-+-+-+-+-+-+-+-+-+-+-+-+-+-+-+-+-+-+-+-+-+-+-+-//-+-+-+-+-+-+-
```

Fragment Packets
```
+-+-+-+-+-+-+-+-+-+-+-+-+-+-+-+-+-+-+-+-+-
|  Unfragmentable  |Fragment|    First     |
|      Part        | Header |   Fragment   |
+-+-+-+-+-+-+-+-+-+-+-+-+-+-+-+-+-+-+-+-+-+-
+-+-+-+-+-+-+-+-+-+-+-+-+-+-+-+-+-+-+-+-+-+-
|  Unfragmentable  |Fragment|    Second    |
|      Part        | Header |   Fragment   |
+-+-+-+-+-+-+-+-+-+-+-+-+-+-+-+-+-+-+-+-+-+-
                         o
                         o
                         o
+-+-+-+-+-+-+-+-+-+-+-+-+-+-+-+-+-+-+-
|  Unfragmentable  |Fragment|   Last   |
|      Part        | Header | Fragment |
+-+-+-+-+-+-+-+-+-+-+-+-+-+-+-+-+-+-+-+-
```

Each fragment packet is composed of the following:

1. The unfragmentable part of the original packet, with the payload length of the original IPv6 header changed to contain the length of this fragment packet only (excluding the length of the IPv6 header itself), and the Next Header field of the last header of the unfragmentable part changed to 44

2. A fragment header containing the following:

   ▪ A next header value that identifies the first header of the fragmentable part of the original packet

   ▪ A fragment offset containing the offset of the fragment, in 8-octet units, relative to the start of the fragmentable part of the original packet

- The fragment offset of the first (leftmost) fragment, set to zero
- An M flag value of zero if the fragment is the last rightmost one, else an M flag value of one
- The identification value generated for the original packet

3. The fragment itself

The lengths of the fragments must be chosen such that the resulting fragment packets fit within the MTU of the path to the packets' destination(s). At the destination, fragment packets are reassembled into their original, unfragmented form, as illustrated:

Reassembled Original Packet:

```
+-+-+-+-+-+-+-+-+-+-+-+-+-+-+-+-+-+-+-//-+-+-+-+-+-+-+-+-+-+-
|  Unfragmentable  |              Fragmentable            |
|      Part        |                  Part                |
+-+-+-+-+-+-+-+-+-+-+-+-+-+-+-+-+-+-+-//-+-+-+-+-+-+-+-+-+-+-
```

The following rules govern reassembly:

- An original packet is reassembled only from fragment packets that have the same source address, destination address, and fragment identification.
- The unfragmentable part of the reassembled packet consists of all headers up to, but not including, the fragment header of the first fragment packet (that is, the packet whose fragment offset is zero), with the following two changes:

  - The Next Header field of the last header of the unfragmentable part is obtained from the Next Header field of the first fragment's fragment header.
  - The payload length of the reassembled packet is computed from the length of the unfragmentable part and the length and offset of the last fragment. For example, a formula for computing the payload length of the reassembled original packet is

    PLorig = PL.first - FL.first - 8 + (8 · FO.last) + FL.last

    where

    PLorig = Payload Length field of reassembled packet

    PL.first = Payload Length field of first fragment packet

    FL.first = Length of fragment following fragment header
        of first fragment packet

> FO.last = Fragment Offset field of fragment header
> of last fragment packet
>
> FL.last = Length of fragment following Fragment header
> of last fragment packet

- The Fragmentable Part of the reassembled packet is constructed from the fragments following the fragment headers in each of the fragment packets. The length of each fragment is computed by subtracting from the packet's payload length the length of the headers between the IPv6 header and fragment itself; its relative position in fragmentable part is computed from its fragment offset value.

The fragment header is not present in the final, reassembled packet. The following error conditions may arise when reassembling fragmented packets:

- If insufficient fragments are received to complete reassembly of a packet within 60 seconds of the reception of the first-arriving fragment of that packet, reassembly of that packet must be abandoned and all the fragments that have been received for that packet must be discarded. If the first fragment (i.e., the one with a fragment offset of zero) has been received, an ICMP Time Exceeded—Fragment Reassembly Time Exceeded message should be sent to the source of that fragment.

- If the length of a fragment, as derived from the fragment packet's Payload Length field, is not a multiple of eight octets and the M flag of that fragment is one, then that fragment must be discarded and an ICMP Parameter Problem, Code 0, message should be sent to the source of the fragment, pointing to the Payload Length field of the fragment packet.

- If the length and offset of a fragment are such that the Payload Length of the packet reassembled from that fragment would exceed 65,535 octets, then that fragment must be discarded and an ICMP Parameter Problem, Code 0, message should be sent to the source of the fragment, pointing to the Fragment Offset field of the fragment packet.

The following conditions are not expected to occur, but are not considered errors if they do:

- The number and content of the headers preceding the fragment header of different fragments of the same original packet differ.

Whatever headers are present, preceding the fragment header in each fragment packet, are processed when the packets arrive, prior to queuing the fragments for reassembly. Only those headers in the Offset zero fragment packet are retained in the reassembled packet.

■ The Next Header values in the fragment headers of different fragments of the same original packet differ. Only the value from the Offset zero fragment packet is used for reassembly.

8.8   IPv6 Destination Options Header

The Destination Options header is used to carry optional information that needs to be examined only by a packet's destination node(s). The Destination Options header is identified by a Next Header value of **60** in the immediately preceding header, and has the following format:

```
+-+-+-+-+-+-+-+-+-+-+-+-+-+-+-+-+-+-+-+-+-+-+-+-+-+-+-+-+-+-+-+-+
|  Next Header  |  Hdr Ext Len  |                               |
+-+-+-+-+-+-+-+-+-+-+-+-+-+-+-+-+                               +
|                                                               |
.                                                               .
.                           Options                             .
.                                                               .
|                                                               |
+-+-+-+-+-+-+-+-+-+-+-+-+-+-+-+-+-+-+-+-+-+-+-+-+-+-+-+-+-+-+-+-+
```

■ *Next Header* is an 8-bit selector that identifies the type of header immediately following the Destination Options header. It uses the same values as the IPv4 Protocol field.

■ *Hdr Ext Len* is an 8-bit unsigned integer with the length of the Destination Options header in 8-octet units, not including the first eight octets.

■ *Options* is a variable-length field, of length such that the complete Destination Options header is an integer multiple of eight octets. Contains one or more TLV-encoded options.

The only destination options defined in this document are the Pad1 and PadN options specified in Sec. 8.2.

Note that there are two possible ways to encode optional destination information in an IPv6 packet: either as an option in the Destination Options header, or as a separate extension header. The Fragment header and the Authentication header are examples of the latter approach. Which approach can be used depends on what action is desired of a destination node that does not understand the optional information.

If the desired action is for the destination node to discard the packet and, only if the packet's destination address is not a multicast address, send an ICMP Unrecognized Type message to the packet's source address, then the information may be encoded either as a separate header or as an option in the Destination Options header whose option type has the value 11 in its highest-order two bits. The choice might depend on such factors as which takes fewer octets, or which yields better alignment or more efficient parsing.

If any other action is desired, the information must be encoded as an option in the Destination Options header whose option type has the value 00, 01, or 10 in its highest-order two bits, specifying the desired action (see Sec. 8.2).

8.9   IPv6 No Next Header

The value 59 in the Next Header field of an IPv6 header or any extension header indicates that there is nothing following that header. If the Payload Length field of the IPv6 header indicates the presence of octets past the end of a header whose Next Header field contains 59, those octets must be ignored, and passed on unchanged if the packet is forwarded.

8.10   IPv6 Packet Size Considerations

IPv6 requires that every link in the Internet have an MTU of 576 octets or greater. On any link that cannot convey a 576-octet packet in one piece, link-specific fragmentation and reassembly must be provided at a layer below IPv6. From each link to which a node is directly attached, the node must be able to accept packets as large as that link's MTU. Links that have a configurable MTU (for example, PPP links must be configured to have an MTU of at least 576 octets), it is recommended that a larger MTU be configured, to accommodate possible encapsulations (i.e., tunneling) without incurring fragmentation.

It is strongly recommended that IPv6 nodes implement Path MTU Discovery, in order to discover and take advantage of paths with MTU greater than 576 octets. However, a minimal IPv6 implementation (e.g., in a boot ROM) may simply restrict itself to sending packets no larger than 576 octets, and omit implementation of Path MTU Discovery.

In order to send a packet larger than a path's MTU, a node may use the IPv6 Fragment header to fragment the packet at the source and have it reassembled at the destination(s). However, the use of such fragmentation is discouraged in any application that is able to adjust its packets to fit the measured path MTU (i.e., down to 576 octets).

A node must be able to accept a fragmented packet that, after reassembly, is as large as 1500 octets, including the IPv6 header. A node is permitted to accept fragmented packets that reassemble to more than 1500 octets. However, a node must not send fragments that reassemble to a size greater than 1500 octets unless it has explicit knowledge that the destination(s) can reassemble a packet of that size.

In response to an IPv6 packet that is sent to an IPv4 destination (i.e., a packet that undergoes translation from IPv6 to IPv4), the originating IPv6 node may receive an ICMP Packet Too Big message reporting a Next-Hop MTU less than 576. In that case, the IPv6 node is not required to reduce the size of subsequent packets to less than 576, but must include a Fragment header in those packets so that the IPv6-to-IPv4 translating router can obtain a suitable identification value to use in resulting IPv4 fragments. This means the payload might have to be reduced to 528 octets (576 minus 40 for the IPv6 header and eight for the Fragment header), and smaller still if additional extension headers are used. The Path MTU Discovery must be performed even in cases where a host "thinks" a destination is attached to the same link as itself.

Unlike IPv4, it is unnecessary in IPv6 to set a Don't Fragment flag in the packet header in order to perform Path MTU Discovery; that is an implicit attribute of every IPv6 packet. Also, those parts of the RFC-1191 procedures that involve use of a table of MTU "plateaus" do not apply to

IPv6 because the IPv6 version of the Datagram Too Big message always identifies the exact MTU to be used.

8.11   IPv6 Flow Labels

The 24-bit Flow Label field in the IPv6 header may be used by a source to label those packets for which it requests special handling by the IPv6 routers, such as non-default quality of service or real-time service. This aspect of IPv6 is, at the time of writing, still experimental and subject to change as the requirements for flow support in the Internet become clearer. Hosts or routers that do not support the functions of the Flow Label field are required to set the field to zero when originating a packet, pass the field on unchanged when forwarding a packet, and ignore the field when receiving a packet.

A *flow* is a sequence of packets sent from a particular source to a particular (unicast or multicast) destination for which the source desires special handling by the intervening routers. The nature of that special handling might be conveyed to the routers by a control protocol, such as a resource reservation protocol, or by information within the flow's packets themselves, e.g., in a hop-by-hop option. The details of such control protocols or options are beyond the scope of this book.

There might be multiple active flows from a source to a destination, as well as traffic that is not associated with any flow. A flow is uniquely identified by the combination of a source address and a non-zero flow label. Packets that do not belong to a flow carry a flow label of zero.

A flow label is assigned to a flow by the flow's source node. New flow labels must be chosen (pseudo-)randomly and uniformly from the range 1 to FFFFFF, hexadecimal. The purpose of the random allocation is to make any set of bits within the Flow Label field suitable for use as a hash key by routers, for looking up the state associated with the flow.

All packets belonging to the same flow must be sent with the same source address, destination address, priority, and flow label. If any of those packets includes a Hop-by-Hop Options header, then they all must be originated with the same Hop-by-Hop Options header contents (excluding the Next Header field of the Hop-by-Hop Options header). If any of those packets includes a Routing header, then they all must be originated with the same contents in all extension headers up to and including the Routing header (excluding the Next Header field in the Routing header). The routers or destinations are permitted, but not required, to verify that these conditions are satisfied. If a violation is detected, it should be reported to the source by an ICMP Parameter Problem message, Code 0, pointing to the high-order octet of the Flow Label field (i.e., offset 1 within the IPv6 packet).

Routers are free to "opportunistically" set up a flow-handling state for any flow, even when no explicit flow-establishment information has been provided to them via a control protocol, a hop-by-hop option, or other means. For example, upon receiving a packet from a particular source with an unknown, non-zero flow label, a router may process its IPv6 header and any necessary extension headers as if the flow label were zero. That processing would include determining the next-hop interface, and possibly other actions, such as updating a hop-by-hop option, advancing the pointer and addresses in a Routing header, or deciding how to queue the packet based on its Priority field. The router may then choose to "remember" the results of those processing steps and cache that information, using the source address plus the flow label as the cache key. Subsequent packets with the same source address and flow label may then be handled by referring to the cached information rather than examining all those fields that, according to the requirements of the previous paragraph, can be assumed unchanged from the first packet seen in the flow.

A cached flow-handling state that is set up opportunistically must be discarded no more than six seconds after it is established, regardless of whether or not packets of the same flow continue to arrive. If another packet with the same source address and flow label arrives after the cached state has been discarded, the packet undergoes full, normal processing as if its flow label were zero, which may result in the re-creation of the cached flow state for that flow.

The lifetime of a flow-handling state that is set up explicitly, for example by a control protocol or a hop-by-hop option, must be specified as part of the specification of the explicit set-up mechanism; it may exceed six seconds. A source must not reuse a flow label for a new flow within the lifetime of any flow-handling state that might have been established for the prior use of that flow label. Since a flow-handling state with a lifetime of six seconds may be established opportunistically for any flow, the minimum interval between the last packet of one flow and the first packet of a new flow using the same flow label is six seconds. Flow labels used for explicitly set-up flows with longer flow-state lifetimes must remain unused for those longer lifetimes before being reused for new flows.

When a node stops and restarts as a result of a crash, it must be careful not to use a flow label that it might have used for an earlier flow whose lifetime may not have expired yet. This can be accomplished by recording flow label usage on stable storage so that it can be remembered across crashes, or by refraining from using any flow labels until the maximum lifetime of any possible previously established flows has expired (at least six seconds; more if explicit flow set-up mechanisms with longer lifetimes have been used). If the minimum time for rebooting the node is known (often more than six seconds), that time can be deducted from the necessary waiting period before starting to allocate flow labels.

There is no requirement that all, or even most, packets belong to flows, i.e., carry non-zero flow labels. This observation is placed here to remind protocol designers and implementers not to assume otherwise. For example, it would be unwise to design a router whose performance would be adequate only if most packets belonged to flows, or to design a header compression scheme that only worked on packets that belonged to flows.

8.12   IPv6 Packet Priority

The 4-bit Priority field in the IPv6 header enables a source to identify the desired delivery priority of its packets, relative to other packets from the same source. The priority values are divided into two ranges: Values 0 through 7 are used to specify the priority of traffic for which the source is providing congestion control, i.e., traffic that "backs off" in response to congestion, such as TCP traffic. Values 8 through 15 are used to specify the priority of traffic that does not back off in response to congestion, e.g., "real-time" packets being sent at a constant rate.

For congestion-controlled traffic, the following priority values are recommended for particular application categories:

- 0—Uncharacterized traffic
- 1—"Filler" traffic, such as news
- 2—Unattended data transfer such as e-mail
- 3—Reserved
- 4—Attended bulk transfer such as FTP and NFS

- 5—Reserved
- 6—Interactive traffic such as TELNET
- 7—Internet control traffic, such as routing protocols and SNMP)

For non-congestion-controlled traffic, the lowest priority value (8) should be used for those packets that the sender is most willing to have discarded under conditions of congestion, such as high-fidelity video traffic, and the highest value (15) should be used for those packets that the sender is least willing to have discarded, such as low-fidelity audio traffic. There is no relative ordering implied between the congestion-controlled priorities and the non-congestion-controlled priorities.

8.13   IPv6 and Upper-Layer Protocols

Any transport or other upper-layer protocol that includes the addresses from the IP header in its checksum computation must be modified for use over IPv6, to include the 128-bit IPv6 addresses instead of 32-bit IPv4 addresses. In particular, the following illustration shows the TCP and UDP "pseudo-header" for IPv6:

```
+-+-+-+-+-+-+-+-+-+-+-+-+-+-+-+-+-+-+-+-+-+-+-+-+-+-+-+-+-+-+-+-+
|                                                               |
+                                                               +
|                                                               |
+                        Source Address                         +
|                                                               |
+                                                               +
|                                                               |
+-+-+-+-+-+-+-+-+-+-+-+-+-+-+-+-+-+-+-+-+-+-+-+-+-+-+-+-+-+-+-+-+
|                                                               |
+                                                               +
|                                                               |
+                     Destination Address                       +
|                                                               |
+                                                               +
|                                                               |
+-+-+-+-+-+-+-+-+-+-+-+-+-+-+-+-+-+-+-+-+-+-+-+-+-+-+-+-+-+-+-+-+
|                        Payload Length                         |
+-+-+-+-+-+-+-+-+-+-+-+-+-+-+-+-+-+-+-+-+-+-+-+-+-+-+-+-+-+-+-+-+
|                          Zero                  | Next Header  |
+-+-+-+-+-+-+-+-+-+-+-+-+-+-+-+-+-+-+-+-+-+-+-+-+-+-+-+-+-+-+-+-+
```

If the packet contains a Routing header, the destination address used in the pseudo-header is that of the final destination. At the originating node, that address will be in the last element of the Routing header; at the recipient(s), that address will be in the Destination Address field of the IPv6 header.

The Next Header value in the pseudo-header identifies the upper-layer protocol (6 for TCP or 17 for UDP). It will differ from the Next Header value in the IPv6 header if there are extension headers between the IPv6 header and the upper-layer header.

The Payload Length value used in the pseudo-header is the length of the upper-layer packet, including the upper-layer header. It will be less than the payload length in the IPv6 header (or in the Jumbo Payload option) if there are extension headers between the IPv6 header and the upper-layer header.

Unlike IPv4, when UDP packets are originated by an IPv6 node, the UDP checksum is not optional. That is, whenever originating a UDP packet, an IPv6 node must compute a UDP checksum over the packet and the pseudo-header. If that computation yields a result of zero, it must be changed to hexadecimal FFFF for placement in the UDP header. IPv6 receivers must discard UDP packets containing a zero checksum, and should log the error.

IPv6 includes the pseudo-header in its checksum computation; this is a change from IPv4 of ICMP, which does not include a pseudo-header in its checksum. The reason for the change is to protect ICMP from misdelivery or corruption of those fields of the IPv6 header on which it depends, which, unlike IPv4, are not covered by an Internet-layer checksum. The Next Header field in the pseudo-header for ICMP contains the value 58, which identifies the IPv6 version of ICMP.

Maximum Packet Lifetime

Unlike IPv4, IPv6 nodes are not required to enforce a maximum packet lifetime. That is the reason the IPv4 Time to Live field was renamed "Hop Limit" in IPv6. In practice, very few, if any, IPv4 implementations conform to the requirement that they limit packet lifetime, so this is not really a change. Any upper-layer protocol that relies on the Internet layer (whether IPv4 or IPv6) to limit packet lifetime ought to be upgraded to provide its own mechanisms for detecting and discarding obsolete packets.

Maximum Upper-Layer Payload Size

When computing the maximum payload size available for upper-layer data, an upper-layer protocol must take into account the larger size of the IPv6 header relative to the IPv4 header. For example, in IPv4, TCP's MSS option is computed as the maximum packet size (a default value or a value learned through Path MTU Discovery) minus 40 octets (20 octets for the minimum-length IPv4 header and 20 octets for the minimum-length TCP header). When using TCP over IPv6, the MSS must be computed as the maximum packet size minus 60 octets, because the minimum-length IPv6 header (i.e., an IPv6 header with no extension headers) is 20 octets longer than a minimum-length IPv4 header.

Formatting Guidelines for Options

This section gives some advice on how to lay out the fields when designing new options to be used in the Hop-by-Hop Options header or the Destination Options header, as described in Sec. 8.2. These guidelines are based on the following assumptions:

- One desirable feature is that any multi-octet fields within the Option Data area of an option be aligned on their natural boundaries, i.e., fields of width n octets should be placed at an integer multiple of n octets from the start of the Hop-by-Hop or Destination Options header, for $n = 1, 2, 4,$ or 8.

- Another desirable feature is that the Hop-by-Hop or Destination Options header take up as little space as possible, subject to the requirement that the header be an integer multiple of eight octets.

- It may be assumed that, when either of the option-bearing headers are present, they carry a very small number of options, usually only one.

These assumptions suggest the following approach to laying out the fields of an option:

Order the fields from smallest to largest, with no interior padding, then derive the alignment requirement for the entire option based on the alignment requirement of the largest field (up to a maximum alignment of eight octets). This approach is illustrated in the following examples:

If an option X required two data fields, one of length 8 octets and one of length 4 octets, it would be laid out as follows:

```
+-+-+-+-+-+-+-+-+-+-+-+-+-+-+-+-+-+-+-+-+-+-+-+-+-+-+-+-+-+-+-+-+
|  Next Header  | Hdr Ext Len=1 | Option Type=X |Opt Data Len=12|
+-+-+-+-+-+-+-+-+-+-+-+-+-+-+-+-+-+-+-+-+-+-+-+-+-+-+-+-+-+-+-+-+
|                         4-Octet Field                         |
+-+-+-+-+-+-+-+-+-+-+-+-+-+-+-+-+-+-+-+-+-+-+-+-+-+-+-+-+-+-+-+-+
|                                                               |
+                         8-Octet Field                         +
|                                                               |
+-+-+-+-+-+-+-+-+-+-+-+-+-+-+-+-+-+-+-+-+-+-+-+-+-+-+-+-+-+-+-+-+
```

Its alignment requirement is $8n+2$, to ensure that the 8-octet field starts at a multiple-of-8 offset from the start of the enclosing header. A complete Hop-by-Hop or Destination Options header containing this one option would look as follows:

```
+-+-+-+-+-+-+-+-+-+-+-+-+-+-+-+-+
| Option Type=X |Opt Data Len=12|
+-+-+-+-+-+-+-+-+-+-+-+-+-+-+-+-+-+-+-+-+-+-+-+-+-+-+-+-+-+-+-+-+
|                         4-Octet Field                         |
+-+-+-+-+-+-+-+-+-+-+-+-+-+-+-+-+-+-+-+-+-+-+-+-+-+-+-+-+-+-+-+-+
|                                                               | +
|                         8-Octet Field                         +
|                                                               |
+-+-+-+-+-+-+-+-+-+-+-+-+-+-+-+-+-+-+-+-+-+-+-+-+-+-+-+-+-+-+-+-+
```

If an option Y required three data fields, one of length 4 octets, one of length 2 octets, and one of length 1 octet, it would be laid out as follows:

```
+-+-+-+-+-+-+-+-+
| Option Type=Y |
+-+-+-+-+-+-+-+-+-+-+-+-+-+-+-+-+-+-+-+-+-+-+-+-+-+-+-+-+-+-+-+-+
|Opt Data Len=7 | 1-Octet Field |          2-Octet Field        |
+-+-+-+-+-+-+-+-+-+-+-+-+-+-+-+-+-+-+-+-+-+-+-+-+-+-+-+-+-+-+-+-+
|                         4-Octet Field                         |
+-+-+-+-+-+-+-+-+-+-+-+-+-+-+-+-+-+-+-+-+-+-+-+-+-+-+-+-+-+-+-+-+
```

Its alignment requirement is $4n+3$, to ensure that the 4-octet field starts at a multiple-of-4 offset from the start of the enclosing header. A complete Hop-by-Hop or Destination Options header containing this one option would look as follows:

```
+-+-+-+-+-+-+-+-+-+-+-+-+-+-+-+-+-+-+-+-+-+-+-+-+-+-+-+-+-+-+-+-+
| Next Header  | Hdr Ext Len=1 | Pad1 Option=0 | Option Type=Y |
+-+-+-+-+-+-+-+-+-+-+-+-+-+-+-+-+-+-+-+-+-+-+-+-+-+-+-+-+-+-+-+-+
|Opt Data Len=7 | 1-Octet Field |        2-Octet Field         |
+-+-+-+-+-+-+-+-+-+-+-+-+-+-+-+-+-+-+-+-+-+-+-+-+-+-+-+-+-+-+-+-+
|                          4-Octet Field                       |
+-+-+-+-+-+-+-+-+-+-+-+-+-+-+-+-+-+-+-+-+-+-+-+-+-+-+-+-+-+-+-+-+
| PadN Option=1 |Opt Data Len=2 |       0       |       0       |
+-+-+-+-+-+-+-+-+-+-+-+-+-+-+-+-+-+-+-+-+-+-+-+-+-+-+-+-+-+-+-+-+
```

A Hop-by-Hop or Destination Options header containing both options X and Y would have one of the two following formats, depending on which option appeared first:

```
+-+-+-+-+-+-+-+-+-+-+-+-+-+-+-+-+-+-+-+-+-+-+-+-+-+-+-+-+-+-+-+-+
| Next Header  | Hdr Ext Len=3 | Option Type=X |Opt Data Len=12|
+-+-+-+-+-+-+-+-+-+-+-+-+-+-+-+-+-+-+-+-+-+-+-+-+-+-+-+-+-+-+-+-+
|                          4-octet field                       |
+-+-+-+-+-+-+-+-+-+-+-+-+-+-+-+-+-+-+-+-+-+-+-+-+-+-+-+-+-+-+-+-+
|                                                              |
+                          8-octet field                       +
|                                                              |
+-+-+-+-+-+-+-+-+-+-+-+-+-+-+-+-+-+-+-+-+-+-+-+-+-+-+-+-+-+-+-+-+
| PadN Option=1 |Opt Data Len=1 |       0       | Option Type=Y |
+-+-+-+-+-+-+-+-+-+-+-+-+-+-+-+-+-+-+-+-+-+-+-+-+-+-+-+-+-+-+-+-+
|Opt Data Len=7 | 1-octet field |        2-octet field         |
+-+-+-+-+-+-+-+-+-+-+-+-+-+-+-+-+-+-+-+-+-+-+-+-+-+-+-+-+-+-+-+-+
|                          4-octet field                       |
+-+-+-+-+-+-+-+-+-+-+-+-+-+-+-+-+-+-+-+-+-+-+-+-+-+-+-+-+-+-+-+-+
| PadN Option=1 |Opt Data Len=2 |       0       |       0       |
+-+-+-+-+-+-+-+-+-+-+-+-+-+-+-+-+-+-+-+-+-+-+-+-+-+-+-+-+-+-+-+-+
+-+-+-+-+-+-+-+-+-+-+-+-+-+-+-+-+-+-+-+-+-+-+-+-+-+-+-+-+-+-+-+-+
| Next Header  | Hdr Ext Len=3 | Pad1 Option=0 | Option Type=Y |
+-+-+-+-+-+-+-+-+-+-+-+-+-+-+-+-+-+-+-+-+-+-+-+-+-+-+-+-+-+-+-+-+
|Opt Data Len=7 | 1-Octet Field |        2-Octet Field         |
+-+-+-+-+-+-+-+-+-+-+-+-+-+-+-+-+-+-+-+-+-+-+-+-+-+-+-+-+-+-+-+-+
|                          4-Octet Field                       |
+-+-+-+-+-+-+-+-+-+-+-+-+-+-+-+-+-+-+-+-+-+-+-+-+-+-+-+-+-+-+-+-+
| PadN Option=1 |Opt Data Len=4 |       0       |       0       |
+-+-+-+-+-+-+-+-+-+-+-+-+-+-+-+-+-+-+-+-+-+-+-+-+-+-+-+-+-+-+-+-+
|       0       |       0       | Option Type=X |Opt Data Len=12|
+-+-+-+-+-+-+-+-+-+-+-+-+-+-+-+-+-+-+-+-+-+-+-+-+-+-+-+-+-+-+-+-+
|                          4-Octet Field                       |
+-+-+-+-+-+-+-+-+-+-+-+-+-+-+-+-+-+-+-+-+-+-+-+-+-+-+-+-+-+-+-+-+
|                                                              |
+                          8-Octet Field                       +
|                                                              |
+-+-+-+-+-+-+-+-+-+-+-+-+-+-+-+-+-+-+-+-+-+-+-+-+-+-+-+-+-+-+-+-+
```

 **8.14 Summary**

IPv6 is more robust than its predecessor. The ability to accommodate many more entities in the addressing scheme is but one of the abilities it has over IPv4.

IPv6 Address Architecture

This chapter defines the addressing architecture of the IP version 6 protocol. The chapter reviews the IPv6 addressing model, and provides text representations of IPv6 addresses, definition of IPv6 unicast addresses, anycast addresses, and multicast addresses, and an IPv6 node's required addresses.

9.1   IPv6 Addressing: A Perspective

IPv6 addresses are 128-bit. There are three types of addresses: unicast, anycast, and multicast. There are no broadcast addresses in IPv6; this type of address is superseded by multicast addresses. Here, address fields are given a specific name, such as *subscriber*. When this name is used with *ID* after the name, as in *subscriber ID*, it refers to the contents of that field. When it is used with the term *prefix* it refers to all of the address up to and including this field.

In IPv6, all zeros and ones are legal values for any field unless specifically excluded. Specifically, prefixes may contain zero-valued fields or end in zeros. For a more detailed review of IPv6 addressing, see section 3.5.

9.2   Address Type Representation

The specific type of an IPv6 address is indicated by the address's leading bits. The variable-length field comprising these leading bits is called the Format Prefix (FP). The initial allocation of these prefixes is as follows:

Allocation of Space	Prefix (binary)	Fraction Address Space
Reserved	0000 0000	$1/256$
Unassigned	0000 0001	$1/256$
Reserved for NSAP Allocation	0000 001	$1/128$
Reserved for IPX Allocation	0000 010	$1/128$
Unassigned	0000 011	$1/128$
Unassigned	0000 1	$1/32$
Unassigned	0001	$1/16$
Unassigned	001	$1/8$

Allocation of Space	Prefix (binary)	Fraction Address Space
Provider-Based Unicast Address	010	$\frac{1}{8}$
Unassigned	011	$\frac{1}{8}$
Reserved for Geographically-Based Unicast Addresses	100	$\frac{1}{8}$
Unassigned	101	$\frac{1}{8}$
Unassigned	110	$\frac{1}{8}$
Unassigned	1110	$\frac{1}{16}$
Unassigned	1111 0	$\frac{1}{32}$
Unassigned	1111 10	$\frac{1}{648}$
Unassigned	1111 110	$\frac{1}{128}$
Unassigned	1111 1110 0	$\frac{1}{512}$
Link Local Use Addresses	1111 1110 10	$\frac{1}{1024}$
Site Local Use Addresses	1111 1110 11	$\frac{1}{1024}$
Multicast Addresses	1111 1111	$\frac{1}{256}$

The unspecified address, the loopback address, and the IPv6 addresses with embedded IPv4 addresses are assigned out of the 0000 0000 format prefix space. This allocation supports the direct provider addresses' allocation, local use addresses, and multicast addresses. Space is reserved for NSAP addresses, IPX addresses, and geographic addresses. The remainder of the address space is unassigned for future use. This can be used for expansion of existing uses or for new uses. Fifteen percent of the address space is initially allocated. The remaining 85% is reserved for future use.

Unicast addresses are distinguished from multicast addresses by the value of the high-order octet of the addresses: a value of FF (11111111) identifies an address as a multicast address; any other value identifies an address as a unicast address. Anycast addresses are taken from the unicast address space, and are not syntactically distinguishable from unicast addresses.

9.3 Unicast Addresses

The IPv6 unicast address is contiguous bit-wise maskable, similar to IPv4 addresses under Classless Interdomain Routing (CIDR). There are several forms of unicast address assignment in IPv6, including the global-provider-based unicast address, the geographic-based unicast address, the NSAP address, the IPX hierarchical address, the site-local-use address, the link-local-use address, and the IPv4-capable host address. Additional address types can be defined in the future.

IPv6 nodes may have considerable or little knowledge of the internal structure of the IPv6 address, depending on what the host does. Remember, a host is not necessarily a computer in the sense that a user does work on it; it could be any valid network device. At a minimum, a node may consider that unicast addresses (including its own) have no internal structure:

```
|                        128 bits                       |
+-+-+-+-+-+-+-+-+-+-+-+-+-+-+-+-+-+-+-+-+-+-+-+-+-
|                      node address                     |
+-+-+-+-+-+-+-+-+-+-+-+-+-+-+-+-+-+-+-+-+-+-+-+-+-
```

A slightly sophisticated host (but still rather simple) may additionally be aware of subnet prefix(es) for the link(s) it is attached to and that different addresses can have different n values:

```
|        n bits |    128-n bits   |
+-+-+-+-+-+-+-+-+-+-+-+-+-+-+-+-+-+
|        subnet prefix            |        interface ID   |
+-+-+-+-+-+-+-+-+-+-+-+-+-+-+-+-+-+-+-+-+-+-+-+-+-+-+-+
```

More sophisticated hosts may be aware of other hierarchical boundaries in the unicast address. Though a very simple router may have no knowledge of the internal structure of IPv6 unicast addresses, routers will more generally have knowledge of one or more of the hierarchical boundaries for the operation of routing protocols. The known boundaries will differ from router to router, depending on what positions the router holds in the routing hierarchy.

An example of a unicast address format that will likely be common on LANs and other environments where IEEE 802 MAC addresses are available is the following, where the 48-bit interface ID is an IEEE-802 MAC address:

```
|   n bits   | 80-n bits   |   48 bits   |
+-+-+-+-+-+-+-++-+-+-+-+-+-+-++-+-+-+-+-+-+-
|      subscriber prefix    |   subnet ID   | Interface ID |
+-+-+-+-+-+-+-+-+-+-+-+-+-+-+-++-+-+-+-+-+-+-+-+-+-+-+-+-
```

The use of IEEE 802 MAC addresses as interface IDs is expected to be very common in environments where nodes have an IEEE 802 MAC address. In other environments, where IEEE 802 MAC addresses are not available, other types of link layer addresses can be used, such as E.164 addresses, for the interface ID.

The inclusion of a unique global interface identifier, such as an IEEE MAC address, makes possible a very simple form of auto-configuration of addresses. A node may discover a subnet ID by listening to Router Advertisement messages sent by a router on its attached link(s), and then fabricating an IPv6 address for itself by using its IEEE MAC address as the interface ID on that subnet.

Another unicast address format example is where a site or organization requires additional layers of internal hierarchy. In this example, the subnet ID is divided into an area ID and a subnet ID. Its format is as follows:

```
|   s bits     |  n bits    |    m bits   | 128-s-n-m bits |
+-+-+-+-+-+-+-+-+-+-+-+-+-+-+-+-+-+-+-+-+-+-+-+-+-+-+-+-+-+-+-+-+
| subscriber prefix | area ID | subnet ID | interface ID        |
+-+-+-+-+-+-+-+-+-+-+-+-+-+-+-+-+-+-+-+-+-+-+-+-+-+-+-+-+-+-+-+-+
```

This technique can be continued to allow a site or organization to add additional layers of internal hierarchy. It may be desirable to use an interface ID smaller than a 48-bit IEEE 802 MAC address to allow more space for the additional layers of internal hierarchy. These could be interface IDs that are administratively created by the site or organization.

The address 0:0:0:0:0:0:0:0 is called the *unspecified address.* It must never be assigned to any node. It indicates the absence of an address. One example of its use is in the Source Address field of any IPv6 datagrams sent by an initializing host before it has learned its own address. The unspecified address must not be used as the destination address of IPv6 datagrams or in IPv6 Routing headers.

The unicast address 0:0:0:0:0:0:0:1 is called the *loopback address.* It may be used by a node to send an IPv6 datagram to itself. It may never be assigned to any interface. The loopback address must not be used as the source address in IPv6 datagrams that are sent outside of a single node. An IPv6 datagram with a destination address of loopback must never be sent outside of a single node.

9.4   IPv6 Addresses and IPv4 Addresses

The IPv6 transition mechanisms include a technique for hosts and routers to dynamically tunnel IPv6 packets over the IPv4 routing infrastructure. IPv6 nodes that utilize this technique are assigned special IPv6 unicast addresses that carry an IPv4 address in the low-order 32-bits. This type of address is termed an *IPv4-compatible IPv6 address* and has the following format:

```
|   80 bits    |  16  |   32 bits    |
+-+-+-+-+-+-+-+-+-+-+-+-+-+-+-+-+-+-+-+
|0000......0000|00      00|IPv4 address |
+-+-+-+-+-+-+-+-+-+-+-+-+-+-+-+-+-+-+-+
```

A second type of IPv6 address that holds an embedded IPv4 address is also defined. This address is used to represent the addresses of IPv4-only nodes (those that *do not* support IPv6) as IPv6 addresses. This type of address is termed an *IPv4-mapped IPv6 address* and has this format:

```
|   80 bits    |  16  |     32 bits  |
+-+-+-+-+-+-+-+-+-+-+-+-+-+-+-+-+-+-+-+
|0000......0000|FFFF   | IPv4 address |
+-+-+-+-+-+-+-+-+-+-+-+-+-+-+-+-+-+-+-+
```

NSAP Addresses

The mapping of NSAP address into IPv6 addresses is as follows:

```
|  7     | 121 bits     |
+-+-+-+-+-+-+-+-+-+-+-+-+-
|0000001 |to be defined |
+-+-+-+-+-+-+-+-+-+-+-+-+-
```

IPX Addresses

The mapping of IPX address into IPv6 addresses is as follows:

```
|   7   |   121 bits    |
+-+-+-+-+-+-+-+-+-+-+-+-+-+
|0000010 | to be defined |
+-+-+-+-+-+-+-+-+-+-+-+-+-+
```

Global Unicast Addresses

The initial assignment plan for these unicast addresses is similar to the assignment of IPv4 addresses under the CIDR scheme. The IPv6 global-provider-based unicast address format is as follows:

```
| 3 | n bits | m bits | o bits | 125-n-m-o bits |
+-+-+-+-+-+-+-+-+-+-+-+-+-+-+-+-+-+-+-+-+-+-+-+-+-+
|010|registry ID|provider ID|subscriber ID|intra-subscriber |
+-+-+-+-+-+-+-+-+-+-+-+-+-+-+-+-+-+-+-+-+-+-+-+-+-+
```

The high-order-bit part of the address is assigned to registries, which assign portions of the address space to providers, which assign portions of the address space to subscribers, etc.

The registry ID identifies the registry, which assigns the provider portion of the address. The term *registry prefix* refers to the high-order part of the address, up to and including the registry ID.

The provider ID identifies a specific provider, which assigns the subscriber portion of the address. The term *provider prefix* refers to the high-order part of the address, up to and including the provider ID.

The subscriber ID distinguishes among multiple subscribers attached to the provider identified by the provider ID. The term *subscriber prefix* refers to the high-order part of the address, up to and including the subscriber ID.

The intra-subscriber portion of the address is defined by an individual subscriber and is organized according to the subscriber's local internet topology. It is likely that many subscribers will choose to divide the intra-subscriber portion of the address into a subnet ID and an interface ID. In this case, the subnet ID identifies a specific physical link and the interface ID identifies a single interface on that subnet.

IPv6 Unicast Addresses

There are two types of local-use unicast addresses defined: link-local and site-local. Link-local is for use on a single link. Site-local is for use in a single site. Link-local addresses have the following format:

```
| 10 | bits   |   n bits   |   118-n bits   |
+-+-+-+-+-+-+-+-+-+-+-+-+-+-+-+-+-+-+-+-+-+-+-
|1111111010   |   0        |   interface ID |
+-+-++-+-+-+-+-+-+-+-+-+-+-+-+-+-+-+-+-+-+-+-+
```

Link-local addresses are designed to be used for addressing on a single link for purposes such as auto-address configuration, neighbor discovery, or when no routers are present. Routers are not permitted to forward any packets with link-local source addresses.

Site-local addresses have the following format:

```
| 10 | bits | n  bits | m bits | 118-n-m bits   |
+-+-+-+-+-+-+-+-+-+-+-+-+-+-+-+-+-+-+-+-+-+-+-+-
|1111111011 |   0      | subnet ID | interface ID |
+-+-+-+-+-+-+-+-+-+-+-+-+-+-+-+-+-+-+-+-+-+-+-+-
```

Site-local addresses may be used for sites or organizations that are not (yet) connected to the global Internet. They do not need to request or "steal" an address prefix from the global Internet address space; IPv6 site-local addresses can be used instead. When the organization connects to the global Internet, it can then form global addresses by replacing the site-local prefix with a subscriber prefix. Routers must not forward any packets with site-local source addresses outside of the site.

9.5   Anycast Addresses

An IPv6 anycast address is an address that is assigned to more than one interface (typically belonging to different nodes), with the property that a packet sent to an anycast address is routed to the nearest interface having that address, according to the routing protocols' measure of distance.

Anycast addresses are allocated from the unicast address space, using any of the defined unicast address formats. Thus, anycast addresses are syntactically indistinguishable from unicast addresses. When a unicast address is assigned to more than one interface, thus turning it into an anycast address, the nodes to which the address is assigned must be explicitly configured to know that it is an anycast address.

For any assigned anycast address, there is a longest address prefix, P, that identifies the topological region in which all interfaces belonging to that anycast address reside. Within the region identified by P, each member of the anycast set must be advertised as a separate entry in the routing system (referred to as a *host route*); outside the region identified by P, the anycast address may be aggregated into the routing advertisement for prefix P.

Note that, in the worst case, the prefix P of an anycast set may be the null prefix, i.e., the members of the set may have no topological locality. In that case, the anycast address must be advertised as a separate routing entry throughout the entire Internet, which presents a severe scaling limit on how many such "global" anycast sets may be supported. Therefore, it is expected that support for global anycast sets may be unavailable or very restricted.

One expected use of anycast addresses is to identify the set of routers belonging to an Internet service provider. Such addresses could be used as intermediate addresses in an IPv6 Routing header, to cause a packet to be delivered via a particular provider or sequence of providers. Some other possible uses are to identify the set of routers attached to a particular subnet, or the set of routers providing entry into a particular routing domain.

There is little experience with widespread, arbitrary use of internet anycast addresses, and some known complications and hazards when using them in their full generality [ANYCST]. Until more experience has been gained and solutions agreed upon for those problems, the following restrictions are imposed on IPv6 anycast addresses:

- An anycast address must not be used as the source address of an IPv6 packet.

- An anycast address must not be assigned to an IPv6 host, that is, it may be assigned to an IPv6 router only.

The subnet-router anycast address is predefined. Its format is as follows:

```
|    n bits      |    128-n bits    |
+-+-+-+-+-+-+-+-+-+-+-+-+-+-+-+-+-+
| subnet prefix  | 00000000000000 |
+-+-+-+-+-+-+-+-+-+-+-+-+-+-+-+-+-+
```

The subnet prefix in an anycast address is the prefix that identifies a specific link. This anycast address is syntactically the same as a unicast address for an interface on the link with the interface identifier set to zero.

Packets sent to the subnet-router anycast address will be delivered to one router on the subnet. All routers are required to support the subnet-router anycast addresses for the subnets for which they have interfaces.

The subnet-router anycast address is intended to be used for applications where a node needs to communicate with one of a set of routers on a remote subnet, for example, when a mobile host needs to communicate with one of the mobile agents on its "home" subnet.

9.6   Multicast Addresses

An IPv6 multicast address is an identifier for a group of nodes. A node may belong to any number of multicast groups. Multicast addresses have the following format:

```
| 8 | 4 | 4 |          112 bits                        |
+-+-+-+-+-+-+-+-+-+-+-+-+-+-+-+-+-+-+-+-+-+-+-+-+-+-+-+-
|11111111|flgs|scop|      group ID                     |
+-+-+-+-+-+-+-+-+-+-+-+-+-+-+-+-+-+-+-+-+-+-+-+-+-+-+-+-
```

- *11111111* at the start of the address identifies the address as being a multicast address.
- *Flags* is a set of four flags:

```
+-+-+-+-+
|0|0|0|T|
+-+-+-+-+
```

- The three high-order flags are reserved, and must be initialized to zero.
- $T = 0$ indicates a permanently assigned ("well-known") multicast address, assigned by the global Internet numbering authority.
- $T = 1$ indicates a non-permanently-assigned multicast address. (It is also referred to as transient.)
- *Scope* is a 4-bit multicast scope value used to limit the scope of the multicast group. The values are as follows:

  0   Reserved

  1   Node-local scope

  2   Link-local scope

  3   Unassigned

  4   Unassigned

  5   Site-local scope

6   Unassigned

7   Unassigned

8   Organization-local scope

9   Unassigned

A   Unassigned

B   Unassigned

C   Unassigned

D   Unassigned

E   Global scope

F   Reserved

■ *Group ID* identifies the multicast group, either permanent or transient, within the given scope. The meaning of a permanently assigned multicast address is independent of the scope value. For example, if the "NTP servers group" is assigned a permanent multicast address with a group ID of 43 (hex), then

■ FF01:0:0:0:0:0:0:43 means all NTP servers on the same node as the sender.

■ FF02:0:0:0:0:0:0:43 means all NTP servers on the same link as the sender.

■ FF05:0:0:0:0:0:0:43 means all NTP servers at the same site as the sender.

■ FF0E:0:0:0:0:0:0:43 means all NTP servers in the internet.

Nonpermanently assigned multicast addresses are meaningful only within a given scope. For example, a group identified by the nonpermanent, site-local multicast address FF15:0:0:0:0:0:0:43 at one site bears no relationship to a group using the same address at a different site, nor to a nonpermanent group using the same group ID with a different scope, nor to a permanent group with the same group ID. Multicast addresses must not be used as source addresses in IPv6 datagrams or appear in any routing header.

Predefined Multicast Addresses

The following well-known multicast addresses are predefined:
FF00:0:0:0:0:0:0:0
FF01:0:0:0:0:0:0:0

FF02:0:0:0:0:0:0:0
FF03:0:0:0:0:0:0:0
FF04:0:0:0:0:0:0:0
FF05:0:0:0:0:0:0:0
FF06:0:0:0:0:0:0:0
FF07:0:0:0:0:0:0:0
FF08:0:0:0:0:0:0:0
FF09:0:0:0:0:0:0:0
FF0A:0:0:0:0:0:0:0
FF0B:0:0:0:0:0:0:0
FF0C:0:0:0:0:0:0:0
FF0D:0:0:0:0:0:0:0
FF0E:0:0:0:0:0:0:0
FF0F:0:0:0:0:0:0:0

These multicast addresses are reserved and will never be assigned to any multicast group.

The following multicast addresses identify the group of all IPv6 nodes, within scope 1 (node-local) or 2 (link-local):

All Nodes Addresses: FF01:0:0:0:0:0:0:1

FF02:0:0:0:0:0:0:1

The following multicast addresses identify the group of all IPv6 routers, within scope 1 (node-local) or 2 (link-local):

All Routers Addresses: FF01:0:0:0:0:0:0:2

FF02:0:0:0:0:0:0:2

The following multicast address identifies the group of all IPv6 DHCP Servers and Relay Agents within scope 2 (link-local):

DHCP Server/Relay-Agent: FF02:0:0:0:0:0:0:C

The following multicast address is computed as a function of a node's unicast and anycast addresses:

Solicited-Node Address: FF02:0:0:0:0:1:XXXX:XXXX

The solicited-node multicast address is formed by taking the low-order 32 bits of the address (unicast or anycast) and appending those bits to the 96-bit prefix FF02:0:0:0:0:1, resulting in a multicast address in the range FF02:0:0:0:0:1:0000:0000 to FF02:0:0:0:0:1:FFFF:FFFF.

For example, the solicited-node multicast address corresponding to the IPv6 address 4037::01:800:200E:8C6C is FF02::1:200E:8C6C. IPv6 addresses that differ only in the high-order bits, due to multiple high-order prefixes associated with different providers, will map to the same solicited-node address, thereby reducing the number of multicast addresses a node must join.

A node is required to compute and support a solicited-node multicast addresses for every unicast and anycast address to which it is assigned.

9.7   Node Address Requirements

A host is required to recognize the following addresses as identifying itself:

■ Its link-local address for each interface

■ Assigned unicast addresses

■ Loopback address

■ All-nodes multicast address

■ Solicited-node multicast address for each of its assigned unicast and anycast addresses

■ Multicast addresses of all other groups to which the host belongs

A router is required to recognize the following addresses as identifying itself:

■ Its link-local address for each interface

■ Assigned unicast addresses

■ Loopback address

■ The subnet-router anycast addresses for the links to which it has interfaces

■ All other anycast addresses with which the router has been configured

■ All-nodes multicast address

■ All-router multicast address

■ Solicited-node multicast address for each of its assigned unicast and anycast addresses

■ Multicast addresses of all other groups to which the router belongs

The only address prefixes that should be predefined in an implementation are the following:

■ Unspecified address

- Loopback address
- Multicast prefix (FF)
- Local-use prefixes (link-local and site-local)
- Predefined multicast addresses
- IPv4-compatible prefixes

Implementations should assume all other addresses are unicast unless specifically configured (e.g., anycast addresses).

CHAPTER 10

Transmission Control Protocol (TCP)

Transmission Control Protocol (TCP) is intended for use as a reliable host-to-host protocol between hosts in packet-switched computer communication networks, and in interconnected systems of such networks. As strategic and tactical computer communication networks increase, it is essential to provide means of interconnecting them and to provide standard interprocess communication protocols that can support a broad range of applications.

10.1   TCP: A Perspective

TCP is a connection-oriented, end-to-end, reliable protocol designed to fit into a layered hierarchy of protocols that support multi-network applications. TCP provides for reliable interprocess communication between pairs of processes in host computers attached to distinct but interconnected computer communication networks. Very few assumptions are made as to the reliability of the communication protocols below the TCP layer. TCP assumes it can obtain a simple, potentially unreliable datagram service from the lower-level protocols. In principle, TCP should be able to operate above a wide spectrum of communication systems, ranging from hard-wired connections to packet-switched or circuit-switched networks.

TCP interfaces on one side to user or application processes and on the other side to a lower-level protocol such as Internet Protocol. The interface between an application process and TCP consists of a set of calls much like the calls an operating system provides to an application process for manipulating files. For example, there are calls to open and close connections and to send and receive data on established connections. It is also expected that the TCP can asynchronously communicate with application programs. Although considerable freedom is permitted to TCP implementers to design interfaces that are appropriate to a particular operating system environment, a minimum functionality is required at the TCP/user interface for any valid implementation.

The interface between TCP and lower-level protocol is essentially unspecified, except that it is assumed there is a mechanism whereby the two levels can asynchronously pass information to each other. Typically, the lower-level protocol is expected to specify this interface. TCP is designed to work in a very general environment of interconnected networks.

10.2   TCP Operation

As noted previously, the primary purpose of TCP is to provide reliable, securable, logical circuit or connection service between pairs of processes. Providing this service on top of a less reliable internet communication system requires facilities in the following areas:

- Basic data transfer
- Reliability
- Flow control
- Multiplexing
- Connections
- Precedence and security

Basic Data Transfer

TCP is able to transfer a continuous stream of octets in each direction between its users by packaging some number of octets into segments for transmission through the internet system. In general, TCPs decide when to block and forward data at its own convenience.

Sometimes users need to be sure that all the data they have submitted to TCP has been transmitted. For this purpose, a PUSH function is defined. To assure that data submitted to TCP is actually transmitted, the sending user indicates that it should be pushed through to the receiving user. A PUSH causes the TCPs to promptly forward and deliver data up to that point to the receiver. The exact push point might not be visible to the receiving user, and the PUSH function does not supply a record boundary marker.

Reliability

TCP must recover from data that is damaged, lost, duplicated, or delivered out of order by the internet communication system. This is achieved by assigning a sequence number to each octet transmitted, and requiring a positive acknowledgment (ACK) from the receiving TCP. If the ACK is not received within a timeout interval, the data is retransmitted. At the receiver, the sequence numbers are used to correctly order segments that are received out of order and to eliminate duplicates. Damage is handled by adding a checksum to each segment transmitted, checking it at the receiver, and discarding damaged segments.

As long as TCPs continue to function properly and the internet system does not become completely partitioned, no transmission errors will affect the correct delivery of data. TCP recovers from internet communication system errors.

Flow Control

TCP provides a means for the receiver to govern the amount of data sent by the sender. This is achieved by returning a "window" with every ACK, indicating a range of acceptable sequence numbers beyond the last segment successfully received. The window indicates an allowed number of octets that the sender may transmit before receiving further permission.

Multiplexing

To allow many processes within a single host to use TCP communication facilities simultaneously, TCP provides a set of addresses or ports within each host. Concatenated with the network and host addresses from the internet communication layer, this forms a socket. A pair of sockets uniquely identifies each connection. That is, a socket may be simultaneously used in multiple connections. The binding of ports to processes is handled independently by each host. However, it proves useful to attach frequently used processes (a *logger* or timesharing service) to fixed sockets that are made known to the public. These services can then be accessed through the known addresses. Establishing and learning the port addresses of other processes may involve more dynamic mechanisms.

Connections

The reliability and flow-control mechanisms described in the previous sections require that TCP initialize and maintain certain status information for each data stream. The combination of this information, including sockets, sequence numbers, and window sizes, is called a *connection*. Each connection is uniquely specified by a pair of sockets identifying its two sides.

When two processes wish to communicate, their TCPs must first establish a connection (initialize the status information on each side). When the communication is complete, the connection is terminated or closed to free the resources for other uses. Since connections must be established between unreliable hosts and over the unreliable internet communication system, a handshake mechanism with clock-based sequence numbers is used to avoid erroneous initialization of connections.

Precedence and Security

The users of TCP may indicate the security and precedence of their communication. Provision is made for default values to be used when these features are not needed.

10.3   TCP and the Host Environment

TCP is assumed to be a module in an operating system or a part of the protocol suite running on a given host. The users access TCP much like they would access the file system. TCP may call on other operating system functions, for example, to manage data structures. The actual interface to the network is assumed to be controlled by a device driver module. TCP does not call on the network device driver directly, but rather calls on the internet datagram protocol module, which may in turn call on the device driver.

The mechanisms of TCP do not preclude implementation in a front-end processor. However, in such an implementation, a host-to-front-end protocol must provide the functionality to support the type of TCP-user interface described in this chapter.

Interfaces and TCP

The TCP/user interface provides for calls made by the user on the TCP to open or close a connection, send or receive data, and to obtain the status of a connection. These calls are like other calls from user programs on the operating system, such as the calls to open, read from, and close a file.

The TCP/internet interface provides calls to send and receive datagrams addressed to TCP modules in hosts anywhere in the internet system. These calls have parameters for passing the address, type of service, precedence, security, and other control information.

TCP Reliability

A stream of data sent on a TCP connection is delivered reliably and in order at the destination. Transmission is made reliable via the use of sequence numbers and acknowledgments. Conceptually, each octet of data is assigned a sequence number. The sequence number of the first octet of data in a segment is transmitted with that segment and is called the *segment sequence number*. Segments also carry an acknowledgment number, which is the sequence number of the next expected data octet of transmissions in the reverse direction. When the TCP transmits a segment containing data, it puts a copy on a retransmission queue and starts a timer; when the acknowledgment for that data is received, the segment is deleted from the queue. If the acknowledgment is not received before the timer runs out, the segment is retransmitted.

An acknowledgment by TCP does not guarantee that the data has been delivered to the end user; it means that the receiving TCP has taken the responsibility to do so. To govern the flow of data between TCPs, a flow-control mechanism is employed. The receiving TCP reports a "window" to the sending TCP. This window specifies the number of octets, starting with the acknowledgment number, that the receiving TCP is currently prepared to receive.

TCP Connection Establishment/Clearing

To identify the separate data streams that a TCP may handle, TCP provides a port identifier. Since port identifiers are selected independently by each TCP, they might not be unique. To provide for unique addresses within each TCP, an internet address is concatenated to identify the TCP with a port identifier, creating a socket that is unique throughout all networks connected together.

A connection is fully specified by the pair of sockets at the ends. A local socket may participate in many connections to different foreign sockets. A connection can be used to carry data in both directions; it is full-duplex.

TCPs are free to associate ports with processes however they choose. However, several basic concepts are necessary in any implementation. There must be well-known sockets that the TCP associates only with the appropriate processes by some means. Processes may own ports, and processes can initiate connections only on the ports they own. Means for implementing ownership is a local issue, but may be done with a Request Port user command, or a method of uniquely allocating a group of ports to a given process, by associating the high-order bits of a port name with a given process.

A connection is specified in the OPEN call by the local port and foreign socket arguments. In return, the TCP supplies a (short) local connection name by which the user refers to the connection in subsequent calls. There are several things to remember about a connection. To store this information, there is a data structure called a *Transmission Control Block* (TCB). One implementation strategy would have the local connection name be a pointer to the TCB for this connection. The OPEN call also specifies whether the connection establishment is to be actively pursued, or passively waited for.

A passive OPEN request means that the process wants to accept incoming connection requests rather than attempting to initiate a connection. Often, the process requesting a passive OPEN will accept a connection request from any caller. In this case, a foreign socket of all zeros is used to denote an unspecified socket. Unspecified foreign sockets are allowed only on passive OPENs. A service process that wished to provide services for unknown other processes would issue a passive OPEN request with an unspecified foreign socket, then a connection could be made with any process that requested a connection to this local socket. It would help if this local socket were known to be associated with this service.

Well-known sockets are a convenient mechanism for associating a socket address with a standard service. For instance, the TELNET-server process is permanently assigned to a particular socket, and other sockets are reserved for File Transfer, Remote Job Entry, Text Generator, Echoer, and Sink processes. A socket address might be reserved for access to a look-up service, which would return the specific socket at which a newly created service would be provided. The concept of a well-known socket is part of the TCP specification, but the assignment of sockets to services is outside this specification.

Processes can issue passive OPENs and wait for matching active OPENs from other processes and be informed by the TCP when connections have been established. Two processes that issue active OPENs to each other at the same time will be correctly connected. This flexibility is critical for the support of distributed computing in which components act asynchronously with respect to each other.

There are two principal cases for matching the sockets in the local passive OPENs and a foreign active OPENs. In the first case, the local passive OPEN has fully specified the foreign socket. In this case, the match must be exact. In the second case, the local passive OPEN has left the foreign socket unspecified. In this case, any foreign socket is acceptable as long as the local sockets match. Other possibilities include partially restricted matches.

If there are several pending passive OPENs (recorded in TCBs) with the same local socket, a foreign active OPEN will be matched to a TCB with the specific foreign socket in the foreign active OPEN, if such a TCB exists, before selecting a TCB with an unspecified foreign socket. The procedure to establish connections utilizes the synchronize (SYN) control flag and involves an exchange of three messages. This exchange has been termed a *three-way handshake*.

A connection is initiated by the rendezvous of an arriving segment containing a SYN and a waiting TCB entry, each created by a user OPEN command. The matching of local and foreign sockets determines when a connection has been initiated. The connection becomes *established* when sequence numbers have been synchronized in both directions. The clearing of a connection also involves the exchange of segments, in this case carrying the FIN control flag.

TCP and Data Communication

The data that flows on a connection may be thought of as a stream of octets. The sending user indicates in each SEND call whether the data in that call (and any preceding calls) should be immediately pushed through to the receiving user by the setting of the PUSH flag.

A sending TCP is allowed to collect data from the sending user and to send that data in segments at its own convenience, until the PUSH function is signaled; then, it must send all unsent data. When a receiving TCP sees the PUSH flag, it must not wait for more data from the sending TCP before passing the data to the receiving process.

There is no necessary relationship between PUSH functions and segment boundaries. The data in any particular segment may be the result of a single SEND call, in whole or part, or of multiple SEND calls.

The purpose of the PUSH function and the PUSH flag is to push data through from the sending user to the receiving user. It does not provide a record of service. There is a coupling between the PUSH function and the use of buffers of data that cross the TCP/user interface. Each time a PUSH flag is associated with data placed into the receiving user's buffer, the buffer is returned to the user for processing even if the buffer is not filled. If data arrives that fills the user's buffer before a PUSH is seen, the data is passed to the user in buffer-size units. TCP also provides a means to communicate to the receiver of data that, at some point further along in the data stream than the receiver, it is currently reading there is urgent data. TCP does not attempt to define what the user specifically does upon being notified of pending urgent data, but the general notion is that the receiving process will take action to process the urgent data quickly.

TCP Precedence and Security

The TCP makes use of the Internet Protocol Type of Service field and security option to provide precedence and security on a per-connection basis to TCP users. Not all TCP modules will necessarily function in a multilevel secure environment; some may be limited to unclassified use only, and others may operate at only one security level and compartment. Consequently, some TCP implementations and services to users may be limited to a subset of the multilevel secure case.

TCP modules that operate in a multilevel secure environment must properly mark outgoing segments with the security, compartment, and precedence. Such TCP modules must also provide their users or higher-level protocols such as TELNET or THP an interface to allow them to specify the desired security level, compartment, and precedence of connections.

10.4   TCP Header Format

TCP segments are sent as internet datagrams. The Internet Protocol header carries several information fields, including the source and destination host addresses. A TCP header follows the IP header, supplying information specific to the TCP protocol. This division allows for the existence of host level protocols other than TCP. The TCP header format is as follows:

```
 0                   1                   2                   3
 0 1 2 3 4 5 6 7 8 9 0 1 2 3 4 5 6 7 8 9 0 1 2 3 4 5 6 7 8 9 0 1
+-+-+-+-+-+-+-+-+-+-+-+-+-+-+-+-+-+-+-+-+-+-+-+-+-+-+-+-+-+-+-+-+
|          Source Port          |       Destination Port        |
+-+-+-+-+-+-+-+-+-+-+-+-+-+-+-+-+-+-+-+-+-+-+-+-+-+-+-+-+-+-+-+-+
|                        Sequence Number                        |
+-+-+-+-+-+-+-+-+-+-+-+-+-+-+-+-+-+-+-+-+-+-+-+-+-+-+-+-+-+-+-+-+
|                     Acknowledgment Number                     |
+-+-+-+-+-+-+-+-+-+-+-+-+-+-+-+-+-+-+-+-+-+-+-+-+-+-+-+-+-+-+-+-+
| Data  |           |U|A|P|R|S|F|                               |
| Offset| Reserved  |R|C|S|S|Y|I|            Window             |
|       |           |G|K|H|T|N|N|                               |
+-+-+-+-+-+-+-+-+-+-+-+-+-+-+-+-+-+-+-+-+-+-+-+-+-+-+-+-+-+-+-+-+
|           Checksum            |        Urgent Pointer         |
+-+-+-+-+-+-+-+-+-+-+-+-+-+-+-+-+-+-+-+-+-+-+-+-+-+-+-+-+-+-+-+-+
|                    Options                    |    Padding    |
+-+-+-+-+-+-+-+-+-+-+-+-+-+-+-+-+-+-+-+-+-+-+-+-+-+-+-+-+-+-+-+-+
|                             data                              |
+-+-+-+-+-+-+-+-+-+-+-+-+-+-+-+-+-+-+-+-+-+-+-+-+-+-+-+-+-+-+-+-+
```

Source Port is the 16-bit source port number.

Destination Port is the 16-bit destination port number.

Sequence Number, 32 bits, is the sequence number of the first data octet in this segment (except when SYN is present). If SYN is present, the sequence number is the initial sequence number (ISN) and the first data octet is ISN+1.

Acknowledgment Number is a 32-bit field. If the ACK control bit is set, this field contains the value of the next sequence number the sender of the segment is expecting to receive. Once a connection is established, this is always sent.

Data Offset, 4 bits, is the number of 32-bit words in the TCP header. This indicates where the data begins. The TCP header (even one including options) is a 32-bit integral number.

Reserved is a 6-bit field reserved for future use. It must be zero.

Control Bits has 6 bits (from left to right):

- URG is Urgent Pointer field significant.
- ACK is Acknowledgment field significant.
- PSH is for the Push function.
- RST resets the connection.
- SYN synchronizes sequence numbers.
- FIN indicates no more data from the sender.

- *Window*, 16 bits, is the number of data octets beginning with the one indicated in the Acknowledgment field which the sender of this segment is willing to accept.

Checksum is a 16-bits field that is the one's complement of the one's complement sum of all 16-bit words in the header and text. If a segment contains an odd number of header and text octets to be checksummed, the last octet is padded on the right with zeros to form a 16-bit word for checksum purposes. The pad is not transmitted as part of the segment. While computing the checksum, the checksum field itself is replaced with zeros.

■ The checksum also covers a 96-bit pseudo-header conceptually prefixed to the TCP header. This pseudo-header contains the source address, destination address, protocol, and TCP length. This gives the TCP protection against misrouted segments. This information is carried in the Internet Protocol and is transferred across the TCP/network interface in the arguments or results of calls by the TCP on the IP.

```
+-+-+-+-+-+-+-+-+-+-+-+-+-+-+-+-+-+-+-+-+
|               Source Address          |
+-+-+-+-+-+-+-+-+-+-+-+-+-+-+-+-+-+-+-+-+
|            Destination Address        |
+-+-+-+-+-+-+-+-+-+-+-+-+-+-+-+-+-+-+-+-+
|  zero  |  PTCL  |      TCP Length      |
+-+-+-+-+-+-+-+-+-+-+-+-+-+-+-+-+-+-+-+-+
```

The TCP Length is the TCP header length plus the data length in octets (this is not an explicitly transmitted quantity, but is computed), and does not count the 12 octets of the pseudo-header.

■ *Urgent Pointer* is a 16-bit field that communicates the current value of the urgent pointer as a positive offset from the sequence number in this segment. The urgent pointer points to the sequence number of the octet following the urgent data. This field is only be interpreted in segments with the URG control bit set.

■ *Options* may occupy space at the end of the TCP header and are a multiple of 8 bits. All options are included in the checksum. An option may begin on any octet boundary. There are two cases for the format of an option:
  1. A single octet of option-kind.
  2. An octet of option-kind, an octet of option-length, and the actual option-data octets. The option-length counts the two octets of option-kind and option-length as well as the option-data octets.

The list of options may be shorter than the data offset field might imply. The content of the header beyond the End-of-Option

option must be header padding (i.e., zero). A TCP must implement all options. Currently defined options include the following (kind indicated in octal):

Kind	Length	Meaning
0	-	End of option list
1	-	No operation
2	4	Maximum segment size

The following option code indicates the end of the option list. This might not coincide with the end of the TCP header according to the Data Offset field. This is used at the end of all options, not the end of each option, and need only be used if the end of the options would not otherwise coincide with the end of the TCP header.

```
End of Option List
        +-+-+-+-+-
        |00000000|
        +-+-+-+-+-
         Kind=0
```

The following No-Operation option code may be used between options, for example, to align the beginning of a subsequent option on a word boundary. There is no guarantee that senders will use this option, so receivers must be prepared to process options even if they do not begin on a word boundary.

```
+-+-+-+-+-
|00000001|
+-+-+-+-+-
 Kind=1
```

■ Maximum Segment Size Option Data is a 16-bit field. If this option is present, it communicates the maximum receive segment size at the TCP that sends this segment. This field must only be sent in the initial connection request (i.e., in segments with the SYN control bit set). If this option is not used, any segment size is allowed. It is shown here:

```
+-+-+-+-+-+-+-+-+-+-+-+-+-+-+-+-+-
|00000010|00000100|   max seg size   |
+-+-+-+-+-+-+-+-+-+-+-+-+-+-+-+-+-
Kind=2   Length=4
```

■ Padding is a variable-length field used to ensure that the TCP header ends and data begins on a 32-bit boundary. The padding is composed of zeros.

10.5   TCP Terminology

It is important to understand some detailed TCP-related terminology. Maintaining a TCP connection requires remembering several variables. These variables are considered to be stored in the Transmission Control Block (TCB). Among the variables stored in the TCB are the local and remote socket numbers, the security and precedence of the connection, pointers to the user's send and receive buffers, and pointers to the retransmit queue and to the current segment. In addition, several variables relating to the send and receive sequence numbers are stored in the TCB.

The following are send-sequence variables:

■ SND.UNA—Send unacknowledged
■ SND.NXT—Send next
■ SND.WND—Send window
■ SND.UP—Send urgent pointer
■ SND.WL1—Segment sequence number used for last window update
■ SND.WL2—Segment acknowledgment number used for last window update
■ ISS—Initial send sequence number

The following are receive-sequence variables:

■ RCV.NXT—Receive next
■ RCV.WND—Receive window
■ RCV.UP—Receive urgent pointer
■ IRS—Initial receive sequence number

The following diagram relates some of these variables to Send Sequence Space, where

- 1 = Old sequence numbers that have been acknowledged
- 2 = Sequence numbers of unacknowledged data
- 3 = Sequence numbers allowed for new data transmission
- 4 = Future sequence numbers that are not yet allowed

```
      1              2              3              4
-+-+-+-+-+|-+-+-+-+-+|-+-+-+-+-+|-+-+-+-+-+
     SND.UNA      SND.NXT      SND.UNA
                               +SND.WND
```

The following is a Receive Sequence Space, where

- 1 = Old sequence numbers that have been acknowledged
- 2 = Sequence numbers allowed for new reception
- 3 = Future sequence numbers that are not yet allowed

```
      1              2              3
-+-+-+-+-+|-+-+-+-+-+|-+-+-+-+-+
     RCV.NXT      RCV.NXT
                  +RCV.WND
```

There are also some variables used frequently in the discussion that take their values from the fields of the current segment:

- SEG.SEQ—Segment sequence number
- SEG.ACK—Segment acknowledgment number
- SEG.LEN—Segment length
- SEG.WND—Segment window
- SEG.UP—Segment urgent pointer
- SEG.PRC—Segment precedence value

A connection progresses through a series of states during its lifetime. The states are as follows:

1. LISTEN
2. SYN-SENT
3. SYN-RECEIVED
4. ESTABLISHED
5. FIN-WAIT-1

6. FIN-WAIT-2

7. CLOSE-WAIT

8. CLOSING

9. LAST-ACK

10. TIME-WAIT

11. The fictional state CLOSED

CLOSED is fictional because it represents the state when there is no TCB, and therefore, no connection. Briefly, the meanings of the states are as follows:

LISTEN   represents waiting for a connection request from any remote TCP and port.

SYN-SENT   represents waiting for a matching connection request after having sent a connection request.

SYN-RECEIVED   represents waiting for a confirming connection request acknowledgment after having both received and sent a connection request.

ESTABLISHED   represents an open connection, i.e., data received can be delivered to the user. This is the normal state for the data-transfer phase of the connection.

FIN-WAIT-1   represents waiting for a connection termination request from the remote TCP, or an acknowledgment of the connection termination request previously sent.

FIN-WAIT-2   represents waiting for a connection termination request from the remote TCP.

CLOSE-WAIT   represents waiting for a connection termination request from the local user.

CLOSING   represents waiting for a connection termination request acknowledgment from the remote TCP.

LAST-ACK   represents waiting for an acknowledgment of the connection termination request previously sent to the remote TCP (which includes an acknowledgment of its connection termination request).

TIME-WAIT represents waiting for enough time to pass to be sure the remote TCP received the acknowledgment of its connection termination request.

CLOSED represents no connection state at all.

A TCP connection progresses from one state to another in response to events. The events are the user calls, OPEN, SEND, RECEIVE, CLOSE, ABORT, and STATUS; the incoming segments, particularly those containing the SYN, ACK, RST, and FIN flags; and timeouts.

The state diagrams illustrated here represent only state changes, together with the causing events and resulting actions, but address neither error conditions nor actions that are not connected with state changes. In a later section, more detail is offered with respect to the reaction of the TCP to events.

TCP Sequence Numbers

A fundamental notion in the design is that every octet of data sent over a TCP connection has a sequence number. Since every octet is sequenced, each of them can be acknowledged. The acknowledgment mechanism employed is cumulative, so that an acknowledgment of sequence number X indicates that all octets up to but not including X have been received. This mechanism allows for straightforward duplicate detection in the presence of retransmission. Numbering octets within a segment is such that the first data octet immediately following the header is the lowest numbered, and the following octets are numbered consecutively.

It is essential to remember that the actual sequence number space is finite, though very large. This space ranges from zero to 2^{32}-1. Since the space is finite, all arithmetic dealing with sequence numbers must be performed modulo 2^{32}. This unsigned arithmetic preserves the relationship of sequence numbers as they cycle from 2^{32}-1 to zero again. There are some subtleties to computer modulo arithmetic, so great care should be taken in programming the comparison of such values. In the following discussion, the symbol =< means "less than or equal to" (modulo 2^{32}).

The typical kinds of sequence number comparisons that the TCP must perform include the following:

■ Determining that an acknowledgment refers to some sequence number sent but not yet acknowledged

- Determining that all sequence numbers occupied by a segment have been acknowledged (e.g., to remove the segment from a retransmission queue)
- Determining that an incoming segment contains sequence numbers that are expected (i.e., that the segment "overlaps" the receive window)

In response to sending data, the TCP will receive acknowledgments. The following comparisons are needed to process the acknowledgments:

- SND.UNA = Oldest unacknowledged sequence number
- SND.NXT = Next sequence number to be sent
- SEG.ACK = Acknowledgment from the receiving TCP (next sequence number expected by the receiving TCP)
- SEG.SEQ = First sequence number of a segment
- SEG.LEN = Number of octets occupied by the data in the segment (counting SYN and FIN)
- SEG.SEQ+SEG.LEN-1 = Last sequence number of a segment

A new acknowledgment (called an "acceptable ACK") is one for which the following inequality holds:

- SND.UNA < SEG.ACK =< SND.NXT

A segment on the retransmission queue is fully acknowledged if the sum of its sequence number and length is less than or equal to the acknowledgment value in the incoming segment. When data is received, the following comparisons are needed:

- RCV.NXT = Next sequence number expected on an incoming segment, and the left or lower edge of the receive window
- RCV.NXT+RCV.WND-1 = Last sequence number expected on an incoming segment, and the right or upper edge of the receive window
- SEG.SEQ = First sequence number occupied by the incoming segment
- SEG.SEQ+SEG.LEN-1 = Last sequence number occupied by the incoming segment

A segment is judged to occupy a portion of valid receive sequence space if
RCV.NXT =< SEG.SEQ < RCV.NXT+RCV.WND
  or
RCV.NXT =< SEG.SEQ+SEG.LEN-1 < RCV.NXT+RCV.WND

The first part of this test checks to see if the beginning of the segment falls in the window, the second part of the test checks to see if the end of the segment falls in the window. If the segment passes either part of the test, it contains data in the window. Actually, it is a little more complicated than this. Due to zero windows and zero length segments, we have four cases for the acceptability of an incoming segment:

Segment Length	Receive Window	Test
0	0	SEG.SEQ = RCV.NXT
0	>0	RCV.NXT =< SEG.SEQ < RCV.NXT+RCV.WND
>0	0	not acceptable
>0	>0	RCV.NXT =< SEG.SEQ < RCV.NXT+RCV.WND or RCV.NXT =< SEG.SEQ+SEG.LEN-1 < RCV.NXT+RCV.WND

When the receive window is zero, no segments should be acceptable except ACK segments. Thus, it is be possible for a TCP to maintain a zero receive window while transmitting data and receiving ACKs. However, even when the receive window is zero, a TCP must process the RST and URG fields of all incoming segments.

We have taken advantage of the numbering scheme to protect certain control information as well. This is achieved by implicitly including some control flags in the sequence space so they can be retransmitted and acknowledged without confusion (i.e., one and only one copy of the control will be acted upon). Control information is not physically carried in the segment data space. Consequently, we must adopt rules for implicitly assigning sequence numbers to control. The SYN and FIN are the only controls requiring this protection, and these controls are used only at connection opening and closing. For sequence-number purposes, SYN is considered to occur before the first actual data octet of the segment in which it occurs, while FIN is considered to occur after the last actual data octet in a segment in which it occurs. The segment length (SEG.LEN) includes both data and sequence space occupying controls. When a SYN is present, then SEG.SEQ is the sequence number of the SYN.

Initial Sequence Number Selection

The protocol places no restriction on a particular connection being used over and over again. A connection is defined by a pair of sockets. New instances of a connection are referred to as *incarnations* of the connection. The problem that arises from this is, how does the TCP identify duplicate segments from previous incarnations of the connection? This becomes apparent if the connection is being opened and closed in quick succession, or if the connection breaks with loss of memory and is then reestablished.

To avoid confusion, segments from one incarnation of a connection must be prevented from being used while the same sequence numbers might still be present in the network from an earlier incarnation. This must be assured, even if a TCP crashes and loses all knowledge of the sequence numbers it has been using. When new connections are created, an initial sequence number (ISN) generator is employed, which selects a new 32-bit ISN. The generator is bound to a (possibly fictitious) 32-bit clock whose low-order bit is incremented roughly every four microseconds. Thus, the ISN cycles approximately every 4.55 hours. Since we assume that segments will stay in the network no more than the Maximum Segment Lifetime (MSL), and that the MSL is less than 4.55 hours, we can reasonably assume that the ISNs will be unique.

For each connection, there is a send sequence number and a receive sequence number. The initial send sequence number (ISS) is chosen by the data-sending TCP, and the initial receive sequence number (IRS) is learned during the connection-establishing procedure. For a connection to be established or initialized, the two TCPs must synchronize on each other's initial sequence numbers. This is done in an exchange of connection-establishing segments carrying a control bit called SYN (for synchronize) and the initial sequence numbers. As a shorthand, segments carrying the SYN bit are also called SYNs. Hence, the solution requires a suitable mechanism for picking an initial sequence number and a slightly involved handshake to exchange the ISNs.

The synchronization requires each side to send its own initial sequence number and to receive a confirmation of it in acknowledgment from the other side. Each side must also receive the other side's initial sequence number and send a confirming acknowledgment:

1. A —> B SYN; my sequence number is X

2. A <— B ACK; your sequence number is X

3. A <— B SYN; my sequence number is Y

4. A —> B ACK; your sequence number is Y

Steps 2 and 3 can be combined in a single message called the three-way (or three-message) handshake. A three-way handshake is necessary because sequence numbers are not tied to a global clock in the network, and TCPs may have different mechanisms for picking the ISNs. The receiver of the first SYN has no way of knowing whether the segment was an old delayed one or not, unless it remembers the last sequence number used on the connection (which is not always possible), and so it must ask the sender to verify this SYN. .

Knowing When to Keep Quiet

To be sure that a TCP does not create a segment that carries a sequence number that may be duplicated by an old segment remaining in the network, the TCP must keep quiet for a maximum segment lifetime (MSL) before assigning any sequence numbers upon starting up or recovering from a crash in which memory of sequence numbers in use was lost. For this specification, the MSL is taken to be two minutes. This is an engineering choice, and may be changed if experience indicates it is desirable to do so. Note that if a TCP is reinitialized in some sense, yet retains its memory of sequence numbers in use, then it need not wait at all; it must only be sure to use sequence numbers larger than those recently used.

TCP Quiet Time Concept

This specification provides that hosts that crash without retaining any knowledge of the last sequence numbers transmitted on each active (i.e., not closed) connection delay emitting any TCP segments for at least the agreed maximum segment lifetime (MSL) in the internet system of which the host is a part. In the paragraphs below, an explanation for this specification is given.

TCP implementers may violate the "quiet time" restriction, but only at the risk of causing some old data to be accepted as new or new data rejected as old duplicated by some receivers in the internet system.

TCPs consume sequence-number space each time a segment is formed and entered into the network output queue at a source host. The duplicate detection and sequencing algorithm in the TCP protocol relies on the unique binding of segment data to sequence space to the extent that sequence numbers will not cycle through all 2^{32} values before the segment data bound to those sequence numbers has been delivered and acknowledged by the receiver, and all duplicate copies of the segments have "drained" from the internet. Without such an assumption, two distinct

TCP segments could conceivably be assigned the same or overlapping sequence numbers, causing confusion at the receiver as to which data is new and which is old. Remember that each segment is bound to as many consecutive sequence numbers as there are octets of data in the segment.

Under normal conditions, TCPs keep track of the next sequence number to emit and the oldest awaiting acknowledgment, to avoid mistakenly using a sequence number over before its first use has been acknowledged. This alone does not guarantee that old duplicate data is drained from the internet, so the sequence space has been made very large to reduce the probability that a wandering duplicate will cause trouble upon arrival. At two megabits per second, it takes 4.5 hours to use up 2^{32} octets of sequence space. Since the maximum segment lifetime in the internet is not likely to exceed a few tens of seconds, this is deemed ample protection for foreseeable internets, even if data rates escalate to tens of megabits per second. At 100 megabits per second, the cycle time is 5.4 minutes, which may be a little short, but still within reason.

The basic duplicate detection and sequencing algorithm in TCP can be defeated, however, if a source TCP does not have any memory of the sequence numbers it last used on a given connection. For example, if the TCP were to start all connections with sequence number 0, then upon crashing and restarting, a TCP might re-form an earlier connection (possibly after half-open connection resolution) and emit packets with sequence numbers identical to or overlapping with packets still in the network, which were emitted on an earlier incarnation of the same connection. In the absence of knowledge about the sequence numbers used on a particular connection, the TCP specification recommends that the source delay for MSL seconds before emitting segments on the connection, to allow time for segments from the earlier connection incarnation to drain from the system.

Even hosts that can remember the time of day and use it to select initial sequence number values are not immune from this problem, even if the time of day is used to select an initial sequence number for each new connection incarnation. Suppose a connection is opened starting with sequence number S. Suppose that this connection is not used much, and that eventually the initial sequence number function (ISN(t)) takes on a value equal to the sequence number, say S1, of the last segment sent by this TCP on a particular connection. Now suppose, at this instant, the host crashes, recovers, and establishes a new incarnation of the connection. The initial sequence number chosen is S1 = ISN(t)—the last-used sequence number on old incarnation of connection! If the recovery occurs quickly enough, any old duplicates in the internet bearing sequence numbers in the neighborhood of S1 may arrive and be treated as new packets by the receiver of the new incarnation of the connection. The problem is that the recovering host might not know how long it crashed or whether there are still old duplicates in the system from earlier connection incarnations.

One way to deal with this problem is to deliberately delay emitting segments for one MSL after recovery from a crash. This is the *quiet time* specification. Hosts that prefer to avoid waiting and are willing to risk possible confusion of old and new packets at a given destination may choose not to wait for the quiet time. Implementors may provide TCP users with the ability to select on a connection-by-connection basis whether to wait after a crash, or may informally implement the quiet time for all connections.

Obviously, even where a user selects to wait, this is not necessary after the host has been up for at least MSL seconds. To summarize: Every segment emitted occupies one or more sequence numbers in the sequence space. The numbers occupied by a segment are busy or in use until MSL seconds have passed. Upon crashing, a block of space-time is occupied by the octets of the last emitted segment. If a new connection is started too soon and uses any of the sequence numbers in the space-time footprint of the last segment of the previous connection incarnation, there is a potential sequence-number overlap area, which could cause confusion at the receiver.

10.6 Establishing a TCP Connection

The three-way handshake is the procedure used to establish a connection. This procedure normally is initiated by one TCP and responded to by another TCP. The procedure also works if two TCP simultaneously initiate the procedure. When a simultaneous attempt occurs, each TCP receives a SYN segment that carries no acknowledgment after it has sent a SYN. Of course, the arrival of an old, duplicate SYN segment can potentially make it appear to the recipient that a simultaneous connection initiation is in progress. Proper use of reset segments can eliminate the ambiguity of these cases. Although examples do not show connection synchronization using data-carrying segments, this is perfectly legitimate, so long as the receiving TCP doesn't deliver the data to the user until it is clear the data is valid; that means the data must be buffered at the receiver until the connection reaches the Established state. The three-way handshake reduces the possibility of false connections. It is the implementation of a trade-off between memory and messages to provide information for this checking.

The simplest three-way handshake is shown in the following table. Each line is numbered for reference purposes. A right arrow (—>) indicates the departure of a TCP segment from TCP A to TCP B, or the arrival of a segment at B from A. A left arrow (<—) indicates the reverse. An ellipsis (...) indicates a segment that is still in the network (delayed). An XXX indicates a segment that is lost or rejected. Comments appear in parentheses. TCP states represent the state after the departure or arrival of the segment (whose contents are shown in the center of each line). Segment contents are shown in abbreviated form, with sequence number, control flags, and ACK field. Other fields, such as window, addresses, lengths, and text have been left out in the interest of clarity.

	TCP A	TCP B
1	CLOSED	LISTEN
2	SYN-SENT —> <SEQ=100><CTL=SYN> —>	SYN-RECEIVED
3	ESTABLISHED <-<SEQ=300><ACK=101><CTL=SYN,ACK><-	SYN-RECEIVED
4	ESTABLISHED -><SEQ=101><ACK=301><CTL=ACK>—>	ESTABLISHED
5	ESTABLISHED-><SEQ=101><ACK=301><CTL=ACK><DATA>->	ESTABLISHED

In line 2, TCP A begins by sending a SYN segment indicating that it will use sequence numbers starting with sequence number 100. In line 3, TCP B sends a SYN and acknowledges the SYN it received from TCP A. Note that the acknowledgment field indicates TCP B is now expecting to hear sequence 101, acknowledging the SYN which occupied sequence 100.

At line 4, TCP A responds with an empty segment containing an ACK for TCP B's SYN. In line 5, TCP A sends some data. Note that the sequence number of the segment in line 5 is the same as in line 4 because the ACK does not occupy sequence number space. (If it did, we would wind up ACKing ACKs!)

Simultaneous initiation is only slightly more complex. Each TCP cycles from CLOSED to SYN-SENT to SYN-RECEIVED to ESTABLISHED, as follows:

	TCP A	TCP B
1	CLOSED	CLOSED
2	SYN-SENT—> <SEQ=100><CTL=SYN>	...
3	SYN-RECEIVED <-<SEQ=300><CTL=SYN><-	SYN-SENT
4	... <SEQ=100><CTL=SYN>->	SYN-RECEIVED
5	SYN-RECEIVED -><SEQ=100><ACK=301><CTL=SYN,ACK>	...
6	ESTABLISHED <-<SEQ=300><ACK=101><CTL=SYN,ACK><-	SYN-RECEIVED
7	... <SEQ=101><ACK=301><CTL=ACK> —>	ESTABLISHED

The principle reason for the three-way handshake is to prevent old, duplicate connection-initiations from causing confusion. To deal with this, a special control message, reset, has been devised. If the receiving TCP is in a nonsynchronized state (i.e., SYN-SENT, SYN-RECEIVED), it returns to LISTEN on receiving an acceptable reset. If the TCP is in one of the synchronized states (ESTABLISHED, FIN-WAIT-1, FIN-WAIT-2, CLOSE-WAIT, CLOSING, LAST-ACK, TIME-WAIT), it aborts the connection and informs its user. Consider the half-open connections shown here:

	TCP A	TCP B
1	CLOSED	LISTEN
2	SYN-SENT-><SEQ=100><CTL=SYN>	...
3	(duplicate) ... <SEQ=90><CTL=SYN>—>	SYN-RECEIVED
4	SYN-SENT <-<SEQ=300><ACK=91><CTL=SYN,ACK><-	SYN-RECEIVED
5	SYN-SENT-><SEQ=91><CTL=RST>->	LISTEN
6	... <SEQ=100><CTL=SYN>->	SYN-RECEIVED
7	SYN-SENT<-<SEQ=400><ACK=101><CTL=SYN,ACK><-	SYN-RECEIVED
8	ESTABLISHED-><SEQ=101><ACK=401><CTL=ACK>->	ESTABLISHED

At line 3, an old duplicate SYN arrives at TCP B. TCP B cannot tell that this is an old duplicate, so it responds normally (line 4). TCP A detects that the ACK field is incorrect and returns a RST (reset) with its SEQ field selected to make the segment believable. TCP B, on receiving the RST, returns to the LISTEN state. When the original SYN (pun intended) finally arrives at line 6, the synchronization proceeds normally. If the

SYN at line 6 had arrived before the RST, a more complex exchange might have occurred, with RSTs sent in both directions.

Half-Open Connections and Other Anomalies

An established connection is said to be *half-open* if one of the TCPs has closed or aborted the connection at its end without the knowledge of the other, or if the two ends of the connection have become desynchronized due to a crash that resulted in loss of memory. Such connections will automatically become reset if an attempt is made to send data in either direction. However, half-open connections are expected to be unusual, and the recovery procedure is mildly involved.

If at site A the connection no longer exists, then an attempt by the user at site B to send any data on it will result in the site B TCP receiving a reset control message. Such a message indicates to the site B TCP that something is wrong, and it is expected to abort the connection.

Assume that two user processes, A and B, are communicating with one another when a crash occurs, causing loss of memory to A's TCP. Depending on the operating system supporting A's TCP, it is likely that some error-recovery mechanism exists. When the TCP is up again, A is likely to start again from the beginning or from a recovery point. As a result, A will probably try to open the connection again, or try to send on the connection it believes open. In the latter case, it receives the error message "Connection not open" from the local (A's) TCP. In an attempt to establish the connection, A's TCP will send a segment containing SYN. After TCP A crashes, the user attempts to reopen the connection. TCP B, in the meantime, thinks the connection is open. Consider the following:

	TCP A	TCP B
1	(CRASH)	(send 300,receive 100)
2	CLOSED	ESTABLISHED
3	SYN-SENT-><SEQ=400><CTL=SYN>->	(??)
4	(!!)<-<SEQ=300><ACK=100><CTL=ACK><-	ESTABLISHED
5	SYN-SENT-><SEQ=100><CTL=RST>->	(Abort!!)
6	SYN-SENT	CLOSED
7	SYN-SENT-><SEQ=400><CTL=SYN>	—>

When the SYN arrives at line 3, TCP B, being in a synchronized state, and the incoming segment outside the window, responds with an acknowledgment indicating what sequence it next expects to hear (ACK 100). TCP A sees that this segment does not acknowledge anything it sent and, being unsynchronized, sends a reset (RST) because it has detected a half-open connection. TCP B aborts at line 5. TCP A will continue to try to establish the connection; the problem is now reduced to the basic 3-way handshake.

An interesting alternative case occurs when TCP A crashes and TCP B tries to send data on what it thinks is a synchronized connection. This is illustrated in the example that follows:

	TCP A	TCP B
1	(CRASH)	(send 300,receive 100)
2	(??) <-<SEQ=300><ACK=100><DATA=10><CTL=ACK><-	ESTABLISHED
3	-><SEQ=100><CTL=RST>->	(ABORT!!)

The data arriving at TCP A from TCP B (line 2) is unacceptable because no such connection exists, so TCP A sends an RST. The RST is acceptable, so TCP B processes it and aborts the connection.

In the following, two TCPs, A and B, with passive connections, wait for SYN:

	TCP A	TCP B
1	LISTEN	LISTEN
2	...<SEQ=Z><CTL=SYN>->	SYN-RECEIVED
3	(??)<-<SEQ=X><ACK=Z+1><CTL=SYN,ACK><-	SYN-RECEIVED
4	-><SEQ=Z+1><CTL=RST>->	(return to LISTEN!)
5	LISTEN	LISTEN

An old duplicate arriving at TCP B (line 2) stirs B into action. A SYN-ACK is returned (line 3), causing TCP A to generate a RST (the ACK in line 3 is not acceptable). TCP B accepts the reset and returns to its passive LISTEN state.

Reset Generation

As a general rule, reset (RST) must be sent whenever a segment arrives that apparently is not intended for the current connection. A reset must not be sent if it is not clear that this is the case. There are three groups of states:

1. If the connection does not exist (CLOSED), then a reset is sent in response to any incoming segment except another reset. In particular, SYNs addressed to a nonexistent connection are rejected by this means. If the incoming segment has an ACK field, the reset takes its sequence number from the ACK field of the segment; otherwise, the reset has sequence number zero, and the ACK field is set to the sum of the sequence number and segment length of the incoming segment. The connection remains in the CLOSED state.

2. If the connection is in any nonsynchronized state (LISTEN, SYN-SENT, or SYN-RECEIVED), and the incoming segment acknowledges something not yet sent (the segment carries an unacceptable ACK), or if an incoming segment has a security level or compartment that does not exactly match the level and compartment requested for the connection, a reset is sent.

   If our SYN has not been acknowledged and the precedence level of the incoming segment is higher than the precedence level requested, either raise the local precedence level (if allowed by the user and the system) or send a reset; or if the precedence level of the incoming segment is lower than the precedence level requested, continue as if the precedence matched exactly. (If the remote TCP cannot raise the precedence level to match ours, this will be detected in the next segment it sends, and the connection will be terminated then.) If our SYN has been acknowledged (perhaps in this incoming segment) the precedence level of the incoming segment must match the local precedence level exactly. If it does not, a reset must be sent.

If the incoming segment has an ACK field, the reset takes its sequence number from the ACK field of the segment; otherwise, the reset has sequence number zero, and the ACK field is set to the sum of the sequence number and segment length of the incoming segment. The connection remains in the same state.

3. If the connection is in a synchronized state (ESTABLISHED, FIN-WAIT-1, FIN-WAIT-2, CLOSE-WAIT, CLOSING, LAST-ACK, or TIME-WAIT), any unacceptable segment (out-of-window-sequence number or unacceptable acknowledgment number) must elicit only an empty acknowledgment segment containing the current send-sequence number and an acknowledgment indicating the next sequence number expected to be received. The connection remains in the same state.

If an incoming segment has a security level, compartment, or precedence that does not exactly match the level, compartment, and precedence requested for the connection, a reset is sent and the connection goes to the CLOSED state. The reset takes its sequence number from the ACK field of the incoming segment.

TCP Reset Processing

In all states except SYN-SENT, all reset (RST) segments are validated by checking their SEQ-fields. A reset is valid if its sequence number is in the window. In the SYN-SENT state (an RST received in response to an initial SYN), the RST is acceptable if the ACK field acknowledges the SYN.

The receiver of an RST first validates it, then changes state. If the receiver was in the LISTEN state, it ignores it. If the receiver was in SYN-RECEIVED state and had previously been in the LISTEN state, then the receiver returns to the LISTEN state; otherwise, the receiver aborts the connection and goes to the CLOSED state. If the receiver was in any other state, it aborts the connection, advises the user, and goes to the CLOSED state.

10.7   Closing a TCP Connection

CLOSE is an operation meaning "I have no more data to send." The notion of closing a full-duplex connection is subject to ambiguous interpretation, of course, since it might not be obvious how to treat the receiving side of the connection. We have chosen to treat CLOSE in a simplex fashion. The user who closes may continue to receive until he or she is told that the other side has closed also. Thus, a program could initiate several SENDs followed by a CLOSE, and then continue to receive until signaled that a RECEIVE failed because the other side has closed.

We assume that the TCP will signal a user, even if no RECEIVEs are outstanding, and that the other side has closed, so the user can terminate gracefully. A TCP will reliably deliver all buffers sent before the connection was closed, so a user who expects no data in return need only wait to hear the connection was closed successfully to know that all data was received at the destination TCP. Users must keep reading connections they close for sending until the TCP says there is no more data.

Essentially, three cases exist:

1. The user initiates by telling the TCP to close the connection.

2. The remote TCP initiates by sending a FIN control signal.

3. Both users close simultaneously.

Case 1: Local User Initiates the Close   In the first case, a FIN segment can be constructed and placed on the outgoing segment queue. No further SENDs from the user will be accepted by the TCP, and it enters the FIN-WAIT-1 state. RECEIVEs are allowed in this state. All segments preceding and including FIN will be retransmitted until acknowledged. When the other TCP has both acknowledged the FIN and sent a FIN of its own, the first TCP can ACK this FIN. Note that a TCP receiving a FIN will ACK but not send its own FIN until its user has closed the connection also.

Case 2: TCP Receives a FIN from the Network   If an unsolicited FIN arrives from the network, the receiving TCP can ACK it and tell the user that the connection is closing. The user will respond with a CLOSE, upon which the TCP can send a FIN to the other TCP after sending any remaining data. The TCP then waits until its own FIN is acknowledged, whereupon it deletes the connection. If an ACK is not forthcoming, after the user timeout, the connection is aborted and the user is told.

Case 3: Both Users Close Simultaneously   A simultaneous CLOSE by users at both ends of a connection causes FIN segments to be exchanged. When all segments preceding the FINs have been processed and acknowledged, each TCP can ACK the FIN it has received. Both will, upon receiving these ACKs, delete the connection:

TCP A		TCP B
1	ESTABLISHED	ESTABLISHED
2	Close FIN-WAIT-1->\<SEQ=100>\<ACK=300>\<CTL=FIN,ACK>->	CLOSE-WAIT
3	FIN-WAIT-2<-\<SEQ=300>\<ACK=101>\<CTL=ACK><-	CLOSE-WAIT
4	Close TIME-WAIT<-\<SEQ=300>\<ACK=101>\<CTL=FIN,ACK><-	LAST-ACK
5	TIME-WAIT->\<SEQ=101>\<ACK=301>\<CTL=ACK>->	CLOSED
6	2 MSL	CLOSED

TCP A		TCP B
1	ESTABLISHED	ESTABLISHED
2	Close	Close
	FIN-WAIT-1->\<SEQ=100>\<ACK=300>\<CTL=FIN,ACK>... FIN-WAIT-1 <—\<SEQ=300>\<ACK=100>\<CTL=FIN,ACK><-	...
	\<SEQ=100>\<ACK=300>\<CTL=FIN,ACK>->	
3	CLOSING->\<SEQ=101>\<ACK=301>\<CTL=ACK>...	CLOSING
	\<SEQ=301>\<ACK=101>\<CTL=ACK>	<—
	\<SEQ=101>\<ACK=301>\<CTL=ACK>	—>
4	TIME-WAIT	TIME-WAIT
	2 MSL	2 MSL
	CLOSED	CLOSED

Precedence and Security The intent is that a connection be allowed only between ports operating with exactly the same security and compartment values and at the higher of the precedence level requested by the two ports. The precedence and security parameters used in TCP are exactly those defined in the Internet Protocol (IP). Throughout this TCP specification, the term *security/compartment* is intended to indicate the security parameters used in IP, including security, compartment, user group, and handling restriction. A connection attempt with mismatched security/compartment values or a lower precedence value must be rejected by sending an RST. Rejecting a connection due to too low a precedence only occurs after an acknowledgment of the SYN has been received. TCP modules that operate only at the default value of precedence will still have to check the precedence of incoming segments and possibly raise the precedence level they use on the connection.

The security parameters may be used even in a non-secure environment. (The values would indicate unclassified data.) Thus, hosts in non-secure environments must be prepared to receive the security parameters, though they need not send them.

10.8 TCP and Data Communication

Once the connection is established, data is communicated by the exchange of segments. Because segments might be lost due to errors (checksum test failure) or network congestion, TCP uses retransmission (after a timeout) to ensure delivery of every segment. Duplicate segments might arrive due to network or TCP retransmission. As discussed in the section on sequence numbers, TCP performs certain tests on the sequence and acknowledgment numbers in the segments to verify their acceptability.

The sender of data keeps track of the next sequence number to use in the variable SND.NXT. The receiver of data keeps track of the next sequence number to expect in the variable RCV.NXT. The sender of data keeps track of the oldest unacknowledged sequence number in the variable SND.UNA. If the data flow is momentarily idle and all data sent has been acknowledged, then the three variables will be equal. When the sender creates a segment and transmits it, the sender advances SND.NXT. When the receiver accepts a segment, it advances RCV.NXT and sends an acknowledgment. When the data sender receives an acknowledgment, it advances SND.UNA. The extent to which the values of these variables differ is a measure of the delay in the communication.

The amount by which the variables are advanced is the length of the data in the segment. Note that once in the ESTABLISHED state, all segments must carry current acknowledgment information. The CLOSE user-call implies a PUSH function, as does the FIN control flag in an incoming segment.

TCP Retransmission Timeout

Because of the variability of the networks that compose an internetwork system and the wide range of uses of TCP connections, the retransmission timeout must be dynamically determined. One procedure for determining a retransmission timeout is given here, as an illustration.

An example retransmission-timeout procedure is to measure the elapsed time between sending a data octet with a particular sequence number and receiving an acknowledgment that covers that sequence number (segments sent do not have to match segments received). This measured elapsed time is the round-trip time (RTT). Next, compute a smoothed round-trip time (SRTT) as follows:

SRTT = (ALPHA * SRTT) + ((1-ALPHA) * RTT)

based on this, compute the retransmission timeout (RTO) as

RTO = min[UBOUND,max[LBOUND,(BETA*SRTT)]]

where UBOUND is an upper bound on the timeout (one minute), LBOUND is a lower bound on the timeout (one second), ALPHA is a smoothing factor (0.8 to 0.9), and BETA is a delay variance factor (1.3 to 2.0).

TCP Communication of Urgent Information

The objective of the TCP urgent mechanism is to allow the sending user to stimulate the receiving user to accept some urgent data and to permit the receiving TCP to indicate to the receiving user when all the currently known urgent data has been received by the user. This mechanism permits a point in the data stream to be designated as the end of urgent information. Whenever this point is in advance of the receive sequence number (RCV.NXT) at the receiving TCP, that TCP must tell the user to go into "urgent mode"; when the receive sequence number catches up to the urgent pointer, the TCP must tell the user to go into "normal mode." If the urgent pointer is updated while the user is in urgent mode, the update will be invisible to the user.

The method employs an urgent field, which is carried in all segments transmitted. The URG control flag indicates that the urgent field is meaningful and must be added to the segment sequence number to yield the urgent pointer. The absence of this flag indicates that there is no urgent data outstanding. To send an urgent indication, the user must also send at least one data octet. If the sending user also indicates a PUSH, timely delivery of the urgent information to the destination process is enhanced.

Managing the Window

The window sent in each segment indicates the range of sequence numbers the sender of the window (the data receiver) is currently prepared to accept. It is assumed that this is related to the currently available data buffer space available for this connection.

Indicating a large window encourages transmissions. If more data arrives than can be accepted, it will be discarded. This will result in excessive retransmissions, adding unnecessarily to the load on the network and the TCPs. Indicating a small window might restrict the transmission of data to the point of introducing a round-trip delay between each new segment transmitted.

The mechanisms provided allow a TCP to advertise a large window and to subsequently advertise a much smaller window without having accepted that much data. This procedure, called *shrinking the window*, is strongly discouraged. The robustness principle dictates that TCPs will not shrink the window themselves, but will be prepared for such behavior on the part of other TCPs.

The sending TCP must be prepared to accept from the user and send at least one octet of new data, even if the send window is zero. The sending TCP must regularly retransmit to the receiving TCP even when the window is zero. Two minutes is recommended for the retransmission interval when the window is zero. This retransmission is essential to guarantee that when either TCP has a zero window, the reopening of the window will be reliably reported to the other.

When the receiving TCP has a zero window and a segment arrives, it must still send an acknowledgment showing its next expected sequence number and current window (zero). The sending TCP packages the data to be transmitted into segments that fit the current window, and may repackage segments on the retransmission queue. Such repackaging is not required, but may be helpful.

In a connection with a one-way data flow, the window information will be carried in acknowledgment segments that all have the same sequence number, so there will be no way to reorder them if they arrive out of order. This is not a serious problem, but it will allow the window information to be, on occasion, temporarily based on old reports from the data receiver. A refinement to avoid this problem is to act on the window information from segments that carry the highest acknowledgment number (that is, segments with acknowledgment number equal to or greater than the highest previously received).

Window management procedure has significant influence on the communication performance. The following comments are suggestions:

- Allocating a small window causes data to be transmitted in many small segments when better performance is achieved using fewer large segments.

- To avoid small windows, the receiver should defer updating a window until the additional allocation is at least X percent of the maximum allocation possible for the connection (where X might be 20 to 40).

- The sender should avoid sending small segments by waiting until the window is large enough before sending data. If the user signals a PUSH function, then the data must be sent even if it is a small segment.

Acknowledgments should not be delayed, or unnecessary retransmissions will result. One strategy would be to send an acknowledgment when a small segment arrives (without updating the window information), and then to send another acknowledgment with new window information when the window is larger. The segment sent to probe a zero window may also begin to break up transmitted data into smaller and smaller segments. If a segment containing a single data octet sent to probe a zero window is accepted, it consumes one octet of the window now available. If the sending TCP simply sends as much as it can whenever the window is nonzero, the transmitted data will be broken into alternating big and small segments. As time goes on, occasional pauses in the receiver making window allocation available will result in breaking the big segments into a small and not-quite-so-big pair. After a while, the data transmission will be in mostly small segments.

TCP implementations need to actively attempt to combine small window allocations into larger windows, since the mechanisms for managing the window tend to lead to many small windows in the simplest-minded implementations.

10.9   TCP Interfaces

There are two interfaces of concern: the user/TCP interface and the TCP/lower-level interface.

User/TCP Interface

The following functional description of user commands to the TCP is, at best, fictional, since every operating system will have different facilities. Consequently, we must warn readers that different TCP implementations may have different user interfaces. However, all TCPs must provide a certain minimum set of services to guarantee that all TCP implementations can support the same protocol hierarchy.

TCP User Commands

The following sections functionally characterize a user/TCP interface. The notation used is similar to most procedure or function calls in high-level languages, but this usage is not meant to rule out trap-type service calls such as SVCs, UUOs, and EMTs.

The user commands described on the following pages specify the basic functions the TCP must perform to support interprocess communication. Individual implementations must define their own exact format, and may provide combinations or subsets of the basic functions in single calls. In particular, some implementations might want to automatically open a connection on the first SEND or RECEIVE issued by the user for a given connection.

In providing interprocess communication facilities, the TCP must not only accept commands, but must also return information to the processes it serves. The latter consists of the following:

1. General information about a connection, such as interrupts, remote close, and binding of unspecified foreign socket.

2. Replies to specific user commands indicating success or various types of failure.

OPEN Command
Format:

```
OPEN (local port, foreign socket, active/passive [, timeout] [, prece-
dence] [, security/compartment] [, options]) -> local connection name
```

We assume that the local TCP is aware of the identity of the processes it serves and will check the authority of the process to use the connection specified. Depending upon the implementation of the TCP, the local network and TCP identifiers for the source address will either be supplied by the TCP or the lower-level protocol (e.g., IP). These considerations are the result of concern about security, to the extent that no TCP will be able to masquerade as another one, and so on. Similarly, no process can masquerade as another without the collusion of the TCP.

If the active/passive flag is set to passive, then this is a call to listen for an incoming connection. A passive open may have either a fully specified foreign socket to wait for a particular connection or an unspecified foreign socket to wait for any call. A fully specified passive call can be made active by the subsequent execution of a SEND.

A Transmission Control Block (TCB) is created and partially filled in with data from the OPEN command parameters. On an active OPEN command, the TCP will begin the procedure to synchronize (establish) the connection at once. The timeout, if present, permits the caller to set up a timeout for all data submitted to TCP. If data is not successfully delivered to the destination within the timeout period, the TCP will abort the connection. The present global default is five minutes.

The TCP or some component of the operating system will verify the user's authority to open a connection with the specified precedence or security/compartment. The absence of precedence or security/compartment specification in the OPEN call indicates the default values must be used.

TCP will accept incoming requests as matching only if the security/compartment information is exactly the same and only if the precedence is equal to or higher than the precedence requested in the OPEN call.

The precedence for the connection is the higher of the values requested in the OPEN call and received from the incoming request, and fixed at that value for the life of the connection. Implementers may want to give the user control of this precedence negotiation. For example, the user might be allowed to specify that the precedence must be exactly matched, or that any attempt to raise the precedence be confirmed by the user.

A local connection name will be returned to the user by the TCP. The local connection name can then be used as a short-hand term for the connection defined by the <local socket, foreign socket> pair.

SEND Command
Format:

```
SEND (local connection name, buffer address, byte count, PUSH flag,
URGENT flag [,timeout])
```

This call causes the data contained in the indicated user buffer to be
sent on the indicated connection. If the connection has not been opened,
the SEND is considered an error. Some implementations allow users to
SEND first, in which case an automatic OPEN would be done. If the calling process is not authorized to use this connection, an error is returned.

If the PUSH flag is set, the data must be transmitted promptly to the
receiver, and the PUSH bit will be set in the last TCP segment created
from the buffer. If the PUSH flag is not set, the data may be combined
with data from subsequent SENDs for transmission efficiency.

If the URGENT flag is set, segments sent to the destination TCP will
have the urgent pointer set. The receiving TCP will signal the urgent
condition to the receiving process if the urgent pointer indicates that
data preceding the urgent pointer has not been consumed by the
receiving process. The purpose of URGENT is to stimulate the receiver
to process the urgent data and to indicate to the receiver when all the
currently known urgent data has been received. The number of times
the sending user's TCP signals URGENT will not necessarily be equal
to the number of times the receiving user will be notified of the presence of urgent data.

If no foreign socket was specified in the OPEN, but the connection is
established (because a LISTENing connection has become specific, due to a
foreign segment arriving for the local socket), then the designated buffer is
sent to the implied foreign socket. Users who make use of OPEN with an
unspecified foreign socket can make use of SEND without ever explicitly
knowing the foreign socket address.

However, if a SEND is attempted before the foreign socket becomes
specified, an error will be returned. Users can use the STATUS call to
determine the status of the connection. In some implementations, the
TCP may notify the user when an unspecified socket is bound.

If a timeout is specified, the current user timeout for this connection
is changed to the new one. In the simplest implementation, SEND would
not return control to the sending process until either the transmission
was complete or the timeout had been exceeded. However, this simple
method is both subject to deadlocks (for example, both sides of the connection might try to do SENDs before doing any RECEIVEs) and offers
poor performance, so it is not recommended. A more sophisticated
implementation would return immediately to allow the process to run
concurrently with network I/O, and, furthermore, to allow multiple

SENDs to be in progress. Multiple SENDs are served in first-come, first-served order, so the TCP will queue those it cannot service immediately.

This scenario implicitly assumes an asynchronous user interface in which a SEND later elicits some kind of signal or pseudo-interrupt from the serving TCP. An alternative is to return a response immediately. For instance, SENDs might return immediate local acknowledgment, even if the segment sent had not been acknowledged by the distant TCP. We could optimistically assume eventual success. If we are wrong, the connection will close anyway, due to the timeout. In implementations of this kind (synchronous), there will still be some asynchronous signals, but these will deal with the connection itself, and not with specific segments or buffers.

In order for the process to distinguish among error or success indications for different SENDs, it might be appropriate for the buffer address to be returned along with the coded response to the SEND request. TCP-to-user signals are discussed later in this chapter, indicating the information that should be returned to the calling process.

RECEIVE Command
Format:

```
RECEIVE (local connection name, buffer address, byte count) -> byte
count, urgent flag, push flag
```

The RECEIVE command allocates a receiving buffer associated with the specified connection. If no OPEN precedes this command or the calling process is not authorized to use this connection, an error is returned. In the simplest implementation, control would not return to the calling program until either the buffer was filled, or some error occurred, but this scheme is highly subject to deadlocks. A more sophisticated implementation would permit several RECEIVEs to be outstanding at once. These would be filled as segments arrive. This strategy permits increased throughput at the cost of a more elaborate scheme (possibly asynchronous) to notify the calling program that a PUSH has been seen or a buffer filled.

If enough data arrives to fill the buffer before a PUSH is seen, the PUSH flag will not be set in response to the RECEIVE. The buffer will be filled with as much data as it can hold. If a PUSH is seen before the buffer is filled, the buffer will be returned partially filled, and PUSH indicated. If there is urgent data, the user will have been informed as soon as it arrived via a TCP-to-user signal. The receiving user should thus be in urgent mode. If the URGENT flag is on, additional urgent data remains. If the URGENT flag is off, this call to RECEIVE has returned all the urgent data, and the user may now leave urgent mode. Note that data following the urgent pointer (non-urgent data) cannot be delivered to the user in the same buffer with preceding urgent data unless the boundary is clearly marked for the user.

To distinguish among several outstanding RECEIVEs and to take care of the case where a buffer is not completely filled, the return code is accompanied by both a buffer pointer and a byte count indicating the actual length of the data received. Alternative implementations of RECEIVE might have the TCP allocate buffer storage, or the TCP might share a ring buffer with the user.

CLOSE Command
Format:

```
CLOSE (local connection name)
```

The CLOSE command causes the connection specified to be closed. If the connection is not open or the calling process is not authorized to use this connection, an error is returned.

Closing a connection is intended to be a graceful operation in the sense that outstanding SENDs will be transmitted and retransmitted, as flow control permits, until all have been serviced. Thus, it should be acceptable to make several SEND calls, followed by a CLOSE, and expect all the data to be sent to the destination. It should also be clear that users should continue to receive on closing connections, since the other side might be trying to transmit the last of its data. Thus, CLOSE means "I have no more to send," but does not mean "I will not receive any more." If the user-level protocol is not well thought out, the closing side might be unable to get rid of all its data before timing out. In this event, CLOSE turns into ABORT, and the closing TCP gives up.

The user may close the connection at any time on his or her own initiative, or in response to various prompts from the TCP (e.g., remote close executed, transmission timeout exceeded, or destination inaccessible). Because closing a connection requires communication with the foreign TCP, connections may remain in the closing state for a short time. Attempts to reopen the connection before the TCP replies to the CLOSE command will result in error responses. CLOSE also implies the PUSH function.

STATUS Command
Format:

```
STATUS (local connection name) -> status data
```

STATUS is an implementation-dependent user command and could be excluded without adverse effect. Information returned typically comes from the TCB associated with the connection. This command returns a data block containing the following information:

- Local socket
- Foreign socket
- Local connection name
- Receive window
- Send window
- Connection state
- Number of buffers awaiting acknowledgment
- Number of buffers pending receipt
- Urgent state
- Precedence
- Security/compartment
- Transmission timeout

Depending on the state of the connection or on the implementation itself, some of this information might not be available or meaningful. If the calling process is not authorized to use this connection, an error is returned. This prevents unauthorized processes from gaining information about a connection.

ABORT Command
Format:

```
ABORT (local connection name)
```

The ABORT command causes all pending SENDs and RECEIVES to be aborted, the TCB to be removed, and a special RESET message to be sent to the TCP on the other side of the connection. Depending on the implementation, users might receive abort indications for each outstanding SEND or RECEIVE, or might simply receive an ABORT acknowledgment.

TCP-to-User Messages

It is assumed that the operating system environment provides a means for the TCP to asynchronously signal the user program. When the TCP does signal a user program, certain information is passed to the user. Often, in the specification, the information will be an error message. In other cases, there will be information relating to the completion of processing a SEND, RECEIVE, or other user call. The following information is provided:

- Local connection name—always
- Response string—always
- Buffer address—send and receive
- Byte count (counts bytes received)—receive
- PUSH flag—receive
- URGENT flag—receive

TCP/Lower-Level Interface

TCP calls on a lower-level protocol module to actually send and receive information over a network. One case is that of the ARPA internetwork system, where the lower-level module is the IP. If the lower-level protocol is IP, it provides arguments for Type of Service and for Time to Live. TCP uses the following settings for these parameters:

- Type of Service = Precedence: routine, delay: normal, throughput: normal, reliability: normal or 00000000
- Time to Live = One minute, or 00111100

The assumed maximum segment lifetime is two minutes. Here, we explicitly ask that a segment be destroyed if it cannot be delivered by the internet system within one minute. If the lower level is IP (or some other protocol that provides this feature) and source routing is used, the interface must allow the route information to be communicated. This is especially important so that the source and destination addresses used in the TCP checksum be the originating source and ultimate destination. It is also important to preserve the return route to answer connection requests.

Any lower-level protocol will have to provide the source address, destination address, and protocol fields, and some way to determine the TCP length, both to provide the functional equivalent service of IP and to be used in the TCP checksum.

10.10   TCP Event Processing

The activity of the TCP can be characterized as responding to events. The events that occur can be cast into three categories: user calls, arriving segments, and timeouts. The processing the TCP does is in response to each of the events. In many cases, the processing required depends on the state of the connection.

The processing depicted in this section is an example of one possible implementation. Other implementations might have slightly different processing sequences. The following are events that occur:

- User Calls:
  - OPEN
  - SEND
  - RECEIVE
  - CLOSE
  - ABORT
  - STATUS
- Arriving Segments:
  - SEGMENT ARRIVES
- Timeouts:
  - USER TIMEOUT
  - RETRANSMISSION TIMEOUT
  - TIME-WAIT TIMEOUT

The model of the TCP/user interface is that user commands receive an immediate return and possibly a delayed response via an event or pseudo-interrupt. In the following descriptions, the term *signal* means to cause a delayed response. Error responses are given as character strings. For example, user commands referencing connections that do not exist receive "error: connection not open." Also, please note in the following that all arithmetic on sequence numbers, acknowledgment numbers, windows, etc., is modulo 2^{32}, the size of the sequence number space. Also note that =< means less than or equal to (modulo 2^{32}).

A natural way to think about processing incoming segments is to imagine that they are first tested for proper sequence number (i.e., that their contents lie in the range of the expected receive window in the sequence number space), and then that they are generally queued and processed in sequence number order. When a segment overlaps other segments that have already been received, the segment is reconstructed to contain just the new data, and the header fields are adjusted to be consistent. If no state change is mentioned, the TCP stays in the same state.

OPEN Call

CLOSED State (TCB Does Not Exist) Create a new TCB to hold connection-state information. Fill in local socket identifier, foreign socket, precedence, security/compartment, and user timeout information. Note that some parts of the foreign socket may be unspecified in a passive OPEN and are to be filled in by the parameters of the incoming SYN segment. Verify that the security and precedence requested are allowed for this user. If not, return "error: precedence not allowed" or "error: security/compartment not allowed." If passive, enter the LISTEN state and return. If active and the foreign socket is unspecified, return "error: foreign socket unspecified." If active and the foreign socket is specified, issue a SYN segment. An initial send sequence number (ISS) is selected. A SYN segment of the form <SEQ=ISS><CTL=SYN> is sent. Set SND.UNA to ISS, SND.NXT to ISS+1, enter SYN-SENT state, and return. If the caller does not have access to the local socket specified, return "error: connection illegal for this process." If there is no room to create a new connection, return "error: insufficient resources."

LISTEN State If active, and the foreign socket is specified, then change the connection from passive to active, and select an ISS. Send a SYN segment, set SND.UNA to ISS, and SND.NXT to ISS+1. Enter the SYN-SENT state. Data associated with SEND may be sent with SYN segment or queued for transmission after entering the ESTABLISHED state. The URGENT bit, if requested in the command, must be sent with the data segments sent as a result of this command. If there is no room to queue the request, respond with "error: insufficient resources." If a foreign socket was not specified, then return "error: foreign socket unspecified."

SYN-SENT, SYN-RECEIVED, ESTABLISHED, FIN-WAIT-1, FIN-WAIT-2, CLOSE-WAIT, CLOSING STATE, LAST-ACK, TIME-WAIT State Return "error: connection already exists."

SEND Call

CLOSED State (TCB Does Not Exist) If the user does not have access to such a connection, return "error: connection illegal for this process." Otherwise, return "error: connection does not exist."

LISTEN State If the foreign socket is specified, change the connection from passive to active, and select an ISS. Send a SYN segment, set SND.UNA to ISS, and SND.NXT to ISS+1. Enter the SYN-SENT state. Data associated with SEND may be sent with the SYN segment or queued for transmission after entering the ESTABLISHED state. The URGENT bit, if requested in the command, must be sent with the data segments sent as a result of this command. If there is no room to queue the request, respond with "error: insufficient resources." If a foreign socket was not specified, return "error: foreign socket unspecified."

SYN-SENT, SYN-RECEIVED State Queue the data for transmission after entering the ESTABLISHED state. If there is no space to queue, respond with "error: insufficient resources."

ESTABLISHED or CLOSE-WAIT State Segmentize the buffer and send it with a piggybacked acknowledgment (acknowledgment value = RCV.NXT). If there is insufficient space to remember this buffer, simply return "error: insufficient resources." If the URGENT flag is set, then set SND.UP <- SND.NXT-1 and set the URGENT pointer in the outgoing segments.

FIN-WAIT-1, FIN-WAIT-2, CLOSING, LAST-ACK, TIME-WAIT State Return "error: connection closing" and do not service request.

RECEIVE Call

CLOSED State (TCB Does Not Exist) If the user does not have access to such a connection, return "error: connection illegal for this process." Otherwise return "error: connection does not exist."

LISTEN, SYN-SENT, SYN-RECEIVED State Queue for processing after entering the ESTABLISHED state. If there is no room to queue this request, respond with "error: insufficient resources."

ESTABLISHED, FIN-WAIT-1, FIN-WAIT-2 State If insufficient incoming segments are queued to satisfy the request, queue the request. If there is no queue space to remember the RECEIVE, respond with "error: insufficient resources." Reassemble queued incoming segments into the receive buffer and return to the user. Mark "push seen" (PUSH) if this is the case. If RCV.UP is in advance of the data currently being passed to the user, notify the user of the presence of urgent data.

When the TCP takes responsibility for delivering data to the user, that fact must be communicated to the sender via an acknowledgment. The formation of such an acknowledgment is described in the following discussion of processing an incoming segment.

CLOSE-WAIT State   Since the remote side has already sent FIN, RECEIVEs must be satisfied by text already on hand, but not yet delivered to the user. If no text is awaiting delivery, the RECEIVE will get an "error: connection closing" response. Otherwise, any remaining text can be used to satisfy the RECEIVE.

CLOSING, LAST-ACK, TIME-WAIT State   Return "error: connection closing."

CLOSE Call

CLOSED State (TCB Does Not Exist)   If the user does not have access to such a connection, return "error: connection illegal for this process." Otherwise, return "error: connection does not exist."

LISTEN State   Any outstanding RECEIVEs are returned with "error: closing" responses. Delete TCB, enter the CLOSED state, and return.

SYN-SENT State   Delete the TCB and return "error: closing" responses to any queued SENDs or RECEIVEs.

SYN-RECEIVED State   If no SENDs have been issued and there is no pending data to send, then form a FIN segment and send it, and enter FIN-WAIT-1 state. Otherwise, queue for processing after entering the ESTABLISHED state.

ESTABLISHED State   Queue this until all preceding SENDs have been segmentized, then form a FIN segment and send it. In any case, enter FIN-WAIT-1 state.

FIN-WAIT-1, FIN-WAIT-2 State   Strictly speaking, this is an error and should receive a "error: connection closing" response. An "ok" response would be acceptable, too, as long as a second FIN is not emitted (the first FIN may be retransmitted, though).

CLOSE-WAIT State Queue this request until all preceding SENDs have been segmentized; then send a FIN segment and enter the CLOSING state.

CLOSING, LAST-ACK, TIME-WAIT State Respond with "error: connection closing."

ABORT Call

CLOSED STATE (TCB does not exist) If the user should not have access to such a connection, return "error: connection illegal for this process." Otherwise return "error: connection does not exist."

LISTEN State Any outstanding RECEIVEs should be returned with "error: connection reset" responses. Delete TCB, enter the CLOSED state, and return.

SYN-SENT State All queued SENDs and RECEIVEs should be given "connection reset" notification. Then, delete the TCB, enter the CLOSED state, and return.

SYN-RECEIVED, ESTABLISHED, FIN-WAIT-1, FIN-WAIT-2, CLOSE-WAIT State
Send the reset segment <SEQ=SND.NXT><CTL=RST>. All queued SENDs and RECEIVEs should be given "connection reset" notification. All segments queued for transmission or retransmission should be flushed (except for the RST just formed). Delete the TCB, enter the CLOSED state, and return.

CLOSING, LAST-ACK, TIME-WAIT State Respond with "ok," delete the TCB, enter the CLOSED state, and return.

STATUS Call

CLOSED State (TCB Does Not Exist) If the user should not have access to such a connection, return "error: connection illegal for this process." Otherwise, return "error: connection does not exist."

LISTEN State   Return "state = LISTEN" and the TCB pointer.

SYN-SENT State   Return "state = SYN-SENT" and the TCB pointer.

SYN-RECEIVED State   Return "state = SYN-RECEIVED" and the TCB pointer.

ESTABLISHED State   Return "state = ESTABLISHED" and the TCB pointer.

FIN-WAIT-1 State   Return "state = FIN-WAIT-1" and the TCB pointer.

FIN-WAIT-2 State   Return "state = FIN-WAIT-2" and the TCB pointer.

CLOSE-WAIT State   Return "state = CLOSE-WAIT" and the TCB pointer.

CLOSING State Return   "state = CLOSING" and the TCB pointer.

LAST-ACK State   Return "state = LAST-ACK" and the TCB pointer.

TIME-WAIT State   Return "state = TIME-WAIT" and the TCB pointer.

SEGMENT ARRIVES   If the state is closed (i.e., TCB does not exist), then all data in the incoming segment is discarded. An incoming segment containing an RST is discarded. An incoming segment not containing an RST causes an RST to be sent in response. The acknowledgment and sequence field values are selected to make the reset sequence acceptable to the TCP that sent the offending segment. If the ACK bit is off, sequence number zero is used, <SEQ=0><ACK=SEG.SEQ+SEG.LEN><CTL=RST,ACK>. If the ACK bit is on, <SEQ=SEG.ACK><CTL=RST>. Return.

If the state is LISTEN, follow these steps:

1. Check for an RST. An incoming RST should be ignored. Return.

2. Check for an ACK. Any acknowledgment is bad if it arrives on a connection still in the LISTEN state. An acceptable reset segment should be formed for any arriving ACK-bearing segment. The RST should be formatted as follows:

   <SEQ=SEG.ACK><CTL=RST>

   Return.

3. Check for a SYN. If the SYN bit is set, check the security. If the security/compartment on the incoming segment does not

exactly match the security/compartment in the TCB, then send a reset and return.

ARRIVES

<SEQ=SEG.ACK><CTL=RST>   If the SEG.PRC is greater than the TCB.PRC, then if allowed by the user and the system, set TCB.PRC<-SEG.PRC. If not allowed, send a reset and return.

<SEQ=SEG.ACK><CTL=RST>   If the SEG.PRC is less than the TCB.PRC, then continue. Set RCV.NXT to SEG.SEQ+1, IRS to SEG.SEQ, and queue any other control or text for processing later. ISS should be selected, and a SYN segment sent of the form <SEQ=ISS><ACK=RCV.NXT><CTL=SYN,ACK>.

SND.NXT Is Set to ISS+1 and SND.UNA to ISS   The connection state should be changed to SYN-RECEIVED. Note that any other incoming control or data (combined with SYN) will be processed in the SYN-RECEIVED state, but processing of SYN and ACK should not be repeated. If the LIS-TEN was not fully specified (i.e., the foreign socket was not fully speci-fied), then the unspecified fields should be filled in now.

Other Text or Control   Any other control or text-bearing segment (not containing SYN) must have an ACK, and thus would be discarded by the ACK processing. An incoming RST segment could not be valid, since it could not have been sent in response to anything sent by this incarnation of the connection. So, you are unlikely to get here, but if you do, drop the segment and return. If the state is SYN-SENT, then first check the ACK bit. If the ACK bit is set to SEG.ACK =< ISS or SEG.ACK > SND.NXT, send a reset unless the RST bit is set. If so, drop the segment, return <SEQ=SEG.ACK><CTL=RST>, and discard the seg-ment. Return.

If SND.UNA =< SEG.ACK =< SND.NXT, then the ACK is acceptable. Check the RST bit.

SEGMENT ARRIVES

If the RST Bit Is Set   If the ACK was acceptable, then signal the user with "error: connection reset," drop the segment, enter the CLOSED state, delete the TCB, and return. Otherwise (no ACK), drop the segment and return.

Check the security and precedence. If the security/compartment in the segment does not exactly match the security/compartment in the TCB, send a reset. If there is an ACK, set

<SEQ=SEG.ACK><CTL=RST>

Otherwise, set

<SEQ=0><ACK=SEG.SEQ+SEG.LEN><CTL=RST,ACK>

If There Is an ACK The precedence in the segment must match the precedence in the TCB, if not, send a reset:

<SEQ=SEG.ACK><CTL=RST>

If There Is no ACK If the precedence in the segment is higher than the precedence in the TCB, then, if allowed by the user and the system, raise the precedence in the TCB to that in the segment. If not allowed to raise the precedence, then send a reset:

<SEQ=0><ACK=SEG.SEQ+SEG.LEN><CTL=RST,ACK>

If the Precedence in the Segment Is Lower Than the Precedence in the TCB If a reset was sent, discard the segment and return.

Check the SYN bit. This step should be reached only if the ACK is OK, or there is no ACK, and if the segment did not contain a RST. If the SYN bit is on and the security/compartment and precedence SEGMENT ARRIVES are acceptable, then RCV.NXT is set to SEG.SEQ+1 and IRS is set to SEG.SEQ.

SND.UNA should be advanced to equal SEG.ACK (if there is an ACK), and any segments on the retransmission queue that are thereby acknowledged should be removed. If SND.UNA > ISS (our SYN has been ACKed), change the connection state to ESTABLISHED, form an ACK segment

<SEQ=SND.NXT><ACK=RCV.NXT><CTL=ACK>

and send it. Data or controls that were queued for transmission may be included. If there are other controls or text in the segment, then continue processing at the sixth step below where the URG bit is checked, otherwise return. Otherwise, enter SYN-RECEIVED, form a SYN,ACK segment

<SEQ=ISS><ACK=RCV.NXT><CTL=SYN,ACK>

and send it. If there are other controls or text in the segment, queue them for processing after the ESTABLISHED state has been reached, and return.

If neither of the SYN or RST bits is set, then drop the segment and return. Otherwise, first check sequence number

SYN-RECEIVED, ESTABLISHED, FIN-WAIT-1, FIN-WAIT-2, CLOSE-WAIT, CLOSING, LAST-ACK, TIME-WAIT

Segments are processed in sequence. Initial tests on arrival are used to discard old duplicates, but further processing is done in SEG.SEQ order. If a segment's contents straddle the boundary between old and new, only the new parts should be processed.

There are four cases for the acceptability test of an incoming segment:

Segment Length	Receive Window	Test
0	0	SEG.SEQ = RCV.NXT
0	>0	RCV.NXT =< SEG.SEQ < RCV.NXT+RCV.WND
>0	0	Not acceptable
>0	>0	RCV.NXT =< SEG.SEQ < RCV.NXT+RCV.WND or RCV.NXT =< SEG.SEQ+SEG.LEN-1 < RCV.NXT+RCV.WND

If the RCV.WND is zero, no segments will be acceptable, but special allowance should be made to accept valid ACKs, URGs, and RSTs. If an incoming segment is not acceptable, an acknowledgment should be sent in reply, unless the RST bit is set. If so, drop the segment, and return the following:

<SEQ=SND.NXT><ACK=RCV.NXT><CTL=ACK>

After sending the acknowledgment, drop the unacceptable segment and return.

In the following, it is assumed that the segment is the idealized segment that begins at RCV.NXT and does not exceed the window. One could tailor actual segments to fit this assumption by trimming off any portions that lie outside the window (including SYN and FIN), and only processing further if the segment then begins at RCV.NXT. Segments with higher beginning sequence numbers may be held for later processing. Check the RST bit, SYN-RECEIVED STATE. If the RST bit is set and this connection was initiated with a passive OPEN (i.e., came from the LISTEN state), then return this connection to LISTEN state and return. The user need not be informed. If this connection was initiated with an active OPEN (i.e., came from SYN-SENT state) then the connection was refused; signal the user "connection refused." In either case, all segments on the retransmission queue should be removed. In the active OPEN case, enter the CLOSED state and delete the TCB, and return.

ESTABLISHED, FIN-WAIT-1, FIN-WAIT-2, CLOSE-WAIT State If the RST bit is set, then any outstanding RECEIVEs and SEND should receive "reset" responses. All segment queues should be flushed. Users should also receive an unsolicited general "connection reset" signal. Enter the CLOSED state, delete the TCB, and return.

CLOSING, LAST-ACK, TIME-WAIT State If the RST bit is set, then enter the CLOSED state, delete the TCB, and return.

SEGMENT ARRIVES Check security and precedence.

SYN-RECEIVED If the security/compartment and precedence in the segment do not exactly match the security/compartment and precedence in the TCB then send a reset, and return.

ESTABLISHED State If the security/compartment and precedence in the segment do not exactly match the security/compartment and precedence in the TCB, send a reset. Any outstanding RECEIVEs and SENDs should receive "reset" responses. All segment queues should be flushed, and users should receive an unsolicited general "connection reset" signal. Enter the CLOSED state, delete the TCB, and return. This check is placed after the sequence check to prevent a segment from an old connection between these ports with a different security or precedence from causing an abort of the current connection. Check the SYN bit.

SYN-RECEIVED, ESTABLISHED, FIN-WAIT-1, FIN-WAIT-2, CLOSE-WAIT, CLOSING, LAST-ACK, TIME-WAIT State If the SYN is in the window, it is an error. Send a reset. Any outstanding RECEIVEs and SENDs should receive "reset" responses, all segment queues should be flushed, and the user should also receive an unsolicited general "connection reset" signal. Enter the CLOSED state, delete the TCB, and return. If the SYN is not in the window, this step would not be reached and an ACK would have been sent in the first step (sequence number check). Check the ACK field. If the ACK bit is off, drop the segment and return if the ACK bit is on.

SYN-RECEIVED State If SND.UNA =< SEG.ACK =< SND.NXT, then enter ESTABLISHED state and continue processing. If the segment acknowledgment is not acceptable, form a reset segment, <SEQ=SEG.ACK><CTL=RST>, and send it.

ESTABLISHED State   If SND.UNA < SEG.ACK =< SND.NXT, then set SND.UNA <- SEG.ACK. Any segments on the retransmission queue that are thereby entirely acknowledged are removed. Users should receive positive acknowledgments for buffers that have been sent and fully acknowledged (i.e., SEND buffer should be returned with "ok" response). If the ACK is a duplicate (SEG.ACK < SND.UNA), it can be ignored. If the ACK acks something not yet sent (SEG.ACK > SND.NXT) then send an ACK, drop the segment, and return.

If SND.UNA < SEG.ACK =< SND.NXT, the send window should be updated. If (SND.WL1 < SEG.SEQ or (SND.WL1 = SEG.SEQ and SND.WL2 =< SEG.ACK)), set SND.WND <- SEG.WND, setSND.WL1 <- SEG.SEQ, and set SND.WL2 <- SEG.ACK. SND.WND is an offset from SND.UNA, checking that SND.WL1 records the sequence number of the last segment used to update SND.WND, and that SND.WL2 records the acknowledgment number of the last segment used to update SND.WND. The check here prevents using old segments to update the window.

FIN-WAIT-1 State   In addition to the processing for the ESTABLISHED state, if our FIN is now acknowledged, then enter FIN-WAIT-2 and continue processing in that state.

FIN-WAIT-2 State   In addition to the processing for the ESTABLISHED state, if the retransmission queue is empty, the user's CLOSE can be acknowledged ("ok"), but do not delete the TCB.

CLOSE-WAIT State   Do the same processing as for the ESTABLISHED state.

CLOSING State   In addition to the processing for the ESTABLISHED state, if the ACK acknowledges our FIN, then enter the TIME-WAIT state. Otherwise, ignore the segment.

LAST-ACK State   The only thing that can arrive in this state is an acknowledgment of our FIN. If our FIN is now acknowledged, delete the TCB, enter the CLOSED state, and return.

TIME-WAIT State   The only thing that can arrive in this state is a retransmission of the remote FIN. Acknowledge it, and restart the 2 MSL timeout. Check the URG bit.

ESTABLISHED, FIN-WAIT-1, FIN-WAIT-2 State   If the URG bit is set, RCV.UP <- max(RCV.UP,SEG.UP), and signal the user that the remote side has urgent data if the urgent pointer (RCV.UP) is in advance of the data consumed. If the user has already been signaled (or is still in the urgent mode) for this continuous sequence of urgent data, do not signal the user again.

CLOSE-WAIT, CLOSING, LAST-ACK State, TIME-WAIT   This should not occur, since a FIN has been received from the remote side. Ignore the URG. Process the segment text.

ESTABLISHED, FIN-WAIT-1, FIN-WAIT-2 State   Once in the ESTABLISHED state, it is possible to deliver segment text to user RECEIVE buffers. Text from segments can be moved into buffers until either the buffer is full or the segment is empty. If the segment empties and carries a PUSH flag, then the user is informed when the buffer is returned, that a PUSH has been received.

When the TCP takes responsibility for delivering the data to the user, it must also acknowledge the receipt of the data. Once the TCP takes responsibility for the data, it advances RCV.NXT over the data accepted, and adjusts RCV.WND as appropriate to the current buffer availability. The total of RCV.NXT and RCV.WND should not be reduced.

Send an acknowledgment of the form:

    <SEQ=SND.NXT><ACK=RCV.NXT><CTL=ACK>

This acknowledgment should be piggybacked on a segment being transmitted, if possible without incurring undue delay.

CLOSE-WAIT, CLOSING, LAST-ACK, TIME-WAIT State   This should not occur, since a FIN has been received from the remote side. Ignore the segment text. Check the FIN bit. Do not process the FIN if the state is CLOSED, LISTEN, or SYN-SENT, since the SEG.SEQ cannot be validated; drop the segment and return.

If the FIN bit is set, signal the user with "connection closing" and return any pending RECEIVEs with same message. Advance RCV.NXT over the FIN, and send an acknowledgment for the FIN. Note that FIN implies PUSH for any segment text not yet delivered to the user.

SYN-RECEIVED, ESTABLISHED State   Enter the CLOSE-WAIT state.

FIN-WAIT-1 State   If our FIN has been ACKed (perhaps in this segment), then enter TIME-WAIT, start the time-wait timer, and turn off the other timers. Otherwise, enter the CLOSING state.

FIN-WAIT-2 State   Enter the TIME-WAIT state. Start the time-wait timer, and turn off the other timers.

CLOSE-WAIT State   Remain in the CLOSE-WAIT state.

CLOSING State   Remain in the CLOSING state.

LAST-ACK State   Remain in the LAST-ACK state.

TIME-WAIT State   Remain in the TIME-WAIT state. Restart the 2 MSL time-wait timeout and return.

USER TIMEOUT   For any state, if the user timeout expires, flush all queues, signal the user with "error: connection aborted due to user timeout in general." For any outstanding calls, delete the TCB, enter the CLOSED state, and return.

RETRANSMISSION TIMEOUT   For any state, if the retransmission timeout expires on a segment in the retransmission queue, send the segment at the front of the retransmission queue again, reinitialize the retransmission timer, and return.

TIME-WAIT TIMEOUT   If the time-wait timeout expires on a connection, delete the TCB, enter the CLOSED state, and return.

10.11   TCP Glossary

ACK   A control bit (acknowledge) occupying no sequence space, which indicates that the acknowledgment field of this segment specifies the next sequence number the sender of this segment is expecting to receive, hence acknowledging receipt of all previous sequence numbers.

ARPANET message   The unit of transmission between a host and an IMP in the ARPANET. The maximum size is about 1012 octets (8096 bits).

ARPANET packet   A unit of transmission used internally in the ARPANET between IMPs. The maximum size is about 126 octets (1008 bits).

connection   A logical communication path identified by a pair of sockets.

datagram   A message sent in a packet-switched computer communications network.

destination address   The destination address, usually the network and host identifiers.

FIN   A control bit (finis) occupying one sequence number, which indicates that the sender will send no more data or control occupying sequence space.

fragment   A portion of a logical unit of data. In particular, an internet fragment is a portion of an internet datagram.

FTP   A file transfer protocol.

header   Control information at the beginning of a message, segment, fragment, packet, or block of data.

host   A computer. In particular, a source or destination of messages from the point of view of the communication network.

Identification   An Internet Protocol field. This identifying value, assigned by the sender, aids in assembling the fragments of a datagram.

IMP   The Interface Message Processor, the packet switch of the ARPANET.

internet address   A source or destination address specific to the host level.

internet datagram   The unit of data exchanged between an internet module and the higher-level protocol, together with the internet header.

internet fragment   A portion of the data of an internet datagram with an internet header.

IP   The Internet Protocol.

IRS   The Initial Receive Sequence number, the first sequence number used by the sender on a connection.

ISN   The Initial Sequence Number, the first sequence number used on a connection (either ISS or IRS). It is selected on a clock-based procedure.

ISS   The Initial Send Sequence number, the first sequence number used by the sender on a connection.

leader   Control information at the beginning of a message or block of data. In particular, in the ARPANET, the control information on an ARPANET message at the host-IMP interface.

left sequence   The next sequence number to be acknowledged by the data receiving TCP (or the lowest currently unacknowledged sequence number); sometimes referred to as the left edge of the send window.

local packet   The unit of transmission within a local network.

module   An implementation, usually in software, of a protocol or other procedure.

MSL   Maximum Segment Lifetime, the time a TCP segment can exist in the internetwork system. Arbitrarily defined to be two minutes.

octet   An eight-bit byte.

options   A field that may contain several options, and each may be several octets in length. Options are used primarily in testing situations; for example, to carry timestamps. Both IP and TCP provide for options fields.

packet   A package of data with a header, which may or may not be logically complete. It is more often a physical packaging than a logical packaging of data.

port   The portion of a socket that specifies which logical input or output channel of a process is associated with the data.

process   A program in execution; a source or destination of data from the point of view of the TCP or other host-to-host protocol.

PUSH   A control bit occupying no sequence space, indicating that this segment contains data that must be pushed through to the receiving user.

RCV.NXT   Receive next sequence number.

RCV.UP   Receive urgent pointer.

RCV.WND   Receive window.

Receive Next Sequence Number   The next sequence number the local TCP is expecting to receive.

receive window   The sequence numbers the local (receiving) TCP is willing to receive. Thus, the local TCP considers that segments overlapping the range RCV.NXT to RCV.NXT + RCV.WND - 1 carry acceptable data or control. Segments containing sequence numbers entirely outside of this range are considered duplicates and discarded.

RST   A control bit (reset) occupying no sequence space, indicating that the receiver should delete the connection without further interaction. The receiver can determine, based on the sequence number and acknowledgment fields of the incoming segment, whether it should honor the RST command or ignore it. In no case does receipt of a segment containing RST give rise to a RST in response.

RTP   Real Time Protocol, a host-to-host protocol for the communication of time-critical information.

SEG.ACK   Segment acknowledgment.

SEG.LEN   Segment length.

SEG.PRC   Segment precedence value.

SEG.SEQ   Segment sequence.

SEG.UP   Segment urgent pointer field.

SEG.WND   Segment window field.

segment   A logical unit of data; in particular, a TCP segment is the unit of data transferred between a pair of TCP modules.

segment acknowledgment   The sequence number in the acknowledgment field of the arriving segment.

segment length   The amount of sequence-number space occupied by a segment, including any controls that occupy sequence space.

segment sequence   The number in the sequence field of the arriving segment.

send sequence   The next sequence number the local (sending) TCP will use on the connection. It is initially selected from an initial sequence number curve (ISN) and is incremented for each octet of data or sequenced control transmitted.

send window   Represents the sequence numbers that the remote (receiving) TCP is willing to receive. It is the value of the window field specified in segments from the remote (data receiving) TCP. The range of new sequence numbers that·may be emitted by a TCP lies between SND.NXT and SND.UNA + SND.WND - 1. (Retransmissions of sequence numbers between SND.UNA and SND.NXT are expected, of course.)

SND.NXT   Send sequence.

SND.UNA   Left sequence.

SND.UP   Send urgent pointer.

SND.WL1   Segment sequence number at last window update.

SND.WL2   Segment acknowledgment number at last window update.

SND.WND   Send window.

socket   An address that specifically includes a port identifier; that is, the concatenation of an internet address with a TCP port.

source address   The source address, usually the network and host identifiers.

SYN   A control bit in the incoming segment, occupying one sequence number, used at the initiation of a connection, to indicate where the sequence numbering will start.

TCB   The Transmission Control Block, the data structure that records the state of a connection.

TCB.PRC   The precedence of the connection.

TCP   Transmission Control Protocol, a host-to-host protocol for reliable communication in internetwork environments.

TOS   Type of Service, an IP field that indicates the type of service for this internet fragment.

URG   A control bit (urgent) occupying no sequence space, used to indicate that the receiving user should be notified to do urgent processing as long as there is data to be consumed with sequence numbers less than the value indicated in the urgent pointer.

Urgent Pointer   A control field meaningful only when the URG bit is on. This field communicates the value of the urgent pointer, which indicates the data octet associated with the sending user's urgent call.

11

User Datagram
Protocol (UDP)

The User Datagram Protocol (UDP) is defined to make a datagram mode of packet-switched computer communication available in the environment of an interconnected set of computer networks. This protocol assumes that IP is used as the underlying protocol.

UDP provides a procedure for application programs to send messages to other programs with a minimum of protocol mechanism. The protocol is transaction oriented, and delivery and duplicate protection are not guaranteed.

11.1 UDP Header Format

The UDP header format is as follows:

```
 0      7 8     15 16    23 24     31
+-+-+-+-+-+-+-+-+-+-+-+-+-+-+-+-+
|     Source     |  Destination   |
|     Port       |     Port       |
+-+-+-+-+-+-+-+-+-+-+-+-+-+-+-+-+
|                |                |
|     Length     |    Checksum    |
+-+-+-+-+-+-+-+-+-+-+-+-+-+-+-+-+
|                                 |
|         Data Octets ...
+-+-+-+-+-+-+-+-+ ...
```

- *Source Port* is an optional field. When meaningful, it indicates the port of the sending process, and may be assumed to be the port to which a reply should be addressed in the absence of any other information. If not used, a value of zero is inserted.

- *Destination Port* has a meaning within the context of a particular internet destination address.

- *Length* is the length, in octets, of this user datagram including this header and the data; the minimum value of the length is eight.

- *Checksum* is the 16-bit one's complement of the one's complement sum of a pseudo-header of information from the IP header, the UDP header, and the data, padded with zero octets at the end (if necessary) to make a multiple of two octets.

The pseudo-header conceptually prefixed to the UDP header contains the source address, the destination address, the protocol, and the UDP length. This information gives protection against misrouted datagrams. This checksum procedure is the same as is used in TCP.

```
 0       7 8      15 16     23 24      31
+-+-+-+-+-+-+-+-+-+-+-+-+-+-+-+-+
|            Source Address            |
+-+-+-+-+-+-+-+-+-+-+-+-+-+-+-+-+
|          Destination Address         |
+-+-+-+-+-+-+-+-+-+-+-+-+-+-+-+-+
|   Zero  |Protocol|   UDP Length      |
+-+-+-+-+-+-+-+-+-+-+-+-+-+-+-+-+
```

If the computed checksum is zero, it is transmitted as all ones (the equivalent in one's complement arithmetic). An all-zero transmitted checksum value means that the transmitter generated no checksum (for debugging or for higher-level protocols that don't care).

A user interface should allow the creation of new receive ports, receive operations on the receive ports that return the data octets and an indication of source port and source address, and an operation that allows a datagram to be sent, specifying the data, source and destination ports, and addresses to be sent.

11.2   IP Interface

The UDP module must be able to determine the source and destination internet addresses and the protocol field from the internet header. One possible UDP/IP interface would return the whole internet datagram, including all of the internet header, in response to a receive operation. Such an interface would also allow the UDP to pass a full internet datagram, complete with header, to the IP to send. The IP would verify certain fields for consistency and compute the internet header checksum.

11.3   Protocol Application

The major uses of this protocol is the Internet Name Server and the Trivial File Transfer. The protocol number is 17 (21 octal) when used in the Internet Protocol.

11.4 Summary

UDP's strength lies in providing the vehicle an application can hook into, thus having a transport protocol. Many custom applications use this protocol in the TCP/IP environment.

UDP applications do not have the support infrastructure in UDP that TCP applications have in TCP. Since this is the case, UDP applications must do much of the network work that TCP applications leave to TCP to perform.

CHAPTER 12

Understanding X

X is a distributed windowing environment. It is a separate protocol from TCP/IP, but in many TCP/IP implementations, X is included as an integral part. X is both a protocol and an application. Figure 12-1 illustrates its relationship to the TCP/IP protocol stack.

Notice where X is located in Fig. 12-1. Notice also the overall positioning of all components in the illustration. It is a protocol with multiple layers defined, as discussed in this chapter. It is also a TCP application.

12.1   A Perspective on X

In the early 1980s, developers at MIT were looking for a way to develop applications in a distributed computed environment. This was cutting-edge technology at the time. During their work, they realized that a distributed windowing system would meet their needs very well.

The MIT group met with individuals at Stanford who had performed similar work, and who gave the MIT group a starting point to begin this endeavor. The group at Stanford working with this technology had dubbed

Fig. 12-1
A perspective on X in
the TCP/IP protocol
stack.

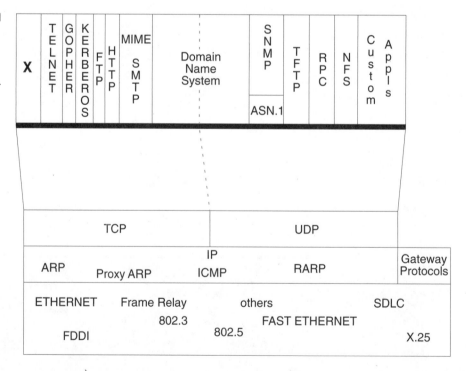

it *W*, for *windowing*. The individuals at MIT renamed it *X*, based upon the reason that it was the next letter in the alphabet. The name stuck.

By the late 1980s, X commanded a considerable market share in UNIX-based environments. One of the factors for its growth was its hardware and software independence. Today, X is a dominant user interface in the UNIX environment and has spread into MS-DOS and VMS environments as well.

12.2   X as a Protocol

X is asynchronous and based on a client/server model. It can manipulate two-dimensional graphics on a bitmapped display. Before examining some of the operational aspects of X, consider the layer of X and its relationship to the TCP/IP protocol suite, as shown in Fig. 12-2.

Figure 12-2 shows the TCP/IP protocol suite, but the focus is upon X. The protocol suite is there to help you understand the relationship of X with TCP/IP. X is not a transport-layer protocol; it uses TCP for a transport protocol.

X can be evaluated two ways. From a TCP/IP perspective, it comprises layers 5, 6, and 7. However, X itself has five layers. X's layer names and functions include the following:

Protocol   This is the lowest layer in X. It hooks into TCP. This layer is comprised of actual X protocol components.

Library   The X library consists of a collection of C-language routines based upon the X protocol. X library routines perform functions such as responding to the pressing of a mouse button.

Toolkit   The X toolkit is a higher level of programming tools. Examples of support provided from this layer are functions that provide programming related to scroll bars and menus.

Interface   The interface is what a user sees. Examples of an interface include SUN's OpenLook, HP's OpenView, OSF's Motif, and NeXT's interface.

Applications   X applications can be defined as client applications that use X and conform to X programming standards, and that interact with the X server.

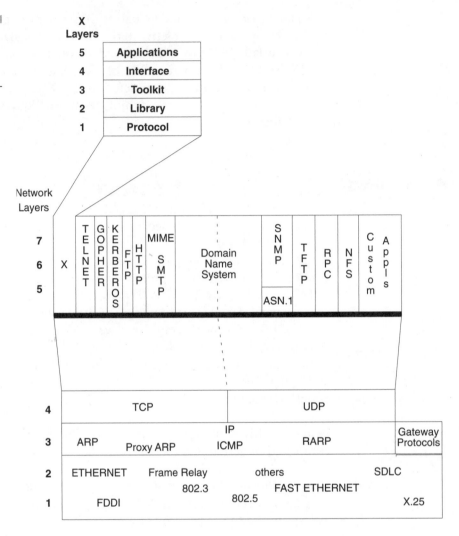

Fig. 12-2
A conceptual view of
X layers in respect to
TCP/IP.

Therefore, X in its totality is a protocol. It does have programs written to it according to the specifications of X protocol.

12.3 X Applications

X clients and X servers do not function in the way other clients and servers do in the TCP/IP environment. Normally, a client initiates something and servers serve, or answer the request of, clients. In X, this concept is skewed.

An X display manager exists in the X environment. Its basic function is starting and keeping the X server operating. The X display manager itself can be started manually or automatically. In respect to X, the display manager (also referred to as *Xdm*) is a client application.

An X display server (also known as *Xds*) is a go-between for hardware components (such as a keyboard or mouse) and X client applications. The Xds operates by catching data as it is entered and directing it to the appropriate X client application.

The correlation of Xdm and Xds can be understood by an example. Consider two windows that are active on a physical display. Each window functions as a client application. With this in mind, the idea of directing data to the appropriate X client application takes on a different meaning. This architectural arrangement is required to maintain order because multiple windows may be on the display (say four or five).

The X display manager and X server control the operations on the display, which is what a user sees. Most entities in an X environment function as X client applications. Examples of this include the Xclock, an Xterm that is an emulator, or even a TN3270 emulation software package used to access a 3270 data stream in an SNA environment. Figure 12-3 shows a TN3270 client application.

A user understands X by what is seen on the display. The following explains what occurs in Fig. 12-3:

- The network uses TCP/IP.
- Three UNIX hosts operate on the network.
- Each host implements X.
- Each host has an X server.
- Each host has X clients.
- A TCP/IP-SNA gateway is present.
- A TN3270 client program exists.
- A TN3270 client provides the required terminal emulation between the TCP/IP network and the SNA network.
- A TN3270 client program communicates through the UNIX X server and the TN3270 server on the gateway.
- The TN3270 client program can be executed from any of the hosts on the network because any UNIX host user (with access permission) can access the UNIX host with the TN3270 client.

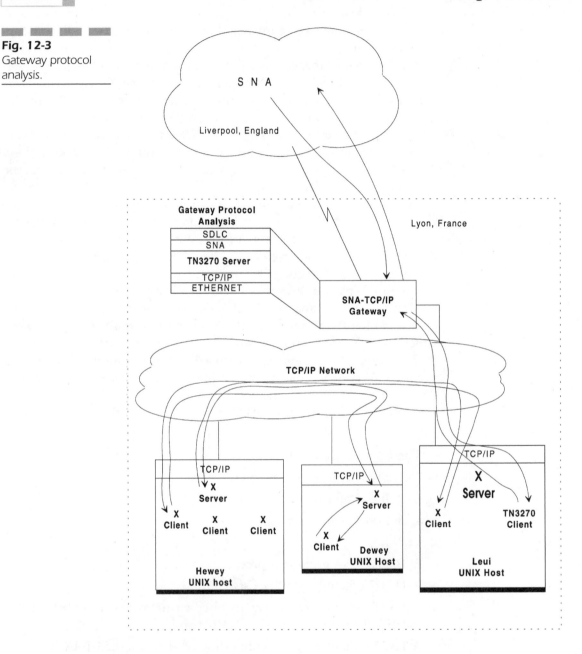

12.4 Understanding X Terminology

One point of confusion is X terminology. Some basic terms and their definitions are presented here:

Access Control List   A list of hosts that are allowed access to each server controlling a display is maintained in the /etc/X*n*.hosts file. The *n* here is the number of displays that hosts can access. This list is also called the *host access list*.

Active Window   This is the window where input is directed.   Background Windows may have a background that consists of a solid color or a pattern of some kind.

Background Window   This is the area that covers the entire screen, against which other windows are displayed. It is also called the *root window*.

Client   Client programs, also known as *X application programs*, include terminal emulation programs, window manager programs, and the clock program. Client programs do not have to run on the same program as the display server.

Display   A display is a screen driven by an X server. The DISPLAY environment variable tells programs which server to connect to unless this is overridden by the

```
-display command-line option.
```

Event   An event is something that must happen prior to an action occurring in response.

Font   This refers to a specific style of text characters.

Font Directory   This refers to the default directory where fonts used with X are stored.

Foreground   This term refers to the pixel value used to draw pictures or text.

Geometry   This is an option that can be used to specify the size and placement of a window on the display screen.

Icon   An icon is a symbol representing a window that, when selected by a mouse, will cause it to take its original form.

Property   This is a general term used to refer to the properties of a window. The basic purpose of properties is to serve as a communication mechanism between clients. For example, windows have properties such as a name, window type, data format, and data within the window.

Server   This term refers to the software and hardware that provides display services for X clients. The server accepts input from the keyboard and the mouse.

Window   This is an area on a display created by a client. For example, the Xclock runs in a window.

Window Manager   The window manager is a client program (application) that permits movement, resizing, and other functions to be performed on a display.

12.5   X Theory of Operation

An X server is responsible for managing a display. Programs that interact with the X server are Xclients. Xclients operate with a program-specific function. The Xclock is considered a client (program) application, for example.

X works differently from MS-Windows because it was designed to operate in a distributed environment with multiple hosts; MS-Windows was not. In practical terms, this means pieces that make an X window environment work can reside physically on different machines. This is not the case with MS-Windows. MS-Windows was originally designed to operate on one machine at a time. Technically, MS-Windows is a shell. It is tied to the operating system; now it has the operating system bundled into it and MS-Windows is referred to as the operating system. It is not incorrect to call MS-Windows a graphical user interface (GUI).

Some X programs are designed to operate in a windowing environment only. This means they will not function correctly in a line-mode environment, which is what UNIX is in native mode.

The X window manager is the major factor in the look and feel of how X is perceived. If SUN's OpenLook window manager is used, the look and feel of the display will take on its characteristics. On the other hand, if the OSF Motif window manager is used, the display will appear differently. The window manager is what a user sees.

The Xds is software that keeps up with input from devices like a keyboard and a mouse. Xds receives messages from an Xclient, and then updates the window on the display to reflect the messages. Display servers can operate on the same machine as the Xclient(s) being used, or they may be located on different machines, or they can even be stored in ROM on special terminals. This is partially how the term *X terminal* was derived.

X itself is about graphics. Its purpose is to support graphics in a distributed window environment. For a long while, X has been dominant in environments where UNIX is the operating system and TCP/IP is the network protocol. X is not confined to UNIX and TCP/IP, but the X windowing system does solve a major problem in UNIX environments.

12.6   Additional Information

More information on X can be found on the Internet. Multiple RFCs exist that directly or indirectly address the topic of X; one is RFC 1013. Information can be obtained from standard-making bodies such as Open Software Foundation (OSF), Unix International (UI), and Open-View. Another good source on X and related topics is books authored by Dr. Sidnie Feit. Her information is informative, accurate, and easy to read. Her books are published by McGraw-Hill.

12.7   Summary

The origins of X can be traced to MIT and Stanford. Originally, it was a distributed windowing environment. It met a real need in UNIX-based environments. During the 1980s, X grew in popularity and gained support from a variety of vendors, including IBM, DEC, and SUN.

X is both a protocol and an application. Technically, it has five layers. X is also an application; it uses TCP as a transport mechanism. User applications can be written with specific program calls against the X protocol components.

X provides a common link between disparate system environments. Because it can operate with a variety of operating systems, it has become popular. X typically comes bundled in many UNIX operating environments. X provides a form of seamlessness to users, but to programmers it is not seamless. X applications must be written and designed to the platform upon which they will run. For example, an X application that operates on the RISC/6000 system will probably not operate on a DEC or SUN platform.

13

A Holistic Approach to TCP/IP Management

Network management carries a variety of connotations. This is true because network management has until recently been segregated into vendor products, so that people with experience in a particular product define network management in terms of that product.

Some people view network management as the means by which technical information is obtained about a network and conclusions made, possibly charting the future of the network or aspects of it. For example, information can be obtained about networks to indicate where additional resources may be required to achieve a desired result. Information can also be obtained to indicate placement of equipment to balance network loads, thus making a network cost-efficient. These and similar decisions are technical by nature and the result of the philosophy of managing networks.

Another philosophy toward network management is based on finance. Some people use network management tools to obtain information to substantiate financial decisions effecting a network. These people usually make decisions that are financially sound at the time, but whose eventual costs exceed initial savings. A trained technical eye can usually discern those corporations who have people making decisions based mostly upon today's costs instead of making smart technical investments.

These two approaches to network management are prevalent in the marketplace. They are not advertised, but observation can reveal one or the other in time. In the best environment, a mix of the two exists.

13.1   What Network Management Means

Network management is a context-dependent term; a solid definition is debatable. "What can be managed in a network?" is not as important a question as, "What should be managed?" Most pieces of a network can be managed, regardless of the category into which a component fits. The degree to which a network component is important within the context of a network generally dictates if it should be managed and if so, to what degree.

For example, many networks consist of the following components:

- Hardware
- Software

■ Core equipment
■ Peripheral equipment
■ Local resources
■ Remote resources
■ Proprietary equipment
■ Nonproprietary equipment

Some networks might have equipment concentrated in a few categories, while others might have pieces of equipment from most of the categories.

Hardware

Networks consist predominately of hardware and software; the question is how much of each. Hardware can be delineated into subcategories such as these:

■ Processors
■ Controllers
■ Interfaces
■ Telecommunication devices
■ Physical links
■ Network devices

These devices perform definable roles. The degree to which they are mixed is usually dictated by the needs the network intends to serve, the capabilities of an individual product, and the vendor. For example, some vendors have processors that can perform functions that other vendor's processors cannot because of the way in which they have been designed.

Software

Software can also be delineated into categories such as

■ Operating systems
■ Network operating systems
■ User applications

■ Software used to control or provide specialized services via network devices

Other categories may be definable, but these exist to some degree in all networks. In some instances, an operating system can supply the abilities required to perform network services, such as a network protocol. In other instances, software that is used to control and provide services for network devices could also perform some type of protocol conversion or aid in the connectivity of multiple platforms that combine to create a network.

Core Equipment

Core equipment can be defined as that equipment crucial to create a network and make it function. Usually, a processor, software, telecommunications equipment, user applications, and physical links comprise the minimum requirements of a network. Variations of what is considered core equipment typically varies by vendor and the size of the network. For example, a network that spans multiple countries and time zones would most likely have different core equipment than a network that meets the needs of a single classroom in a geographically isolated location.

Peripheral Equipment

Peripheral equipment varies greatly, typically by vendor. For example, IBM peripheral equipment might be required to aid in the functionality of terminals or printers. At the opposite extreme, an AppleTalk network could utilize a device to make printer-sharing possible. Defining peripheral equipment in a network is difficult to do outside the confines of a vendor implementation.

Local Resources

Most networks identify equipment as being either local or remote. The former is usually in the same physical location as the network. Local and remote resources might differ in how they are managed and the degree to which they need network management. Local resources tend to be easier to manage in a network, simply because they are at or near the major portion of the network.

Remote Resources

Remote resources are not within the confines of the identified location of a network. *Remote* has multiple connotations. A remote resource might be across the street or halfway around the world. Conservatively, *remote* means not in the locale of the user (whether the user is human or a program).

Proprietary Equipment

It is possible to construct a network with one vendor's equipment. This is becoming more unusual, but it can be done. From a management perspective, this is increasingly important. Some vendors who provide network equipment have their own way of managing hardware and software. These vendors design their management schemes around what components can and cannot be managed.

Many companies today claim to have compatible equipment. This can be misleading. When someone says something is compatible, I immediately wonder to what degree. Compatibility is a growing concern in the network industry today. Just because someone says something is compatible does not mean it is 100% replaceable.

To a considerable degree, incompatibility is unimportant. A considerable amount of modern networking equipment can be classified as compatible according to someone's definition of compatibility. One area where compatibility is not clear, however, is that of network management. In a broad sense, it is easy for a device or software to be compatible with network management tools, but the question still applies: "To what degree?"

Despite popular opinion, there is an argument for maintaining proprietary vendor equipment. It should be noted, however, that even using a single vendor for equipment does not guarantee that an entire network can be managed. Startling as it sounds, this is true.

Nonproprietary Equipment

Probably the most difficult equipment to manage is a network in a hodgepodge of vendor equipment. Nonproprietary equipment is not a bad thing, but it is difficult to manage. This is true because network management, just like any kind of management, implies agreement upon fundamental issues.

The argument here is simple. The more diverse a network is in terms of vendor equipment, the more likely it is that a degree of unmanageability will exist. A simpler way to say this is one size does not fit all.

The idea of nonproprietary equipment in a network permeates the industry, but it is difficult to define exactly what *nonproprietary* means. In today's market, vendors shun the notion of anything being proprietary. However, the idea of nonproprietary carries equal adverse connotations. For example, if a thing is nonproprietary, then what vendor is responsible for it? Suddenly, the notion of being nonproprietary becomes less popular.

These ideas bring us full circle to the question of what can be managed. No exhaustive listing exists to explain those devices that can be managed and those that cannot. At present, partial listings and human experience governs the industry. A better way to examine the question is by way of network protocol.

13.2 Poll-Driven Management

Poll-driven management can be easily understood by examining its function via analogy. Consider a cook, kitchen, and three ovens. Each oven has a loaf of bread cooking, but not yet ready to be removed. One method of determining when a loaf of bread in a given oven is ready is by visual examination of each loaf; the cook examines each loaf of bread every so often to see if it is ready to be removed.

This method is similar to polling. In this example, the cook polls ovens 1, 2, and 3, making a mental note of the condition of each loaf of bread. The condition of each loaf is compared against a reference point of "ready." At some time, each piece of bread is deemed ready and removed from the oven.

This analogy might seem crude, but it is a good example of how polling operates, not only in management schemes, but also in other areas such as data link protocols or other technological implementations. Through this analogy, some observations can be made:

1. The cook exerts tremendous energy and time opening and closing the oven door, or peering through a window in each oven. If you do not think this is the case, expand the analogy to include 100 ovens, 100 loaves of bread, and one cook.

2. Each loaf of bread might be at a different state of readiness. For example, just because the bread in oven 1 appears to be like that

in oven 3 does not mean that it is identical. The difference between the two might be small, but significant.

3. The ovens are not the same. They might be made by the same vendor, manufactured to meet the same specifications, and checked to be the same temperature, but differences nevertheless exist, if for no other reason than their locations.

The point is that each oven might differ by an insignificant amount, but multiplied by a considerable number of ovens, it becomes significant.

13.3   Event-Driven Management

"Event-driven" anything is, to put it bluntly, interrupt-driven. Interrupts in any environment are, by definition, an interruption. An analogy to explain event-driven management comes by way of another example from ordinary life.

Consider how many times an individual visits an emergency medical facility. Normally, people only visit these facilities when emergency care is needed. In other words, they visit on an as-needed basis.

This as-needed basis is problematic by design. The inherent problem of event (on-demand) design is that the number of events inversely effects the ability, at some point, to meet an event's request. The ability to meet the emergency needs of a given community is inversely proportional to the number of individual demands placed on emergency medical clinics. Imagine one person going to an emergency clinic requiring care. Next, imagine ten people needing immediate care in the same clinic. Then, imagine 100 people in the same clinic requiring care. Now, imagine 1000 people entering an emergency clinic requiring immediate attention. I think the point is made.

The same problem exists for network management that utilizes this philosophy. The greater the number of devices in a given network, the greater the requirements to meet the needs of those devices. At some point, there is a point of no return. It is one thing to have a designated host manage another host, ten hosts, or even 100 hosts, but consider a host attempting to manage 1000 hosts. The implications abound.

First, the implication of having a pipeline big enough to move data from the hosts requiring management to the management host either accommodates multiple hosts to be managed or a single one. If the latter, then the implications of increasing the number of hosts managed quickly reaches a saturation point. Furthermore, the nature of event-driven management is similar to having a room of some hundred or thousand individuals listening to a mathematics lecture; the number of possible individuals who have questions increases with each individual. At some point, multiple individuals will have questions simultaneously. How is this to be handled?

13.4   How TCP/IP Is Managed

TCP/IP networks are dominant today. Its protocols might not be known to the end-user, but to those who must cope with the operational aspect of network management, TCP/IP's complexity is known. Complexity in anything is, at best, difficult to manage. Ironically, TCP/IP's complexity is part of the foundational power it delivers in a network protocol.

TCP/IP is defined and maintained by the Internet Engineering Task Force, which oversees the Request For Comments (RFCs) that define how TCP/IP is managed. Although particular vendors have products that enable management of certain products, TCP/IP is generally managed today by Simple Network Management Protocol (SNMP). This management application is defined by the RFCs that specify TCP/IP's architecture and operations. SNMP is offered by different vendors, but operationally works about the same regardless of vendor.

SNMP's implementation is built around a distributed system that employs multiple small programs that communicate with a centralized management program. Hence, many different programs use SNMP as the underlying protocol to manage TCP/IP networks.

TCP/IP network management requires information that is common to other networks as well. TCP/IP networks require tools that make accurate assessments about the network. Some information that is typically obtained by tools for TCP/IP network management include:

- Resource utilization requirements
- Link bandwidth
- Response time
- Resource status
- Application-specific information
- Resource utilization requirements

Resource utilization requirements may come under the guise of a different name, but usually the information is similar. Fundamental to the notion of resource utilization is defining what is a resource. Technical hair-splitting usually begins at this point. In large networks, a resource can be a link, software application, or CPU. Smaller networks might identify a resource in more general terms, identifying applications, CPU, memory, and other requirements.

Regardless of the size of the network, all networks have resource-utilization requirements. If you think this is not the case, try loading a program onto a system that requires more disk space or memory than available. When the number of users exceeds the amount of disk space required for a system to operate, most work comes to a halt. I have experienced this first-hand and learned the importance of understanding resource-utilization requirements.

Obtaining resource-utilization requirements is part science and part art. The precision for measuring resource requirements becomes a blur in most networks. The best resource-utilization requirement tools should be used with the thought in mind that a definite line exists between mathematics and reality.

Link Bandwidth

Link bandwidth is best calculated after obtaining two pieces of information: details of the actual link and an overall view of the surrounding network components. Once this information is acquired, synthesis of both is best.

Link bandwidth is tricky to calculate; many factors enter the picture. In test-bed environments, where isolated conditions can be maintained, figures can be tabulated. In the real world, factors enter into the equation that seldom reach the eyes of those who write specifications, unless they have come from the real world of implementation.

A realistic question about link bandwidth usually overrides other information. "Where is the bottleneck"? is always a sobering question. Sometimes, the bandwidth of the link is insignificant because a bottleneck in another part of the network causes congestion. This is where the reality/cost equation kicks in.

A point exists in most implementations where adding additional bandwidth to the link is moot. It is similar to driving the congested freeways of Dallas at rush hour: What does it matter if you have a Corvette or a Honda when traffic is moving ten miles per hour?

Some networks today have bandwidth capabilities that outstrip the capability of the components that make up the network. When this occurs, a decision must be made; normally, the reality/cost equation dominates.

Response Time

Response time is another topic many network administrators prefer to avoid. Response time in a network is usually proportional to the number of users currently using the network, the location of the network components, and the complexity of the network. Response time can be elusive.

In most networks, response time refers to the amount of time it takes for input from a keyboard to reach the application and a response returned. Some networks, like SNA, have this as a straightforward measurement. It is best to realize because of customization possibilities in SNA (namely NetView), that response-time measurement can be site-specific and relative at best. Response time can reflect measurements differently according to the network.

Resource Status

Resource status is nebulous. Some networks use different terms to identify a resource. Some networks identify an application, processor, or even link as a resource. Other networks identify processors as servers, therefore indirectly identifying them as a resource.

Explaining a resource status is a matter of degrees. One measurement of status might be "active" or "inactive." Another measurement might define a resource's status as "80% with regard to maximum usage". Many variances exist. Normally, they reflect vendors' interpretations of the idea.

Application-Specific Information

Application-specific information can be ascertained by management programs. Applications generally are the products of the environment they operate in, and reflect that environment's characteristics to some degree. An application can be interactive in nature, programmatic, or both.

Application-specific information generally falls into two categories: percentage of utilization with regard to the number of users, and percentage of utilization with regard to the amount of system resources such as CPU, memory, and disk space.

13.5   Examples of TCP/IP Management

One example of TCP/IP network management is TCP/IP management into SNA environments. Too often, network management is a post-implementation discussion, and it should not be. Managing TCP/IP-based LANs from NetView, managing TCP/IP LAN hosts from NetView, and managing network devices from NetView and TCP/IP LANs are all critical considerations of integration.

Managing TCP/IP LANs from NetView involves considering aspects beyond the obvious. Managing network devices used in the integration process is equally important. A brief checklist for managing TCP/IP LANs from NetView (and network devices) follows:

- Can NetView manage network devices used to integrate a TCP/IP network into the SNA?

- *How* does NetView manage network devices used for TCP/IP and SNA integration?

- Is Network Management Vector Transport (NMVT) Protocol support used on behalf of the vendor supplying the network devices being used?

- What level of management detail is available from NetView about network devices?

- How much customization is required for NetView to be able to manage network devices?

- Can NetView "see through" network devices to manage hosts on a TCP/IP LAN?

- What level of information can NetView ascertain about TCP/IP LAN hosts?

- What network management tools are available on network devices used in integration?

- Can network devices be managed remotely?

- What level of information is provided by network management on the network device?

- Do the network devices support TCP/IP network management, typically SNMP?

- What level of information do the network devices provide to SNMP?

- How do TCP/IP-based hosts appear to NetView?

- If a link failure occurs, can a TCP/IP host store management information, then forward it to NetView upon restoration of the link?

These questions provide a starting point for those with TCP/IP LANs integrated into SNA. Individual sites differ; your list might include additional questions or be shorter. Regardless, some basic questions remain. An area of focus is devices used in the integration process.

Managing Network Devices

Integrating TCP/IP-based LANs is typically achieved by network devices. In this case, a common device is a TCP/IP-to-SNA gateway. When this type of device is used, consumers should ask whether NetView can manage this device.

From a NetView perspective, TCP/IP-to-SNA gateways used in integration should support two basic functions. First, they should support alerts, specifically generic alerts. Generic alerts can be user- (vendor-) customized to report conditions to NetView. These type of alerts can report time and date stamps, and identification of the device sending an alert, all in terms NetView can understand. Generic alerts can also support other states a device might be in and pass them to NetView.

Second, a TCP/IP-to-SNA gateway should support Response Time Monitor (RTM). Response time is a requirement in many sites, and this aspect of NetView management support should not be overlooked by the consumer. Response time is a measuring tool used to define how much time it takes for a response to be sent back to the terminal after a user presses an AID key, causing data to be sent to the target SNA host.

A TCP/IP-to-SNA gateway supporting these NetView management abilities would appear like Fig. 13-1. Figure 13-1 shows a gateway positioned between SNA and the TCP/IP network. Additionally, gateways should support TCP/IP management.

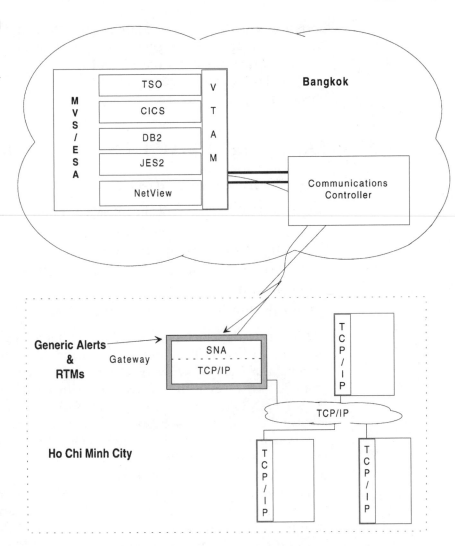

Figure 13-1
Network manage-
ment and gateways.

Managing TCP/IP Hosts From NetView

Network management in SNA is accomplished via NetView. Network management in a TCP/IP network is accomplished via SNMP. Network devices such as TCP/IP-to-SNA gateways can be managed via NetView if they have that support built into the architecture. In many instances, TCP/IP LANs are integrated into SNA via a gateway.

Once the question is answered concerning NetView management and gateways, the next question is whether NetView can manage TCP/IP-based hosts resident on the LAN. Managing individual TCP/IP hosts from NetView presupposes the gateway used in integration provides a pass-through capability and that these LAN-based hosts can communicate with NetView. This is a pivotal question to ask vendors who provide gateways for integration.

Managing TCP/IP LAN Hosts from NetView

Some critical questions need consideration:

■ Can all TCP/IP LAN hosts be managed from NetView?

■ What type of management information is ascertainable from TCP/IP LAN hosts from NetView?

■ If a link failure occurs, do the TCP/IP LAN hosts have the ability to store management data?

■ Once the link is re-established, is the data forwarded to NetView?

Answers to these questions should be enlightening. Figure 13-2 shows an example where TCP/IP hosts can be managed from NetView.

In light of Fig. 13-2, be aware of some typical vendor responses. To the first question a "yes" is the answer many times. However, does this mean TCP/IP LAN hosts can be viewed from NetView as SNA devices, or does it mean they will be viewed from NetView as some aberration?

A potential purchaser of a gateway should explore what level of SNA management support is provided. For example, are NetView alerts supported? If so, to what detail? These inquiries should be made to clarify exactly how a TCP/IP host appears to a NetView operator.

If a gateway is used between the SNA network and a TCP/IP LAN, critical questions are in order. For example, can NetView view TCP/IP hosts on the LAN beyond the gateway, or is only the gateway visible from NetView? The answer to this question is important. If NetView cannot manage TCP/IP hosts on the LAN *through* a gateway, only gateway management is available, not TCP/IP host management. On the other hand, if the TCP/IP hosts are manageable from NetView through the gateway, two other questions need to be asked: at what cost and what degree of detail is available to NetView from these hosts?

Further clarification should be made concerning what network management means. This is another gray area, where many vendors define management their own way. If TCP/IP hosts can be managed from NetView, does this mean they can have new releases of software downloaded to them and rebooted? Or does it mean that only status informa-

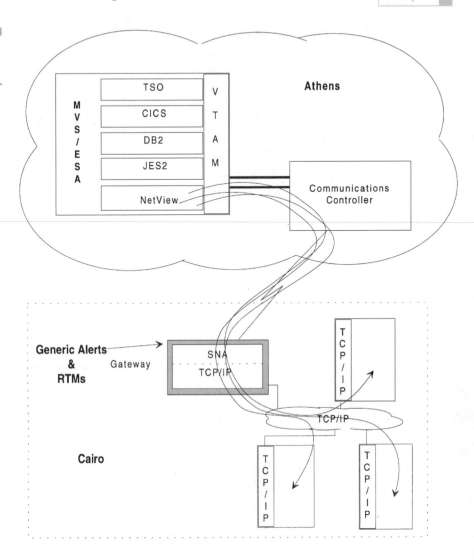

Figure 13-2
Remote network
management.

tion is available, such as a link failure or link recovery? Does it mean performance statistics can be gathered? A number of other valid questions exist, and this area of integration should not be overlooked. Sooner or later, it will become important.

TCP/IP and SNA Gateways

If TCP/IP networks are integrated into SNA via a gateway, they sit between two networks. Hence, it is only fair to ask how a network device is managed from a TCP/IP-based LAN. Conceptually, TCP/IP-to-SNA gateway management, from a network device perspective, would appear as in Fig. 13-3.

SNMP is the popular choice for TCP/IP network management today. It does not work like NetView, nor does it have the same look and feel. SNMP is used to manage devices such as hosts, network devices, and other nodes. SNMP is a system whereby events can be reported to network administrators, who can ascertain the status of a particular device.

Figure 13-3
SNMP gateway management.

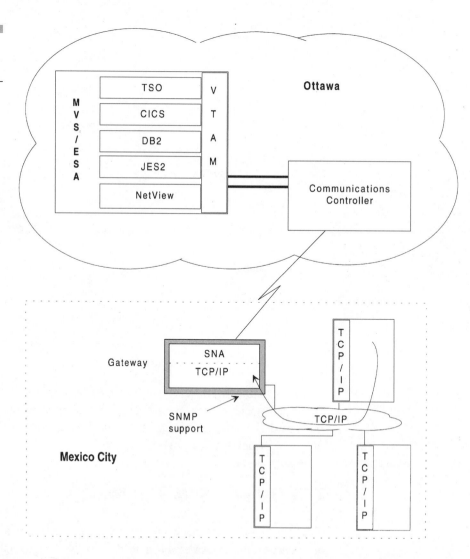

Managing TCP/IP-based LAN hosts from NetView includes managing network devices used in the integration process, typically TCP/IP-to-SNA gateways. It also includes managing individual TCP/IP network hosts. Comprehensive network management also includes being able to manage network devices from TCP/IP-based LANs. This means network devices used in the integration process should support NetView (SNA) management and TCP/IP management. If so, true heterogeneous network management can be realized.

13.6 Summary

Managing a network today is difficult at best. Managing heterogeneous networks requires significant artistic talent as well as an ability to implement the skills acquired through learning the science of management. Network management difficulty is usually proportional to the size of the network (in number of users), its location, the complexity of the tasks being performed in the network, and site-specific issues.

In little more than a decade, network management has exploded from gathering statistical information generally used by the elite to gathering information to be digested by technical and management personnel alike. Not only have networks themselves changed radically in the past decade, but the management tools used to understand them have as well.

Network management is best implemented with consideration given to long-term consequences. Along with this thought, it is best to remember that different network protocols usually have inherently different methods of management. Furthermore, just because a network is integrated does not necessarily mean the sum will be more easily managed than the parts.

14

TCP/IP TELNET Application

TELNET is popular in the TCP/IP protocol suite because it is the premier way to perform a remote logon. This chapter explains TELNET generally, raw TELNET, TELNET usage, valid TELNET commands, and hints for using TELNET. If you are new to some aspects of TCP/IP and have not asked in-depth questions, this chapter explains TELNET for you.

14.1 TELNET Application Orientation

TELNET is an application that provides logon capabilities to remote systems. It consists of two parts, as shown in Fig. 14-1.

Figure 14-1
TELNET as an application.

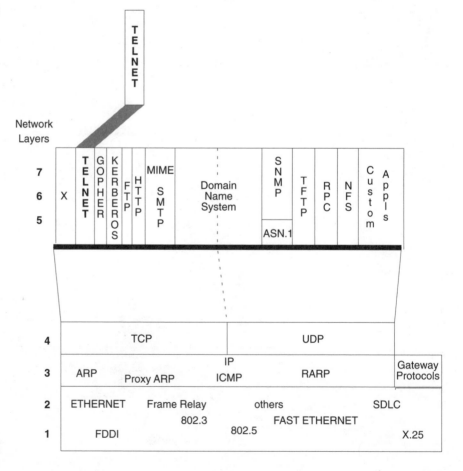

Remote logon capabilities are important because sharing resources is a fundamental idea behind a distributed environment. To share resources, they must be accessible. TELNET makes remote logon possible, thus making accessibility possible. Many times, resource sharing implies multiple computers and related devices connected together in such a manner that access from multiple sources is possible. In order for a user on any given computer to access resources on another computer, there must be a remote logon capability.

For example, consider three computers in a corporation. Assume that each computer has a primary purpose. Computer 1 is a machine dedicated for sales; it has sales information on it that anyone on the network can use. Computer 2 is a machine used to archive employee files; it has a large database running on it. Computer 3 has remote users connected to it, performing interactive tasks. Users are indirectly, or logically, attached to computer 3, and they do most of their work on this machine. Because of the network, however, users on computer 3 also have access to computers 1 and 2.

Consider Fig. 14-2, which shows a network with TCP/IP on each system, all connected to a common medium. If a user on computer 3 wanted to access a training program on computer 1, the user would enter

```
TELNET 1 at the command prompt on computer 3.
```

Once the user on computer 3 enters the TELNET 1 command, the TELNET client gets invoked on computer 3. After the TELNET client is invoked, it examines a file commonly known (particularly on UNIX-based computers) as /etc/hosts searching for a device named 1. Assuming it finds it, it then examines its corresponding internet address as the computer is known to all TCP/IP software. Next, sparing much detail, the TELNET server located on machine 1 answers the client's request (from computer 3) for a logical connection to be established. When this is done, the logon prompt from computer 1 is displayed on the user's terminal connected to computer 3.

A TELNET client, once invoked, can perform another function that is quite common in many networks: it interacts with what is called a *Domain Name Resolver*. The Domain Name Resolver works through a database that maintains device names and internet addresses on networks. When employed, TELNET clients use this instead of the typical /etc/hosts file to find the address for the target host. The reason behind this is efficiency. In large networks, the Domain Name System is used. The consequence if a Domain Name System is not used in a network is that all /etc/hosts files and other pertinent files must be updated whenever new hosts are added to the network.

Figure 14-2
Remote connectivity
to a network.

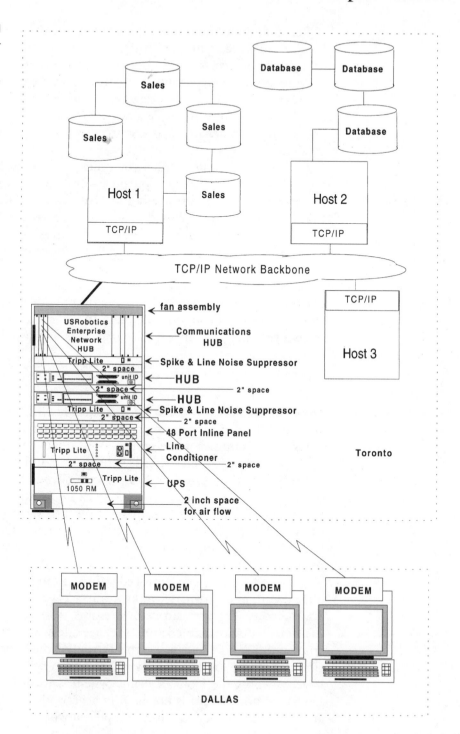

14.2 TELNET Application Characteristics

TELNET is an application and a protocol. From an application perspective, TELNET provides the ability for a user to invoke it and perform a remote logon with another host, as shown in Fig. 14-3.

In Fig. 14-3, the TELNET client on host 1 requests a logical connection with the TELNET server on host 2. The important thing to note is that both the TELNET client and server are part of the TCP/IP stack on each network device.

Figure 14-3
TELNET operation.

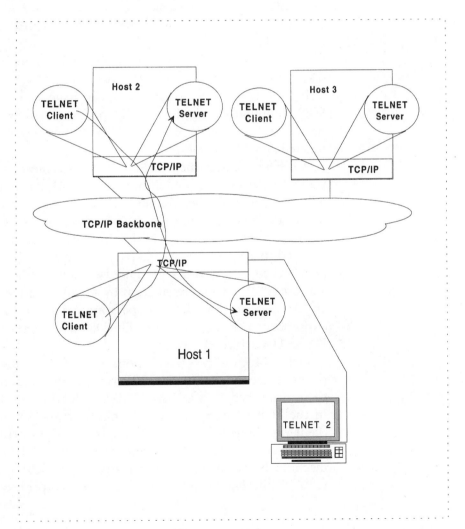

Raw TELNET

Figure 14-3 depicts three hosts with a TCP/IP stack, each with a native TELNET client and server. It also shows a user invoking a TELNET client from the TCP/IP protocol stack on host 1 by entering telnet at an operating system prompt. In many technical circles, this use of the TEL-NET client is referred to as a *raw TELNET,* specifically referring to the TELNET client. The TELNET client is considered native because it is inherent to the TCP/IP protocol stack.

Since TELNET is also a protocol, additional explanation is in order. So far, the focus has been upon TELNET as an application, but the TELNET protocol can be used to create a program called a *TN3270 client application.* Operational differences exist between a raw TELNET client and a TN3270 client application.

A raw TELNET client is popular for logons to hosts that use ASCII for data representation. For example, users working on a Convex computer (using the UNIX operating system) can issue a raw TELNET against a SUN computer (using the UNIX operating system). Both systems use ASCII for data representation. This is typical where TCP/IP is implemented and TELNET is used.

A different requirement exists when a user needs to log on to a computer that uses Extended Binary Coded Decimal Interchange Code (EBCDIC) for data representation, while the source computer uses ASCII for data representation. In today's heterogeneous networking environments, a variety of hosts with different operating systems may be attached to any given network, particularly TCP/IP. Assume three hosts are attached to a network. Assume two of these hosts are UNIX-based and use ASCII data representation by default. Assume one of the hosts is an IBM system using the VM operating system, and it has TCP/IP operating on it. The VM host uses EBCDIC data representation by default. Figure 14-4 depicts this example.

In Fig. 14-5, the two UNIX hosts and one VM host are connected to the same network and use TCP/IP as the network protocol. The TELNET client is invoked on a UNIX (ASCII-based) host, and the TELNET server in the TCP/IP stack on the VM machine answers the request of the client. Additionally, the TCP/IP stack on the VM machine performs *protocol conversion:* that TCP/IP protocol is converted to SNA on the VM machine. It also performs data translation (ASCII-to-EBCDIC and vice versa). While Fig. 14-5 shows the VM machine performing the two func-

Figure 14-4
EBCDIC- and ASCII-
oriented hosts.

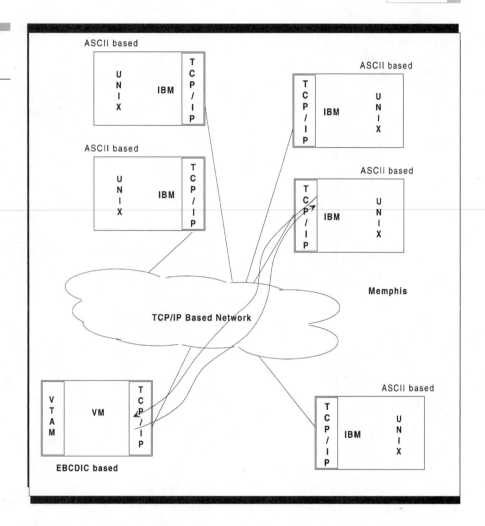

tions of protocol conversion and data translation, with the appropriate product, data translation can be performed on the UNIX host. Over time, this will realize savings when measured by CPU cycles. This is done with a TN3270 client.

TN3270 Client

A TELNET protocol is defined, called TN. An individual with the required knowledge who wishes to design a program based on the TN protocol

Figure 14-5
TELNET and data formatting.

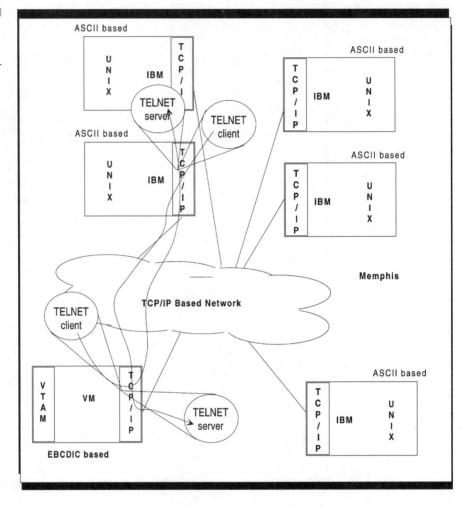

can do so. The most common program written using TN protocol is
an emulator application providing data translation services between
ASCII and EBCDIC and vice versa. This program (application) is called
a *TN3270 client*.

TELNET (within a native TCP/IP protocol stack) has ASCII-based
data, so it does not natively fit into SNA, which is dominated by
EBCDIC. In the SNA world, EBCDIC goes a step further and defines
data streams. A few specific data streams exist, but the dominant one is
the 3270 data stream. It is used with terminals interactively.

Because of the difference in ASCII and EBCDIC, converting ASCII
into EBCDIC, specifically into a 3270 or a 5250 data stream, is required.

But where will this process take place? With the data stream dilemma between TCP/IP networks and SNA networks, this fundamental issue must be resolved. So, how do users on a TCP/IP-based network have ASCII data converted into EBCDIC? Two possible solutions exist:

1. A raw TELNET client can be used to establish a logical connection between a UNIX or other non-EBCDIC host and an EBCDIC-based host. If this is the case, then ASCII-to-EBCDIC translation will occur on the EBCDIC host (with the exception of when a gateway is used between the two and translation services are provided).

2. A TN3270 client application can be used like a raw TELNET to gain entry into the SNA environment, but a TN3270 client application performs data translation. This means it sends an EBCDIC 3270 or 5250 data stream to the destination host. In this case, the TN3270 client application translates ASCII data into EBCDIC.

Figure 14-6 shows two UNIX hosts and a VM host connected to a network. On one UNIX host, a TN3270 client application exists. The TN3270 client is shown establishing a logical connection with the TELNET server native to the TCP/IP protocol on the VM machine. The data stream leaving the ASCII-based host is EBCDIC. This works because data format occurs at layer six within a network, which is the host in this case (hardware and software). By the time the data gets down to the interface card connecting it to the network, the data is represented by voltages or light pulses, whichever the network is based upon.

Having a TN3270 client means data conversion occurs where the client resides. Thus, on inbound data streams toward an EBCDIC-based host, only protocol conversion must be performed. Sometimes TN3270 applications are not needed and provide little benefit to the end user. Both a raw TELNET and a TN3270 client provide the user with remote logon capability. Both are client applications. The difference is merely where data translation is performed.

14.3   TELNET Application Usage

As mentioned previously, TELNET consists of a client and a server. A client always initiates a logical connection and a server always answers

Fig. 14-6
TELNET
communications.

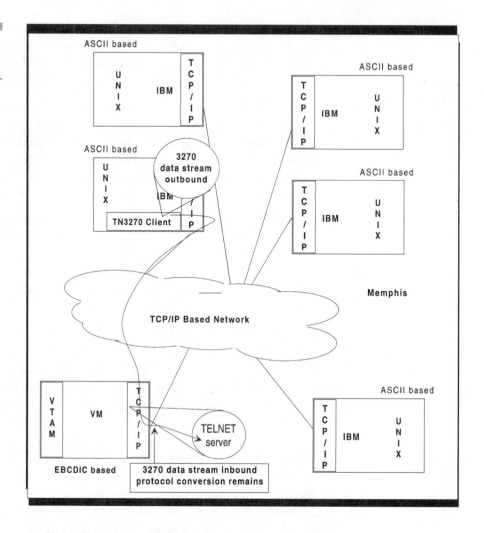

the client's request. To use TELNET, a command must be entered to
invoke the TELNET client. The command to invoke the TELNET client
from the TCP/IP suite is >TELNET. Assuming TCP/IP has been installed
properly and normal setup occurred, entering the >TELNET command
invokes the TELNET client from the TCP/IP protocol stack.

If the TELNET command is entered without a target host name, alias,
or internet address, the following prompt appears:

```
telnet>
```

This command is generated from the TELNET client on that host.
When this prompt appears, valid TELNET client commands can be
entered against it.

14.4 TELNET Application Commands

Valid TELNET client commands can be entered at the TELNET client prompt. To get a list of valid commands, the user can enter a question mark (?), and the list will be displayed. Valid TELNET client commands include the following:

- close–This closes a current connection if one is established.
- display–This command will display the operating parameters in use for TELNET. Because these parameters can be changed, they are site-dependent.
- mode–This command indicates whether entry can be made line-by-line or one character at a time.
- open–This command is required prior to the target host name in order for the session to be established.
- quit–This command is entered to exit the >telnet> prompt, thus exiting TELNET.
- send–In certain instances, special characters might need to be transmitted. This provides the means to accommodate some of these characters.
- set–This command is used to set certain parameters to be enforced during a TELNET session.
- status–This command provides information regarding the connection and any operating parameters in force for the TELNET session.
- toggle–This command is used to toggle (change) operating parameters.
- z–This command will suspend the >telnet> prompt.
- ?–This command prints valid TELNET commands that can be entered against the >telnet> prompt.

14.5 TELNET Application Hints

Using TELNET is easy. After using it a time or two, it becomes quite second nature. It is easiest to learn TELNET by understanding its basic operation, commands, and host logon procedures.

Since TELNET is part of the TCP/IP protocol suite, it works with other components in the suite. For example, if a user attempts to establish a remote logon with a target host, and after a period of time a response such as "host unreachable" is displayed on the terminal, the problem is not necessarily related to TELNET. In this example, the "host unreachable" message comes from the Internet Control Message Protocol (ICMP) component. This is an integral part of the IP layer. It provides messages responding to different conditions. Here, a destination host is not reachable by the TELNET client. The obvious question is, why? In this example, a few possible reasons would be viable. It could be the host is unreachable because of a break in the physical cable connecting the hosts together. It could be that the host is located on another segment of the network and, for some reason, is inaccessible at the moment. Other possibilities exist as well.

When messages such as these appear, they are most often generated from the ICMP portion of the TCP/IP suite. It would be helpful to familiarize yourself with common messages and understand their meanings. It can prove to be a valuable troubleshooting tool.

14.6 Summary

TELNET is popular today. It was also popular a decade ago, when I started using it. It provides remote logon capabilities and operates on the client/server method of communication. Each TELNET in a TCP/IP suite consist of a client and server. A client is used to initiate a request. Servers always respond to the request of a client.

TELNET is both an application and a protocol. It is an application in its native sense. It is also a protocol in its native sense; the two are not diametrically opposed. The TN protocol can be used to write a TN3270 client application.

A TN3270 client performs data translation on the host where it resides. A TN3270 client session requires communication with a TELNET server. This may be a TELNET server in a TCP/IP stack on an EBCDIC host, or it may be a customized program called a TN3270 server.

SNMP Overview

Simple Network Management Protocol (SNMP) is anything but simple! Conceptually, it is simple, but its implementations is complex. To say that SNMP is abstract is an understatement. Granted, some of its terms and concepts are rather straightforward, but other SNMP terms and concepts create an elaborate, abstract management structure that challenges the best minds. Presentation of the information in this book, however, is truly simplified. I had to simplify it to understand it myself!

I remember asking someone a question trying to get a grip on the basic overview. I merely wanted an idea of the overall picture, but the answer I received felt as if the individual turned an intellectual firehose on me.

This chapter, and the next, provides a foundation for those new to this network management tool, those seeking a different angle to the technology, and those who need a simpler explanation of the *Simple* Network Management Protocol.

15.1   SNMP Origins and Dates

The origins of SNMP can be easily traced to 1988. That year, some significant events happened. One need merely reference original Request For Comments (RFCs) to get insight to origins and significant dates in the development of SNMP.

A focal point in the history of RFCs hinges upon RFC 1052, issued in February, 1988. It attempted to narrow the scope of network management methods. At the time, SNMP was included along with Common Management Information Protocol (CMIP) implemented over TCP, which resulted in Common Management Information Protocol Over TCP, called *CMOT*. CMOT was intended for use in OSI environments.

In August of 1988, RFCs 1065 and 1066 appeared. RFC 1065 explains the Structure of Management Information (SMI). SMI explains how managed objects are defined. This RFC was a significant contribution to the initial works related to SNMP because of the content it provided to direct further development. RFC 1066 explains the Management Information Base (MIB) architecture. The MIB is explained in greater detail later, but suffice it to say the MIB is used to specify objects that are managed and how they can be accessed.

April 1989 brought the release of two RFCs: 1095 and 1098. RFC 1095 defined the CMOT protocol, which is used to access objects defined by previous RFCs, such as those explaining the SMI and MIB. RFC 1098 explains SNMP as it works in the access of objects explained in the RFCs previously mentioned. Other RFCs relating directly or indirectly to SNMP were released in the months that followed.

In my opinion, the next significant occurrence relating to SNMP happened in 1990. In May of that year, RFCs 1065, 1066, and 1098 were replaced by RFC 1155, 1156, and 1157, respectively. These three RFCs reflect the SMI, MIB, and SNMP, respectively. Although these RFCs were significant, I think the most significant RFC was 1158. RFC 1158 explains MIB-II definitions. The impact of this RFC was not felt in the marketplace for some years to come. More information is provided about MIB-II later.

RFC 1212, released in March of 1991, explained MIB definitions. RFC 1213, also released that March, is another MIB-II RFC. Other RFCs that contribute to SNMP in different areas include RFCs 1418, 1419, and 1420. These RFCs explain SNMP and OSI, AppleTalk, and Internet Packet Exchange (IPX), respectively.

In the years since 1991, many other RFCs have been introduced that relate to SNMP. RFCs 1441, 1442, and 1443 were released in May 1993. Each of these addresses SNMP version 2. Many permutations on SNMP were released in 1994 and 1995. The RFCs listed here and others related to SNMP directly and indirectly can be accessed via the Internet. The trick is knowing where they are. An RFC listing is also available on CD-ROM. The following is the source:

InfoMagic
11950 North Highway 89
Flagstaff, AZ 86004
www.infomagic.com

This company also has the following titles on CD-ROM:

- BSDisc CD-ROM
- CICA Windows
- Games for Daze
- Internet Tools
- LINUX Developer's Toolkit
- Moo-Tiff for LINUX
- Moo-Tiff for FreeBSD
- Mother of Perl
- Source Code
- TCL/TK

■ TEX
■ USENET
■ World Wide Catalog

15.2 Perspective

Before exploring SNMP itself, it helps to become acquainted with its
related protocol environment. Operationally, it is related to TCP/IP.
TCP/IP itself has been around for over two decades, and, as mentioned
earlier, SNMP dates back to around 1988.

Protocol Environment

Before examining multiple dates that affected SNMP, some consideration
should be given to understanding SNMP in context with TCP/IP. Figure
15-1 illustrates this concept. Figure 15-1 might be disconcerting for first-
time viewers. A dilemma exists for those who desire to understand
SNMP internal operations in its operational environment. To understand SNMP operations, you must understand TCP/IP operation.

If you do not understand TCP/IP internals to some degree, then trying to understand SNMP is like trying to learn calculus without being
competent in linear algebra. It can be done, but it is painful. Therefore, a
brief explanation of SNMP and TCP/IP operation is provided here.

TCP/IP is a collection of programs. These programs are protocols
defining the operation of various functions. TCP/IP is a client-server
architecture; clients invoke a process and servers answer clients requests.
In even simpler terms, clients initiate something and servers respond to
meet the needs of clients.

TCP/IP is basically a three-layer protocol. Its specifications begin at
layer three compared to the OSI, include layer four, and layers five, six,
and seven are generally viewed together. This is not precise because at
the session layer (layer five), logical connections are made between end-
users (whether human or program). Nevertheless, the layer functions are
routing, transport, and application. In a practical sense, this could be
applied to network protocol, but for those who split technical hairs, I
understand your concerns.

Users "see" TCP/IP operations through the eyes of the client-server
relationship or through the X-windowing environment. X itself is com-

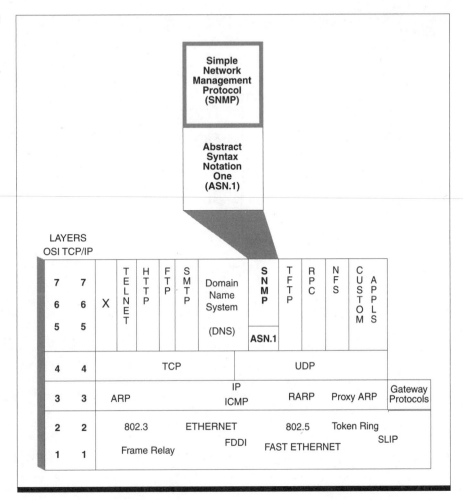

Figure 15-1
SNMP, TCP/IP, and
OSI layer correlation.

plex. Consider a conceptual view of X in relation to TCP/IP and its layers in Fig. 15-2.

Figures 15-1 and 15-2 provide enough information to substantiate the contention regarding protocol complexity. Like X, SNMP is complex and functions at layers five, six, and seven. Unlike X, it is not a windowing environment. Also unlike X, SNMP uses User Datagram Protocol (UDP), whereas the former uses Transmission Control Protocol (TCP).

Understanding SNMP

SNMP operation requires the integration of multiple components. Normally, these components are designed into a given device. Understanding

Figure 15-2
X in perspective to
TCP/IP.

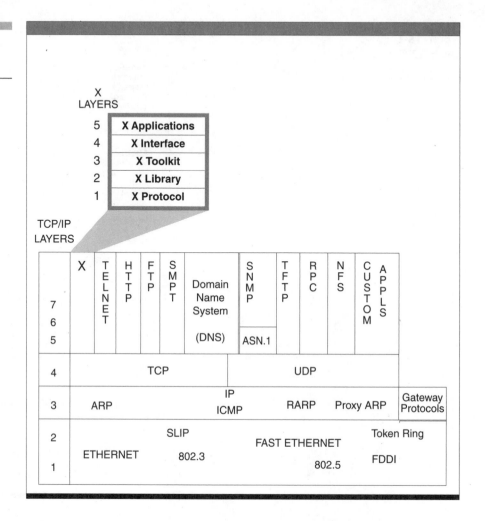

Figure 15-2
X in perspective to
TCP/IP.

overall SNMP operation requires some degree of knowledge of its structure. In order to make sense of the structure, terms and concepts must
be understood. The focus of the next section is a functional overview
of SNMP.

15.3   SNMP Functional Overview

SNMP uses *agents* and *application managers* (or simply *managers*). A user
agent can reside on any node that supports SNMP. Each agent maintains
status information about the node on which it operates.

Consider Fig. 15-3. This figure illustrates multiple devices and multiple networks. An application manager resides on one host, and agents are shown on hosts and devices.

The term *element* is a generic reference to a node. As Fig. 15-3 illustrates, multiple network elements exist, and each has its own agent. Typically, one node is designated as a network management node. Some people refer to this node as the *network manager*. It has an application that communicates with network elements to obtain their status. Communication between network element agents and the application manager occurs via messages. Different message types exist, as shown in Fig. 15-4.

Information gathered about these agents is stored in a Management Information Base (MIB). A MIB is a database containing information about a particular element. Examples of MIB-specific element information include the following:

- Statistical information regarding segments transferred to and from the manager application
- A community name
- An interface type

MIB information structure is defined by the Structure of Management Information (SMI) language. SMI is a language used to define a data structure and the methods for identifying an element for the manager application. This information identifies object variables in the MIB. A minimum of object descriptions defined by SMI includes the following :

ACCESS Object access-control is maintained via this description.

DEFINITION This provides a textual description of an object.

NAMES This term is synonymous with object identifiers. It refers to a sequence of integers.

OBJECT DESCRIPTOR This is a text name ascribed to identify the object.

OBJECT IDENTIFIER This is a numeric ID used to identify the object.

STATUS This describes the level of object support for the status.

SNMP implementations use Abstract Syntax Notation 1 (ASN.1) for defining the data structure in network elements. Because this language is based on a data type definition, it can be used to define practically

Figure 15-3
SNMP agents.

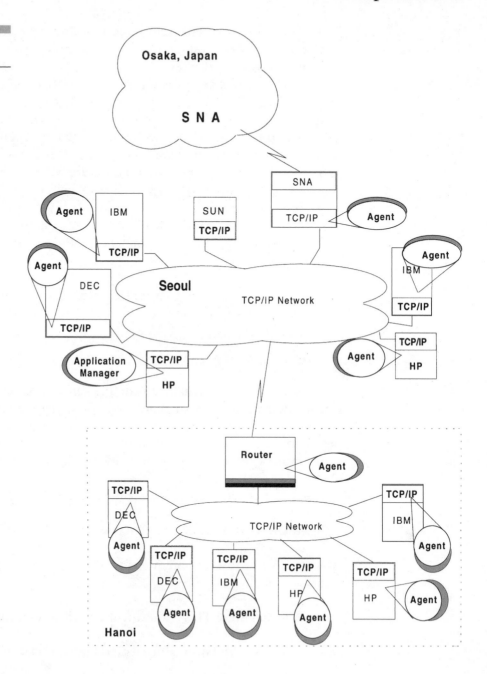

Figure 15-4
Message types.

GET REQUEST	This type request is used by the network manager to communicate with an element to request or list about that particular network element
GET RESPONSE	This is a reply to GET REQUEST, SET REQUEST, and GET NEXT REQUEST
GET NEXT REQUEST	This request is used to sequentially read information about an element.
SET REQUEST	This request enables variable values to be set in an element.
GET NEXT REQUEST	This type of message is designed to report information such as link status, whether a neighbor responds or a message is received, and the status of the element.

is based on a data type definition, it can be used to define practically any element on a network.

SNMP is event-oriented. An event is generated when a change occurs to an element. SNMP operation is such that approximately every 10 to 15 minutes, the manager application communicates with each network element regarding its individual MIB data.

15.4 Structure of Management Information (SMI)

SMI defines data structures in networked systems. It is also used to identify schemes used in the network. Definitions of these data types are coded in ASN.1, which also provides a method for inventing new objects.

SMI requires that the following information must be maintained about each object:

OBJECT DESCRIPTOR This is an object's textual name, such as sysDescri, ifSpeed, and icmpOutEchos.

OBJECT IDENTIFIER This is a numeric ID ascribed to an object. The structure for this is the same as that used in the Domain Name System (DNS).

SYNTAX This defines the structure of an object.

DEFINITION This is a textual description of the object.

ACCESS This identifies how an object can be accessed. For example, it could be read-only or read-write.

STATUS This specifies the level of support for an object, such as mandatory, optional, or obsolete.

Object identifiers, also called *SNMP names,* are implemented just as those in DNS, in dotted-decimal notation. Individual programs implement these in a series of one-byte integers, and each integer represents a level on a tree. The following list includes object identifiers and basic MIB structure, with the significance of each explained:

system This is a general system description. It includes the vendor's name and product ID.

interfaces This reflects the type of interfaces on a system, including the type of media, operating status, and counters.

at This is the map of network addresses to physical addresses.

ip This includes routing information and datagram statistics.

icmp This group keeps statistics on ICMP messages sent to and received by the system.

tcp This group reflects information on the total connections in the system and the status of current connections.

udp This group keeps statistics on UDP datagrams sent and received. It also includes a table of current UDP listeners.

egp This group maintains information on egp messages passed between egp systems. This table also includes information on egp neighbors.

CMOT In this use, CMOT is not used or defined.

transmission This is simply a node position on a tree. Beyond this, it is not like other groups.

snmp This is the group that keeps track of numbers of incoming and outgoing messages.

Figure 15-5 shows that iso is at the top of the design structure. One of the subtrees under iso is org. One of the subtrees under org is dod. One of the subtrees under dod is internet.

Figure 15-5

Conceptual view of MIB-II.

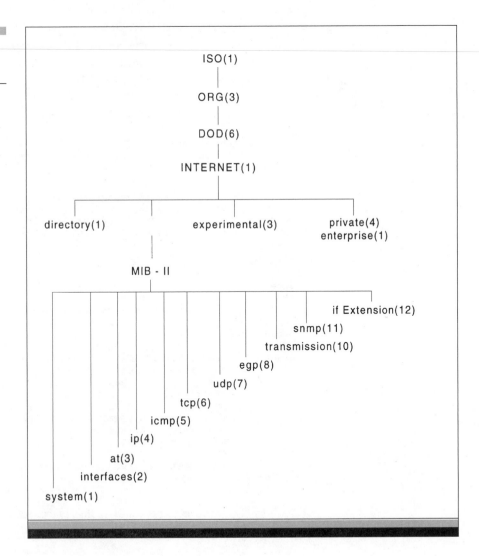

Figure 15-5 also shows numbers associated with a given subtree. This number is the object identifier, an integer assigned to that level. For example, the object identifier associated with the management node under internet is 1.3.6.1.2.

Figure 15-5 shows *private*, then *enterprise*. The enterprise is where companies can define their extensions to the MIB. This is important for those buying equipment who intend to manage it via SNMP.

Beneath mib(1) are 11 nodes. These nodes are called *object groups*. Each of these groups has related objects beneath it. For example, the system group under mib(1) includes sysDescr(1), sysObjectID(2), and sysUpTime(3). To better understand this, consider an example.

An agent of a network element that reports its system description would use the object identifier 1.3.6.1.2.1.1.1.0, that is

```
(iso.org.dod.internet.mgmt.mib.system.sysDescr.0)
```

```
This information is interpreted to mean the data that follows is
about the system description. The zero after sysDescr identifies a
specific occurrence of data.
```

```
The SMI definition of network management information is helpful to
understand how the data is organized. The following information is
an excerpt from RFC 1155. Here, the SMI definition of network man-
agement information is presented in ASN.1 syntax:
```

```
RFC1155-SMI DEFINITIONS ::= BEGIN
EXPORTS — EVERYTHING
    internet, directory, mgmt,
    experimental, private, enterprises,
    OBJECT-TYPE, ObjectName, ObjectSyntax, SimpleSyntax,
    ApplicationSyntax, NetworkAddress, IpAddress, Counter,
    Gauge, TimeTicks, Opaque;
                — the path to the root
    internet        OBJECT IDENTIFIER ::= { iso org(3) dod(6) 1 }
                    directory       OBJECT IDENTIFIER ::= { internet 1 }
                    mgmt            OBJECT IDENTIFIER ::= { internet 2 }
                    experimental    OBJECT IDENTIFIER ::= { internet 3 }
                    private         OBJECT IDENTIFIER ::= { internet 4 }
                    enterprises     OBJECT IDENTIFIER ::= { private 1  }
                    — definition of object types
                    OBJECT-TYPE MACRO ::=
                    BEGIN
                    TYPE NOTATION ::=       "SYNTAX" type (TYPE
    ObjectSyntax)
                                            "ACCESS" Access
                                            "STATUS" Status
                    VALUE NOTATION ::= value (VALUE ObjectName)
                        Access ::= "read-only"
                                                | "read-write"
                                                | "write-only"
                                                | "not-accessible"
```

```
Status ::= "mandatory"
                        | "optional"
                        | "obsolete"
END
— names of objects in the MIB
ObjectName ::=
OBJECT IDENTIFIER
— syntax of objects in the MIB
ObjectSyntax ::=
        CHOICE {
                simple
                SimpleSyntax,
— note that simple SEQUENCEs are not directly
— mentioned here to keep things simple (i.e.,
— prevent mis-use). However, application-wide
— types which are IMPLICITly encoded simple
— SEQUENCEs may appear in the following

CHOICE
                application-wide
                        ApplicationSyntax
                        }
SimpleSyntax ::=
        CHOICE {
        number
        INTEGER,
                string                    OCTET
STRING,
                object
        OBJECT IDENTIFIER,
                empty
        NULL
                        }
ApplicationSyntax ::=
        CHOICE {
        address
        NetworkAddress,
        counter
        Counter,
        gauge
        Gauge,
        ticks
        TimeTicks,
                arbitrary
                Opaque
— other application-wide types, as they are
— defined, will be added here
        }
— application-wide types
NetworkAddress ::=
        CHOICE {
        internet
        IpAddress
        }
IpAddress ::=
        [APPLICATION 0]— in network-byte
order
        IMPLICIT OCTET STRING (SIZE (4)
Counter ::=
        [APPLICATION 1]
```

```
                    IMPLICIT INTEGER (0..4294967295)
        Gauge ::=
                    [APPLICATION 2]
                    IMPLICIT INTEGER (0..4294967295)
        TimeTicks ::=
                    [APPLICATION 3]
                    IMPLICIT INTEGER (0..4294967295)
        Opaque ::=
                    [APPLICATION 4]— arbitrary ASN.1
value,
                    IMPLICIT OCTET STRING — "double-
wrapped"
                END
```
Many more details exist in this RFC. For more information, consult this and related RFCs.

15.5   Summary

SNMP is a complex network management scheme. It is powerful and capable of managing an unlimited number of devices in a given network. Currently, SNMP is the dominant network management tool for TCP/IP-based networks.

SNMP exists as a result of the RFCs mentioned in this chapter. These RFCs and newer ones contribute to SNMP development. Numerous vendors support SNMP. SNMP is one of the open standards that seems to have agreement from a wide variety of vendors in the U.S. and worldwide. The following chapter has additional information about SNMP.

SNMP Details

Chapter 15 explained some of the terms and concepts SNMP uses. This chapter includes more information about SNMP, building on the information presented in the previous chapter.

16.1   SNMP Protocol

SNMP architecture is built around the need for one or more network management stations to have administrative authority over a group of network elements. This arrangement is called a *community*.

The notion of a community works because of how SNMP is designed and the role of network administrators. One role of a network administrator is to assign a name to groups of managed elements, thus identifying them as communities. The name used to identify these elements is the name used in messages that pass between a manager application and agents. In a sense, a community name functions like a password. It controls the degree of network access to managed elements. Hence, the name *public* has wide access and is considered the default for managed elements.

Another facet of SNMP is its use of UDP for a transport-level protocol. UDP does not perform retransmissions, nor is it considered reliable. As a result, SNMP is responsible for reliability.

SNMP is a network management protocol that acts as an application protocol, where variables of an agent's MIB can be inspected or altered. Communication among protocol entities (elements) is accomplished by message exchange. Each message is entirely and independently represented within a single UDP datagram using the basic encoding rules of ASN.1. A message consists of a version identifier, SNMP community name, and a protocol data unit (PDU).

An element receives messages at UDP port 161 on the host with which it is associated for all messages except those that report the Trap-PDU. Messages that report traps are received on UDP port 162. An implementation of this protocol need not accept messages whose length exceeds 484 octets. However, recommendations suggest implementing support for larger datagrams when reasonable.

According to SNMP RFCs, five PDUs must be supported for SNMP operation:

- GetRequest
- GetNextRequest

- GetResponse
- SetRequest
- Trap

According to RFC 1067, the following applies to SNMP definitions:

```
        SNMP DEFINITIONS ::= BEGIN
IMPORTS
        ObjectName, ObjectSyntax, NetworkAddress, IpAddress,
TimeTicks
        FROM RFC1065-SMI;
        — top-level message
        Message ::=
        SEQUENCE       {
                version — version-1 for this RFC
        INTEGER {
                version-1(0)
                       },
                community      — community name
        OCTET STRING,
                data    — e.g., PDUs if trivial
        ANY     — authentication is being used
                       }
                — protocol data units
        PDUs ::=
        CHOICE {
                get-request
                        GetRequest-PDU,
                get-next-request
                        GetNextRequest-PDU,              get-response
                        GetResponse-PDU,
                set-request
                        SetRequest-PDU,
                trap
                        Trap-PDU
                }
        — the individual PDUs and commonly used
        — data types will be defined later
        END
```

The actions of a protocol entity (element) that implements SNMP need consideration. The term *transport address* reflects the address associated with the transport-level protocol. With UDP, a transport address consists of an IP address along with a UDP port. Other transport services may be used to support the SNMP. When this is the case, the definition of a transport address should be made appropriately.

The top-level actions of an element that generates a message are as follows:

1. Construct the appropriate PDU, e.g., the GetRequest PDU, as an ASN.1 object.

2. Pass the ASN.1 object along with a community name, source transport address, and destination transport address to the service that implements the desired authentication scheme. This authentication service returns another ASN.1 object.

3. Construct an ASN.1 message object, using the community name and the resulting ASN.1 object.

4. Serialize the new ASN.1 object, using the basic encoding rules of ASN.1, then send it using a transport service to a peer protocol entity.

Top-level actions of an element that receives a message include the following:

1. Perform a rudimentary parse of the incoming datagram to build an ASN.1 object corresponding to an ASN.1 message object. If the parse fails, the datagram is discarded and no other actions are performed.

2. Verify the SNMP message version number. If a mismatch exists, the datagram is discarded and no other actions are performed.

3. Pass the community name, user data found in the ASN.1 message object, and the datagram's source and destination transport addresses to the service implementing the desired authentication scheme. The receiving entity returns another ASN.1 object or signals an authentication failure. If the latter occurs, the element notes this failure and may generate a trap. It then discards the datagram and no other actions are performed.

4. Perform a parse on the ASN.1 object returned from the authentication service. This is performed in order to build an ASN.1 object corresponding to an ASN.1 PDU object. If a parse fails, the datagram is discarded. If the parse does not fail, the named SNMP community is used, the appropriate profile selected, and the PDU is processed accordingly. If a message is returned as a result of this processing, the source transport address is identical to the destination transport address to which the original request message was sent.

Requests and response packets flow through the network as UDP packets. The SNMP message format contains a version number, community name, response identifier, and one of five types of PDUs. The following information provides further details about the packets:

- *PDU Type* is GetRequest(0), GetNextRequest(1), GetResponse(2), or SetRequest(3).
- *ResponseID* is a number corresponding to a request.
- *Error Status* is noError(0), tooBig(1), noSuchName(2), badValue(3), readOnly(4), or genError(5).
- *Error Index* indicates which variable in a list caused an exception
- *Variable Bindings* is a list of object identifiers and values for an object.
- *Trap PDU* is Trap(4).
- *Enterprise* contains the value of sysObjectID.
- *Address* is the internet address of the object generating the trap.
- *Generic Trap* is coldStart(0), warmStart(1), linkDown(2), linkUp(3), authenticationFailure(4), egpNeighborLoss(5), or enterpriseSpecific(6).
- *Specific Trap* explains enterprise-specific traps.
- *Timestamp* is the elapsed time since an agent was initialized.
- *Variable Data* holds optional data that pertains to the trap.

SNMP is based on event reporting. Because of the structure of agents and managers, SNMP can retrieve network configuration and statistical data from managed objects. These managed objects may be bridges, hosts, links, or a variety of other things. Also, a managed object can include an abstract thing such as a protocol.

SNMP structure includes a MIB. The current MIB has over 100 objects identified. These objects are included in the groups within the MIB itself.

16.2 Frequently Used ASN.1 Construction

The following reflects a commonly used ASN.1 construction according to RFC 1067:

```
Request/response information
        RequestID ::=
        INTEGER
```

```
ErrorStatus ::=
INTEGER {
noError(0),
tooBig(1),
noSuchName(2),
badValue(3),
readOnly(4)
genErr(5)
}
ErrorIndex ::=
INTEGER
Variable bindings
VarBind ::=
SEQUENCE {
name
ObjectName,
value
ObjectSyntax
}
VarBindList ::=
SEQUENCE OF
VarBind
```

RequestIDs distinguish among outstanding requests. By use of the RequestID, an SNMP application entity can correlate incoming responses with outstanding requests. Where unreliable datagram service is used, the RequestID also provides a means of identifying messages duplicated by the network.

A non-zero instance of error-status is used to indicate that an exception occurred while processing a request. In exception cases, the error-index may provide additional information by indicating which variable caused the exception.

The term *variable* refers to an instance of a managed object. A variable binding, or VarBind, refers to the pairing of the name of a variable to the variable's value. A VarBindList is a simple list of variable names and corresponding values. Some PDUs are concerned only with the name of a variable and not its value. GetRequest is one such PDU. Here, the value portion of the binding is ignored by an element. However, the value portion must have a valid ASN.1 syntax. The ASN.1 value NULL is used for the value portion of such bindings.

The GetRequest PDU

The form of the GetRequest PDU is as follows:

```
GetRequest-PDU ::=
[0]
```

```
IMPLICIT SEQUENCE {
       request-id
                 RequestID,
       error-status- always 0
                 ErrorStatus,
       error-index- always 0
                 ErrorIndex,
       variable-bindings
                 VarBindList
       }
```

The GetRequest PDU is generated by an element only when requested by an SNMP application entity. When a GetRequest PDU is received, the receiving element responds according to any applicable rule in the following list:

1. If any object named in the variable-bindings field does not exactly match the name of some object available for GET operations in the relevant MIB view, the receiving element sends the originator of the received message a GetResponse PDU of identical form. This occurs except when the value of the error-status field is noSuchName and the value of the error-index field is the index of that object-name component in the received message.

2. If any object named in the variable-bindings field is an aggregate type (as defined in the SMI), the receiving entity sends the originator of the received message a GetResponse PDU of identical form. This occurs unless the value of the error-status field is noSuchName and the value of the error-index field is the index of that object-name component in the received message.

3. If the size of the GetResponse PDU exceeds a local limitation, the receiving element sends the originator a GetResponse PDU of identical form. This occurs unless the value of the error-status field is tooBig and the value of the error-index field is zero.

4. If an object's value named in the variable-bindings field cannot be retrieved, the receiving element sends the message originator a GetResponse PDU of identical form. This is the case unless the value of the error-status field is genErr and the value of the error-index field is the index of that object-name component in the received message.

If none of these rules apply, the receiving element sends the originator a GetResponse PDU. This applies for each object named in the variable-bindings field of the received message, the corresponding component of the GetResponse PDU that represents the name, and the value of that variable. The value of the error-status field of the GetResponse PDU is noError and the value of the error-index field is zero. The value of the RequestID field of the GetResponse PDU is that of the received message.

The GetNextRequest PDU

The form of the GetNextRequest PDU is identical to the GetRequest PDU, except for the indication of the PDU type. In ASN.1 language, it appears as follows:

```
GetNextRequest-PDU ::=
[1]
IMPLICIT SEQUENCE {
        request-id
                RequestID,
        error-status — always 0
                ErrorStatus,
        error-index — always 0
                ErrorIndex,
        variable-bindings
                VarBindList
        }
```

The GetNextRequest PDU is generated by an element at the request of its SNMP application entity. When the GetNextRequest PDU is received, the receiving element responds according to any applicable rule in the following list:

1. If an object name in the variable-bindings field does not lexicographically precede the name of some object available for GET operations in the relevant MIB view, the receiving entity sends a GetResponse PDU of identical form unless the value of the error-status field is noSuchName and the value of the error-index field is the index of that name component in the received message.

2. If the size of the GetResponse PDU exceeds a local limitation, the receiving entity sends the originator a GetResponse PDU of identical form. This occurs unless the value of the error-status field is tooBig and the value of the error-index field is zero.

3. If an object's value named in the variable-bindings field is the lexicographical successor to the named object and cannot be retrieved, the receiving entity sends the originator a GetResponse PDU of identical form. This occurs unless the value of the error-status field is genErr and the value of the error-index field is the

index of that object-name component in the received message.

If none of these rules apply, the receiving protocol entity sends the GetResponse PDU to the sending entity. The value of the error-status field of the GetResponse PDU is noError, and the value of the error-index field is zero. The value of the request-id field of the GetResponse PDU is that of the received message.

An important use of the GetNextRequest PDU is traversal of conceptual tables of information within the MIB. The semantics of this type of SNMP message and the protocol-specific mechanisms for identifying individual instances of object types in the MIB affords access to related objects in the MIB as if they enjoyed a tabular organization. By the SNMP exchange in the following example, an SNMP application entity might extract the destination address and next-hop gateway for each entry in the routing table of a particular network element.

Suppose that the routing table has three entries:

Destination	NextHop	Metric
10.0.0.99	89.1.1.42	5
9.1.2.3	99.0.0.3	3
10.0.0.51	89.1.1.42	5

A management station sends an SNMP agent a GetNextRequest PDU containing the indicated OBJECT IDENTIFIER values as the requested variable names:

```
GetNextRequest (ipRouteDest, ipRouteNextHop, ipRouteMetric1)
```

The SNMP agent responds with a GetResponse PDU:

```
GetResponse    ((ipRouteDest.9.1.2.3 =  "9.1.2.3"),
(ipRouteNextHop.9.1.2.3 = "99.0.0.3"),
(ipRouteMetric1.9.1.2.3 = 3))
```

The management station continues with the following:

```
GetNextRequest (ipRouteDest.9.1.2.3,
ipRouteNextHop.9.1.2.3,
ipRouteMetric1.9.1.2.3)
```

The SNMP agent responds:

```
GetResponse     ((ipRouteDest.10.0.0.51 = "10.0.0.51"),
(ipRouteNextHop.10.0.0.51 = "89.1.1.42"),
(ipRouteMetric1.10.0.0.51 = 5))
```

The management station continues with the following:

```
GetNextRequest (ipRouteDest.10.0.0.51,
ipRouteNextHop.10.0.0.51,
ipRouteMetric1.10.0.0.51)
```

The SNMP agent responds:

```
GetResponse     ((ipRouteDest.10.0.0.99 = "10.0.0.99"),
(ipRouteNextHop.10.0.0.99 = "89.1.1.42"),
(ipRouteMetric1.10.0.0.99 = 5))
```

The management station continues with the following:

```
GetNextRequest (ipRouteDest.10.0.0.99,
ipRouteNextHop.10.0.0.99,
ipRouteMetric1.10.0.0.99)
```

With no further entries in the table, the SNMP agent returns those objects that are next in the lexicographical ordering of the known object names. This response signals the end of the routing table to the management station.

The GetResponse PDU

The form of the GetResponse PDU is identical to that of the GetRequest PDU except for the indication of the PDU type. It appears in the ASN.1 language as follows:

```
GetResponse-PDU ::=
[2]
IMPLICIT SEQUENCE {
        request-id
                RequestID,
        error-status
                ErrorStatus,             error-index
                ErrorIndex,
```

```
variable-bindings
        VarBindList
}
```

The GetResponse PDU is generated by a protocol entity upon receipt of the GetRequest PDU, GetNextRequest PDU, or SetRequest PDU. Upon receipt of the GetResponse PDU, the receiving protocol entity presents its contents to its SNMP application entity.

The SetRequest PDU

The form of the SetRequest PDU is identical to that of the GetRequest PDU except for the indication of the PDU type. In ASN.1 language, it appears as follows:

```
SetRequest-PDU ::=
[3]
IMPLICIT SEQUENCE {
        request-id
                RequestID,
        error-status    — always 0
                ErrorStatus,
        error-index     — always 0
                ErrorIndex,
        variable-bindings
                VarBindList
        }
```

The SetRequest PDU is generated by a protocol entity at the request of its SNMP application entity. Upon receipt of the SetRequest PDU, the receiving entity responds according to any applicable rule in the following list:

1. If any object named in the variable-bindings field is not available for SET operations in the relevant MIB view, the receiving entity sends a GetResponse PDU of identical form. This occurs unless the value of the error-status field is noSuchName and the value of the error-index field is the index of that object-name component in the received message.

2. If any object named in the variable-bindings field does not manifest a type, length, and value consistent with that required, the receiving entity sends a GetResponse PDU of identical form.

3. If the GetResponse type message generated exceeds a local limitation, the receiving entity sends a GetResponse PDU of

identical form. This occurs unless the value of the error-status field is tooBig and the value of the error-index field is zero.

4. If any object named in the variable-bindings field cannot be altered, the receiving entity sends the originator a GetResponse PDU message of identical form.

The Trap PDU

The form of the Trap PDU is as follows:

```
Trap-PDU ::=
[4]
IMPLICIT SEQUENCE {
enterprise — type of object generating
— trap, see sysObjectID in
[2]
OBJECT IDENTIFIER,
agent-addr        — address of object generating
NetworkAddress,— trap
generic-trap    — generic trap type
INTEGER {
        coldStart(0),
        warmStart(1),
        linkDown(2),
        linkUp(3),                    authenticationFailure(4),
        egpNeighborLoss(5),
        enterpriseSpecific(6)
        },
        specific-trap  — specific code, present even
INTEGER,        — if generic-trap is not
                — enterpriseSpecific
        time-stamp      — time elapsed between the last
        TimeTicks,    — (re)initialization of the network
                — entity and the generation of the trap
        variable-bindings    — "interesting" information
        VarBindList
        }
```

The Trap PDU is generated by a protocol entity only at the request of the SNMP application entity. The means by which an SNMP application entity selects the destination addresses of the SNMP application entities is implementation-specific.

Upon receipt of the Trap PDU, the receiving protocol entity presents its contents to its SNMP application entity. The significance of the variable-bindings component of the Trap PDU is implementation-specific. Interpretations of the value of the generic-trap field are as follows:

The coldStart Trap   A coldStart(0) trap signifies that the sending entity is reinitializing itself such that the agent's configuration or the entity's implementation may be altered.

The warmStart Trap   A warmStart(1) trap signifies that the sending entity is reinitializing itself such that neither the agent configuration nor the protocol entity implementation is altered.

The linkDown Trap   A linkDown(2) trap signifies that the sending entity recognizes a failure in one of the communication links represented in the agent's configuration. The Trap PDU of type linkDown contains, as the first element of its variable-bindings, the name and value of the ifIndex instance for the effected interface.

The linkUp Trap   A linkUp(3) trap signifies that the sending entity recognizes that one of the communication links represented in the agent's configuration has come up. The Trap PDU of type linkUp contains, as the first element of its variable-bindings, the name and value of the ifIndex instance for the effected interface.

The authenticationFailure Trap   An authenticationFailure(4) trap signifies that the sending entity is the addressee of a protocol message that is not properly authenticated. While implementations of the SNMP must be capable of generating this trap, they must also be capable of suppressing the emission of such traps via an implementation-specific mechanism.

The egpNeighborLoss Trap   An egpNeighborLoss(5) trap signifies that an EGP neighbor for whom the sending entity was an EGP peer has been marked down, and the peer relationship no longer obtains. The Trap PDU of type egpNeighborLoss contains, as the first element of its variable-bindings, the name and value of the egpNeighAddr instance for the effected neighbor.

The enterpriseSpecific Trap   A enterpriseSpecific(6) trap signifies that the sending entity recognizes that some enterprise-specific event has occurred. The specific-trap field identifies the particular trap that occurred.

16.3 SNMP MIB

The Management Information Base (MIB) is a database where definitions of managed objects are kept. MIB-II is the focus here. It includes most things in the first MIB, with the exception of CMOT. One addition was made to MIB-II and is discussed in this section.

A MIB, in general, contains managed-object information such as

Figure 16-1
MIB-II structure.

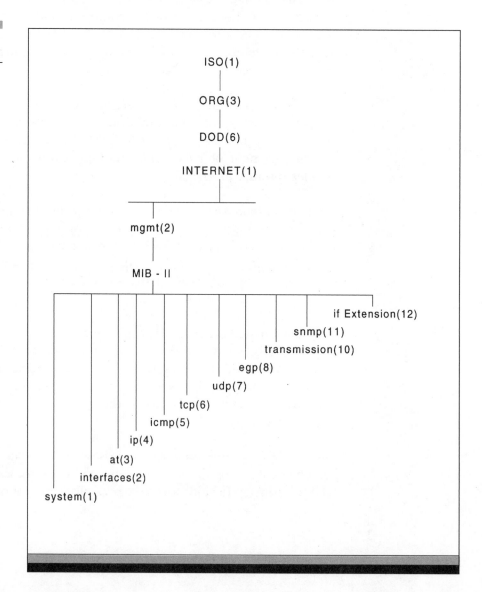

- Device configuration
- Device performance
- Device status
- Device system

A MIB's contents may change frequently to stay up-to-date. Technically, MIB definitions are not part of the SNMP protocol itself. Instead, a MIB works in conjunction with SNMP. Figure 16-1 illustrates MIB-II.

MIB-II includes the following groups:

- *SYSTEM* is required for all devices. It includes critical information such as what the device is and where it is. This group also contains the following information within it, sometimes referred to as the system subtree:

  - sysdescr contains information such as the operating system, hardware, and network software included in this group.

  - sysobjectID is an identifier used by a manufacturer.

  - sysuptime reflects the elapsed time since the system was last initialized.

  - syscontact reflects an individual's name and the number where they can be reached.

  - sysname is the name assigned to this element by a network administrator. This is typically the DNS name.

  - syslocation indicates the actual location of the device.

  - sysservices is a number reflecting a layer number for which the node performs services.

- *INTERFACE* reflects interfaces used to connect the device to the network. Interface information such as configuration, availability, and error messages is included. The list of interfaces continues to grow, and includes the following:

  - ATM
  - 802.3
  - 802.5
  - ETHERNET
  - X.25
  - SMDS
  - ISDN
  - DS3

- FDDI
- Frame Relay
- LAPB

- *AT* contains routing information. Currently, it is being phased out because of restrictions and limitations imposed by it upon multi-protocol environments. The diverse needs of upper- and lower-layer protocols are better addressed via tables defined as needed. Used in MIB-I implementations, this entry reflects an interface, lower-layer address, and network (upper-layer) protocol.

- *IP* includes information related to the TCP/IP protocol, such as configuration and management. Incoming and outgoing statistics are included, as well as the TTL, reassembly timeouts, an IP routing table, and counts of different kinds. The IP subtree includes the following groups:

  - ipforwarding
  - ipdefaultttl
  - ipinreceives
  - ipinhdrerrors
  - ipinaddrerrors
  - ipforwdatagrams
  - ipinunknownprotos
  - ipindiscards
  - ipindelivers
  - ipoutrequests
  - ipoutdiscards
  - ipoutnoroutes
  - ipreasmtimeout
  - ipreasmreqds
  - ipreasmoks
  - ipreasmfails
  - fragoks
  - ipfragfails
  - ipfragcreates
  - ipaddrtable
  - iproutetable

- ipnettomediatable
- iproutingdiscards

- *ICMP* (Internet Control Message Protocol) sends messages to the source. It is a requirement wherever IP is implemented. Most of these messages send the number of incoming or outgoing messages. The ICMP subtree includes

  - icmpinmsgs
  - icmpinerrors
  - icmpindestunreachs
  - icmpintimeexcds
  - icmpinparmrpobsicmpinsrcquenchs
  - icmpinredirects
  - icmpinechos
  - icmpinechoreps
  - icmpintimestamps
  - icmpintimestampreps
  - icmpinaddrmasks
  - icmpinaddrmaskreps
  - icmpoutmsgs
  - icmpouterrors
  - icmpoutdestunreachs
  - icmpouttimeexcds
  - icmpoutparmrpobs
  - icmpoutsrcquenchs
  - icmpoutredirects
  - icmpoutechos
  - icmpoutechoreps
  - icmpouttimestamps
  - icmpouttimestampreps
  - icmpoutaddrmasks
  - icmpoutaddrmaskreps

- *TCP* permits SNMP managers to check up on TCP configuration values. Incoming, outgoing, and statistical information is obtainable, as well as a view of active TCP sessions, thus giving the applications in use. Additional information can be found in RFC 1122 about requirements for Internet hosts. The TCP subtree includes the following:

- tcprtoalgorithm
- tcprtomin
- tcprtomax
- tcpmaxconn
- tcpactiveopens
- tcppassiveopens
- tcpattemptfails
- tcpestabresets
- tcpcurrestab
- tcpinsegs
- tcpoutsegs
- tcpretranssegs
- tcpconntable
- tcpiners
- tcpoutrsts

- *UDP* lists those services that are actively listening. This group keeps track of traffic and errors passed. The UDP subtree includes udpindatagrams, idpnoports, udpinerrors, udpoutdatagrams, and udptable.

- *EGP* enables a router to talk to another router. Incoming and outgoing traffic can be monitored. The EGP subtree includes egpinmsgs, egpinerrors, egpuotmsgs, egpouterrors, egpneightable, and egpas. Additional information about EGP is available in RFC 904.

- *TRANSMISSION* is actually not a group; it is a position in the naming tree. Its subtree, however, does include most popular technologies used today in lower-layer protocols, for example, ATM, SMDS, DS3, and 802.X. It is more accurate to think of this "group" as having multiple MIBs, those needed to accurately reflect the technology at this level.

- *SNMP* can track errors and incoming and outgoing messages, and records of the SETs, GETs, and TRAPs it uses. Its subtree includes
  - snmpinpkts
  - snmpoutpkts
  - snmpinbadversions
  - snmpinbadcommunitynames
  - snmpinbadcommunityuses
  - snmpinasnpaseerrs
  - snmpintoobigs
  - snmpinnosuchnames
  - snmpinnosuchnames
  - snmpinbadvaluessnmpinreadonlys
  - snmpinfenerrs
  - snmpintotalreqvars
  - snmpintotalsetvars
  - snmpgetrequests
  - snmpingetnexts
  - snmpingetresponses
  - snmpintraps
  - snmpouttoobigs
  - snmpoutnosuchnames
  - snmpoutbadvalues
  - snmpoutgenerrs
  - snmpoutgetrequests
  - snmpoutgetnexts
  - snmpoutsetrequests
  - snmpoutsetresponses
  - snmpouttraps
  - snmpenableauthentraps
- The SNMP subtree does not use items 7 and 23. However, it does use all the rest of the 28 definitions. This group of subtree

components is required for those devices claiming SNMP support.

■ *IFEXTENSIONS* contains information that relates closely to the interface group. The information in this group contains extensions of variables used by the interface group. Two critical documents are associated with this group: RFC 1299, "Extensions to the Generic Interface MIB," and its update, RFC 1573, "The Evolution of the Interfaces Group of MIB-II."

The MIB explanations given here do not include the details of each subtree. This level of information is best derived from original RFCs because the amount of information related to each subtree is considerable.

16.4   SNMP Operation

Figure 16-2 illustrates communication between SNMP agents and a manager. Figure 16-2 shows one manager, and each host has an agent. Regardless of the location, the one manager can manage three networks. In Fig. 16-2, each physical location is considered a different community. Each location has a different host, but all have information used by the manager application in order to manage the networks.

Network elements contain a subset of objects in a MIB that pertain to specific elements; this is called a *MIB view*. MIB view names do not necessarily have to belong to a single object subtree or object name space. The access mode of an element refers to the ability to read, write, or read/write. The combination of an element mode with a MIB is referred to as a *community profile*.

An *access policy* is the pairing of a community and the community profile. This policy is used by the application entities architecturally defined to utilize administrative relationships between SNMP elements.

16.5   The Role of ASN.1 and X

ASN.1 is the OSI language for describing abstract syntax. In essence, it reflects network elements (better known generically as hosts). It is a powerful, flexible language used to describe datatypes and to define structures for managed objects. It is an integral part of SNMP.

Figure 16-2
The management-
agent relationship.

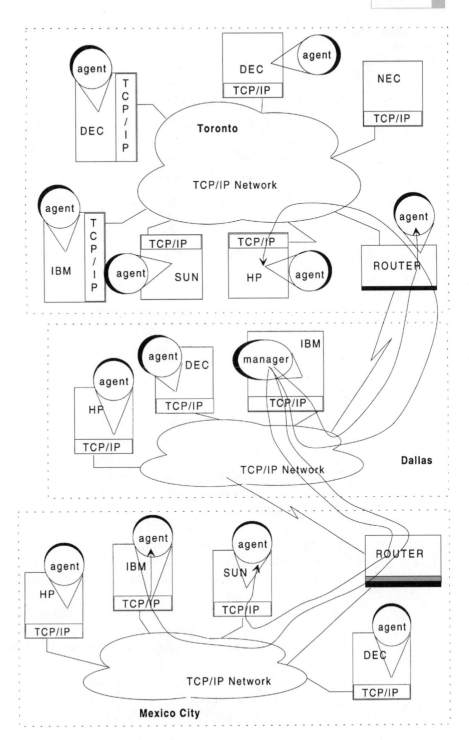

Figure 16-3
SNMP operation.

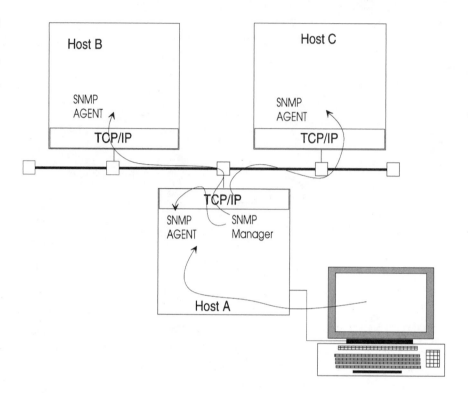

Figure 16-3
SNMP operation.

ASN.1 describes not only data structures, but also management information about the network element. ASN.1 is a formal language and is defined in terms of grammar. Because the nature of ASN.1 is flexible, it can accommodate numerous types of network elements. For example, this language can be used to report information about practically any network device.

When SNMP is used in real environments, it is typically seen where X is employed. X provides a distributed windowing environment whereby multiple hosts or items may be viewed simultaneously. Since this windowing environment is viewed in real-time, an effective management environment can be achieved.

Figure 16-3 shows host A with an SNMP manager used to monitor the TCP/IP network. Host A can view all hosts on the network, including itself. It can view itself because of how SNMP works, while at the same time viewing other participating hosts on the network.

Figure 16-3 shows a central point of network management. Implicit here is the semblance of remote network management. As long as a person has access to the TCP/IP network, he or she can TELNET to the host used to perform network management, so a sense of remote network management can be achieved.

16.6   Summary

SNMP is complex. It is robust and capable of providing a means for managing diverse network devices. SMI, ASN.1, a MIB, and other aspects (including the SNMP protocol itself) work together to make SNMP a popular management tool.

More information can be found in the RFCs listed in this chapter. Also, I strongly encourage you to consider reading Dr. Sidnie Feit's book, *SNMP: A Guide to Network Management.*

17

Managing TCP/IP with Various Products

Many TCP/IP management products are available today. The major ones are evolving into versatile products. Many network management tools, including some indispensable TCP/IP management tools, are a combination of software and hardware. The focus of this chapter, however, is on TCP/IP network management software.

17.1 Managing TCP/IP with NetView

NetView has evolved since the mid-1980s. Its roots reach back into the 1970s, when some of its components were introduced. It is important to have a general understanding of NetView before focusing on the way TCP/IP-based networks are managed with it. Figure 17-1 represents a conceptual view of some NetView components. Years ago, NetView had fewer components. It is evolving to meet market needs, and thus incorporates multiple protocol support.

A few simple examples of NetView component operation help to provide insight into its management abilities. These examples do not highlight TCP/IP, but they show the power behind NetView, which has caused it to evolve into a program that can manage TCP/IP networks and components. Consider Fig. 17-2.

Figure 17-2 illustrates how all terminals can be managed from NetView. Information such as response time can be determined. In this example, NetView examines the controllers, provides information by department, and analyzes the performance of each department. Other detailed information can be obtained also.

Figure 17-3 portrays how NetView would acquire information about network hardware. It shows multiple sites with many pieces of hardware. The site in Winnipeg includes the host where NetView is located, enterprise controllers, network interface cards (NICs), front-end controllers (FEPs), line interface couplers (LICs), and a transmission group identified as connecting the Winnipeg location with Mexico City. From the perspective of NetView, all of these components can be monitored. Details about each piece of hardware can be collected and presented for interpretation.

Now, consider Fig. 17-4. It is another illustration of how NetView can manage multiple sites from a single location, but this time these sites are TCP/IP. For years, NetView has been IBM's premier network management system. Now it can easily manage TCP/IP networks, as Fig. 17-4 illustrates.

Managing TCP/IP with Various Products

359

Figure 17-1
A conceptual view of
NetView.

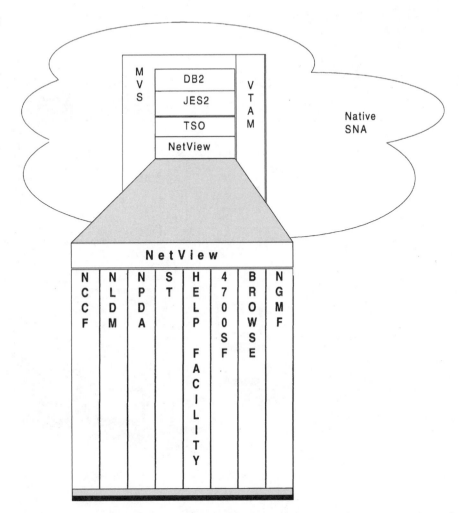

17.2 Managing TCP/IP with NetView/6000

NetView/6000, specifically AIX NetView/6000, is part of the family of
NetView products. NetView/6000 was designed to operate on the
RISC/6000. This machine uses UNIX as an operating environment.
NetView/6000 is a component that requires other components to enable
UNIX workstation management; namely, a RISC/6000 from a NetView
program in an SNA environment. Figure 17-5 portrays this idea.

Figure 17-5 shows multiple components on a RISC/6000 with numerous resources. The size of this diagram does not reflect a certain

Figure 17-2
Managing terminals.

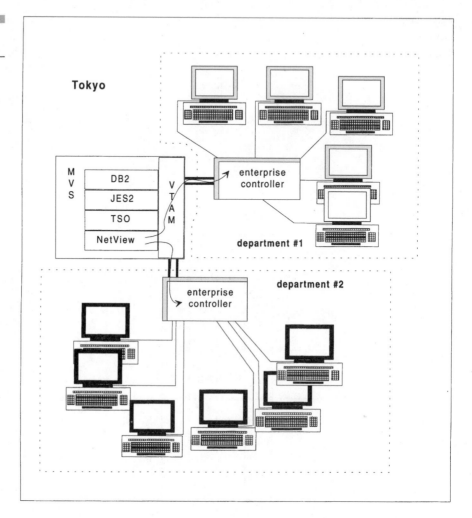

amount of other resources, which may be inferred. The drawing here shows the components used in AIX NetView/6000 operation.

17.3 Managing TCP/IP with SunNet Manager

SunNet Manager is TCP/IP- and UNIX-oriented. It requires SUN's OpenWindows program, which in turn requires X windows, and X

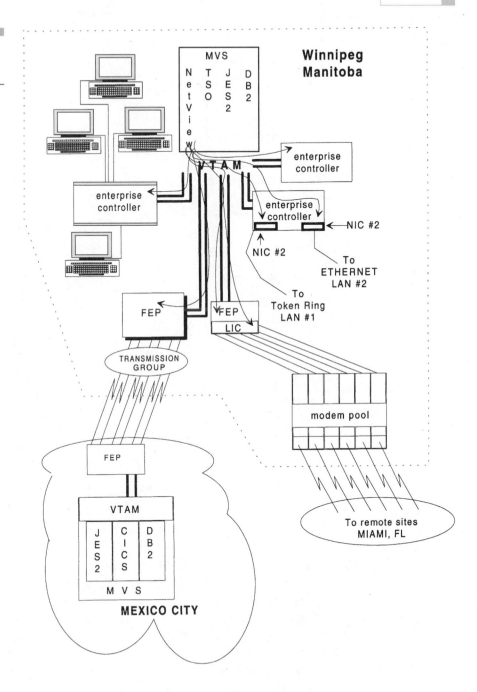

Figure 17-3
Managing hardware
from NetView.

Figure 17-4
Managing TCP/IP
Networks from SNA.

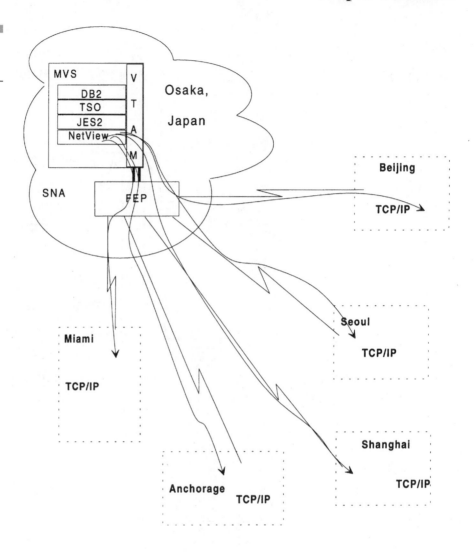

windows is part of a TCP/IP protocol stack. Other specifics are required to install and use SunNet Manager; these are outside the scope of this chapter.

Before exploring the details of how this management program operates, certain observations can be made. First, SUN is a UNIX-oriented system. By default, it operates in a TCP/IP network easily. Conceptually, SunNet Manager would appear as in Fig. 17-6.

Second, because of the structure of SNMP (the dominant TCP/IP management protocol) and SunNet Manager's design, both can work in harmony to leverage SunNet Manager's strength and SNMP capabilities.

They can coexist; multiple SNMP-based systems can participate in communication with a system where SunNet Manager is running. Figure 17-7 presents this idea, showing multiple hosts functioning in a TCP/IP network, with the SunNet Manager working in cooperation with SNMP agents on participating hosts.

Figure 17-5
An overview of
NetView/6000.

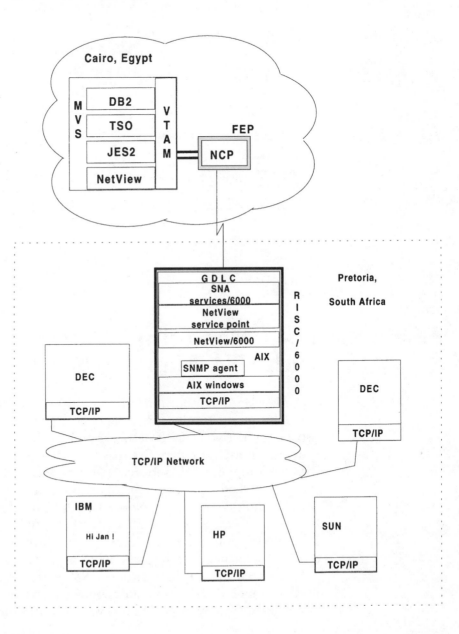

Figure 17-6
A conceptual view of
SunNet Manager.

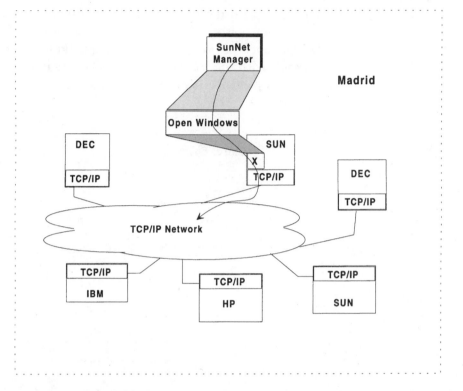

17.4 Managing TCP/IP and SNA with SNA Manager/6000

SNA Manager/6000 is an interesting twist on network management for IBM. This product, and other components, permits a RISC/6000 administrator to manage TCP/IP networks and a sub-area SNA network. This provides significant benefits to sites that have TCP/IP networks, a RISC/6000, and multiple sub-area SNA networks.

Figure 17-8 provides a conceptual overview of how SNA Manager/6000 fits into multiple networks. Notice that TCP/IP networks are managed by the same host where the SNA Manager software is running. In Fig. 17-8, multiple sub-area SNA networks exist. Through SNA Manager/6000, each can be managed from the RISC/6000 where it resides.

In contrast to TCP/IP networks, SNA network management can be illustrated by Fig. 17-9. This is an example of how management information in an SNA network is obtained when the RISC/6000 with SNA Manager/6000 is not directly attached to that network. In Fig. 17-9, specific PU and LU information can be obtained about a given controller. The host managing SNA can manage TCP/IP networks as well. SNA Manager/6000 depicts this under a windowing system. The resulting information is color-coded and portrayed via graphics. Ease of use is built into SNA Manager/6000.

SNA Manager/6000 presents information about SNA networks through a UNIX-based system, namely the RISC/6000, whose administrators are traditionally UNIX- and TCP/IP-oriented. By presenting network information in such a way, these administrators adapt more easily to the SNA environment, which seems formidable to others.

Design of SNA Manager/6000 was well thought-out. By taking typically cryptic information about an SNA network and presenting it to generally non-SNA-oriented personnel through the typical TCP/IP perspective, much was gained. The primary advantage of this program is that it permits SNA to be easily managed from a TCP/IP-based network.

Figure 17-7
SunNet Manager's cooperation with SNMP agents on participating hosts.

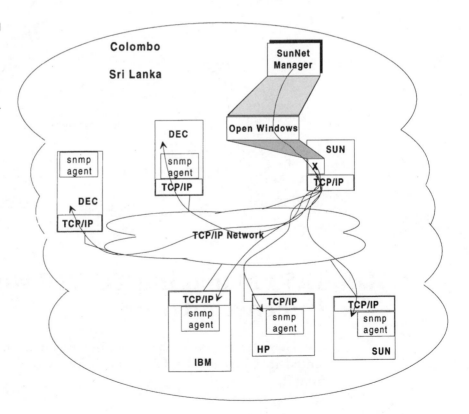

Figure 17-8
A conceptual view of
SNA Manager/6000.

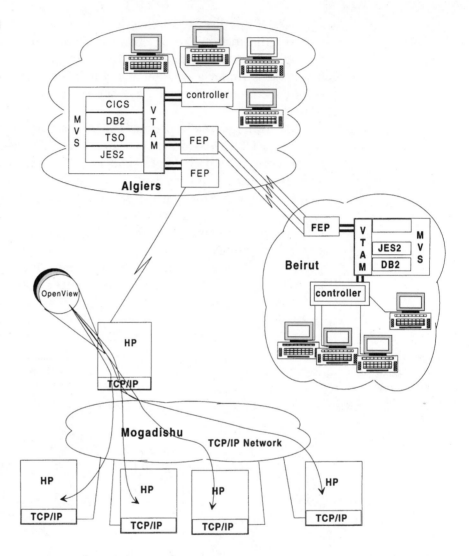

Figure 17-8
A conceptual view of
SNA Manager/6000.

17.5   Managing TCP/IP with HP's Internet Advisor

The Hewlett Packard Internet Advisor was used in the network presented in Chap. 6. The following is an example of data obtained from it:

Protocol	DLL or SAP	Frames	Bytes	DLL Errors	Average Length
21 Jul 8 97 Netware IPX	E0	4558	39	30	9
21 Jul 8 97 10:38:53 IP DOD	0800	4498	13	0	10
21 Jul 8 97 10:38:53 Novell IPX	8137	650	10	0	13
21 Jul 8 97 10:38:53 NetBIOS	F0	1167	7	1	6

Notice the date, time, upper-layer protocol type, DLL or SAP identifier, number of frames, number of bytes, the DLL errors that have occurred, and the average length.

The Internet Analyzer also provides these details as a snapshot of the network at this given point in time:

Ethernet Expert Analyzer

Date	Time	Protocol	#B/Sec, #/Stations, Warnings, Alerts, Utilization, Health, #F/Missed
Jul 8 97	10:34	Network Total	0,5,0,0,"","",0
Jul 8 97	10:34	IP	0,0,0,0
Jul 8 97	10:34	Novell	0,1,0,0
Jul 8 97	10:34	Other Protocols	0,2,"",""
Jul 8 97	10:34	MAC Level	0,2,0,0

The HP Internet Advisor is capable of measuring *any* protocol, upper or lower, in your network. This is just a small sample of the information obtained and used in this network.

17.6 Managing TCP/IP with OpenView

OpenView is a network management product offered by Hewlett Packard (HP) that is typically used with TCP/IP networks. In addition to managing other networks such as SNA and NetWare, OpenView can easily manage TCP/IP networks. Figure 17-10 illustrates the concept behind managing such diverse environments. It shows a TCP/IP-based network

Figure 17-9
An example of infor-
mation obtained
from subarea SNA via
SNA Manager/6000.

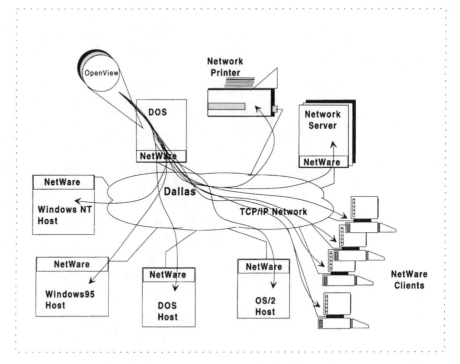

with OpenView implemented, and two SNA sub-area networks. All three networks can be managed from the OpenView host.

OpenView can operate on PCs with Windows or on UNIX-based systems. Viewed conceptually, it appears like Fig. 17-11.

Figure 17-12 illustrates how OpenView is used to manage a NetWare network. This example shows OpenView in a typical NetWare environment.

Another way to view OpenView as a network management product is in light of its surrounding environment, as in Fig. 17-13. This illustration is a high-level view of OpenView showing highlights of the individual OpenView components. Notice the dotted line separating the X protocol and TCP/IP. In reality, this separation does not exist. X is a five-layer protocol itself, and utilizes TCP for a transport mechanism. OpenView uses the X protocol to provide the windowing capability in a UNIX environment. A component not shown in Fig. 17-13 is the database used by OpenView components. This database works with OpenView as an ordinary application in the UNIX environment.

Figure 17-10
SunNet Manager's
cooperation with
SNMP agents.

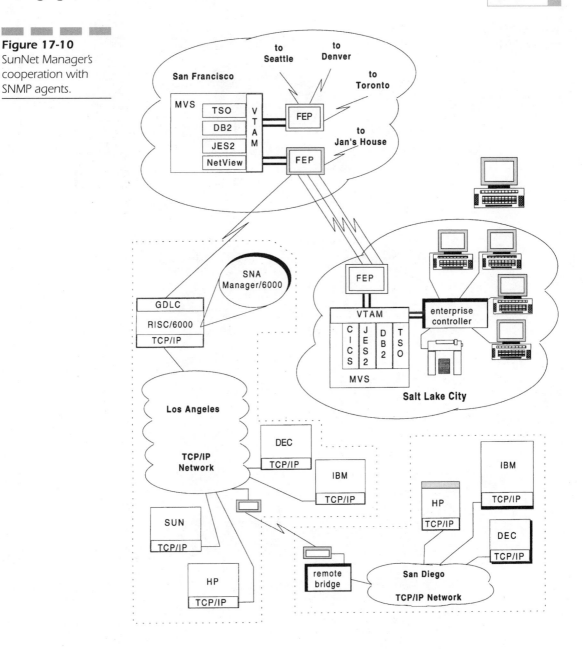

Figure 17-11
A conceptual overview of Open-View.

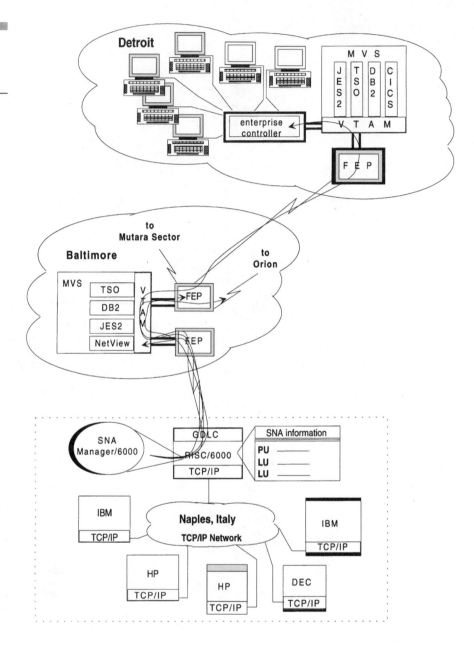

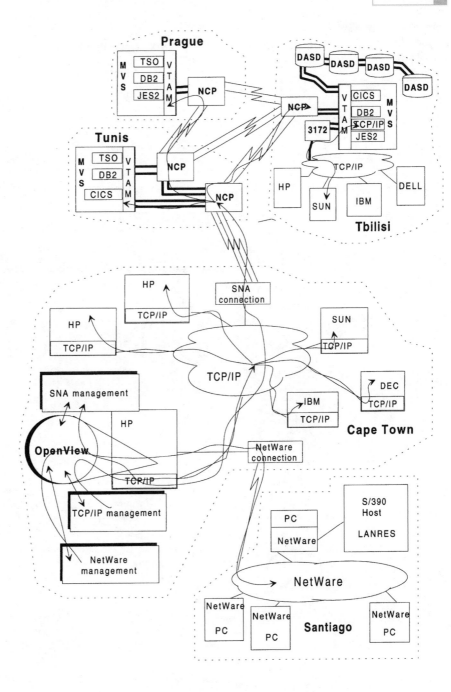

Figure 17-12
OpenView management in an SNA environment.

Figure 17-13
A high-level overview
of OpenView.

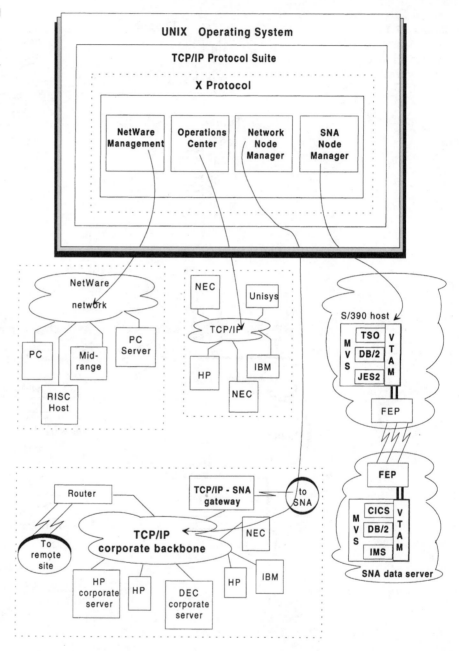

Understanding TCP/IP Management with OpenView

Network management is not new among those who have worked with corporate networks. The topic itself, and the products used by those responsible for it, has grown along with the integration of multi-vendor networks. Today, network management is not a luxury; it is central in most conversations about networks of any size at all. Like many aspects of networking technology, network management tends to get identified after the fact. Even the definition of network management varies.

For some, network management means ascertaining an identifiable measure of response time. Other meanings vary from outage notification of a given device in a remote location to CPU utilization at a given point in time. Beyond these ideas, network management implies a broad range of "management" abilities, such as the ability to download newer revisions of software, or to measure the throughput capability of an identified line.

OpenView and Network Management

Network management is similar to other network functions. For example, AppleTalk network users can perform file transfers as well as NetWare users. What differs is how this occurs. Even if multiple vendors offer the same protocol, how each vendor implements it is still a consideration.

Furthermore, the question of what can be managed is applicable among different vendor networks. Delineation of the upper- and lower-layer protocols used throughout a network must be identified for meaningful information to be obtained via network management. Beyond network management viewed from a protocol, there is application management. When custom applications are exploited across heterogeneous networks, this requires attention. Such applications might use User Datagram Protocol (UDP), Advanced Peer-to-Peer Communication (APPC), Multi-Protocol Transport Networking (MPTN), or other techniques.

Narrowing the scope of network management can be accomplished by determining if hardware management tools are needed, or if software-based tools can be used. The latter can yield significant information about hardware matters quite well.

OpenView's origins are tied to TCP/IP (from a development standpoint). This makes sense, because Hewlett Packard has been traditionally UNIX- and TCP/IP-oriented, and has also focused on research and development. OpenView has evolved into a component package capable of managing SNA- and NetWare-based networks as well as TCP/IP-based networks, be they LANs or WANs. This diversity provides flexibility and robustness for multi-platform network management. Figure 17-14 illustrates this idea.

Figure 17-14 shows OpenView managing three distinct network protocols: SNA, TCP/IP, and NetWare. This configuration, or variations of it, is typical. However, it is important to have a practical understanding of OpenView, particularly in the way the naming structure works.

OpenView Details

Network Node Manager (*NNM*) is probably the most-used acronym and reference to OpenView. NNM is the UNIX-based software part of OpenView that is used to manage TCP/IP-based networks. The NNM operates in conjunction with a database stored on a UNIX-based system. This database has continued to be Ingres in HP-UX environments. The Ingres database is where NNM stores its information about the networks it manages.

HP's NNM can also operate on SUN's Solaris operating system. In addition, in the summer of 1995, HP released version 4.0 of NNM, which now supports Oracle-based databases as well. HP has addressed other environments by supporting other operating software.

NNM can also operate under MS-Windows. This product is called OpenView Windows Node Manager (WNM). WNM is DOS-based and works quite well on PCs to monitor LAN statistical information. The original design intent behind this product seems to be that it would be used to manage a single-segment LAN-based environment. This is a conservative observation; the product may well be capable of going beyond the single-segment environment. An aspect of this product that makes it quite popular is its fairly broad support for third-party vendor applications.

Another twist on OpenView is the Distributed Manager, called OpenView DM. It focuses on the more complex environments and supports these environments by its broad base of Application Program Interfaces (APIs). Some refer to this product as simply DM. Regardless of how it is referred to, it is an aspect of OpenView that is more complex, and less "plug-n-play," than other implementations of it. To use DM presupposes a certain knowledge of Common Management Information Protocol (CMIP). However, the number of professionals who know CMIP well enough to use it is quite limited.

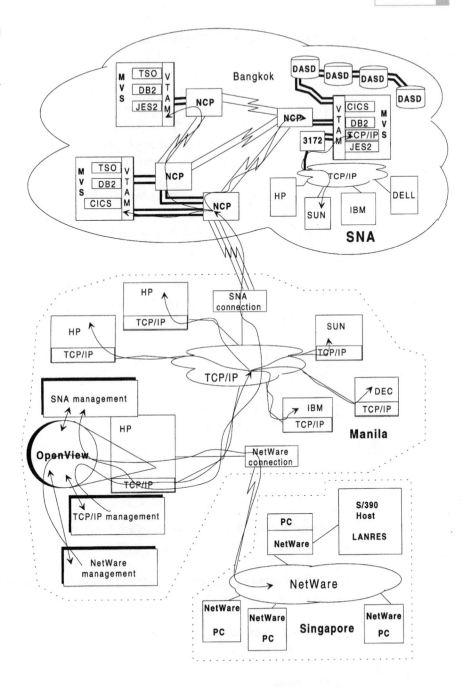

Figure 17-14
OpenView in a
diverse environment.

OpenView Products

Another way of examining OpenView is to consider it as a family of products. The OpenView name is a general reference to the individual components that may or may not be part of any given installation. These components can be separated into those that apply to systems management and those that apply to network management. The following components put OpenView in focus:

Distributed Management Platform This is known as DM and has been explained earlier in this chapter. This component operates in the traditional HP UNIX-based host and can manage TCP/IP- and OSI-based networks.

History Analyzer This component is used to view information previously collected about a network. Normally, this type of information reveals network utilization and capacity.

Interconnect Manager This component focuses on managing HP bridges, hubs, and routers.

NetMetrix This part of OpenView is responsible for internetwork monitoring and exploiting RMON.

Network Node Manager for NetWare Servers This part of OpenView provides the ability to manage NetWare servers. By default, it supports the IPX protocol.

Network Node Manager for NetWare Stations This aspect of OpenView is used to interact with NetWare clients.

Probe Manager This performs traffic-monitoring functions for NetWare clients as implemented on PCs.

Resource Manager This aspect of OpenView is used to obtain real-time management functions.

SNA Node Manager (SNM) This part of OpenView is used to manage SNA environments from UNIX-based LANs or WANs. OpenView's SNM views SNA through a MIB-II view. This concept alone is interesting. Just a few years ago, it was anathema to consider managing SNA networks with the philosophical approach that SNMP employs. However, things change.

Other application programs include the following:

AdminDesk Administrators and users utilize this product with servers and workstations for control and other administrative functions.

OperationsCenter This is used in automated responses. It focuses on distributed UNIX servers.

OpenSpool This is used in various aspects to control print servers.

Other programs are available and used to varying degrees depending on the need and size of the WAN or LAN.

17.7 OpenView Architecture

OpenView can best be understood when examined in light of the overall environment where it operates. Consider Fig. 17-15.

Figure 17-15 illustrates an architectural overview of OpenView. It is a simplistic view, but it conveys some basic component interaction. One important note is the presupposition of OpenView's operation with a

Figure 17-15
OpenView architecture and components.

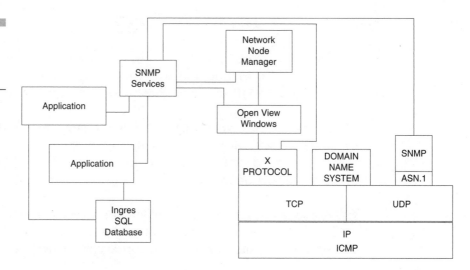

TCP/IP stack. Notice that OpenView utilizes X protocol. X, in turn, is part of most common TCP/IP protocol stacks. X uses SNMP services also. Notice further that X uses TCP as a transport protocol.

Figure 17-16 depicts multiple hosts on two networks in different locations connected via a router. The host where OpenView is loaded is an exploded view, designed to show the components in perspective. This illustration depicts a network in Tokyo and one in Kobe, Japan. The host in Tokyo has a UNIX-based host with OpenView. Figure 17-16 shows other applications operating on the same host as OpenView; this is important. This host, although used for network management, is not *dedicated*. Certain site-specific factors are used to determine the requirements needed to run OpenView.

Figure 17-16 implies the OpenView host can manage the host in Kobe, and it can. The routers connecting these networks together can also be managed. They are viewed as network devices. Another implication of this design is that both networks are TCP/IP-based.

Network Node Manager (NNM) is based on SNMP services. This is partially because of HP's product design and partially because of the way NNM interacts with TCP/IP, the network protocol. NNM has operational parallels to SNMP as it operates in TCP/IP environments. Because NNM operates the way it does, an understanding of certain SNMP and NNM terms is required. The following section explains some of these terms.

17.8 Terminology

Here is a concise list of terms whose meaning is critical to understanding NNM operation and structure. The list includes terms specifically related to SNMP, and those directly or indirectly related to NNM.

Abstract Syntax Notation.1 ASN.1 is the language used to describe network elements. It is a datatype definition language. This language is not only used to define the structure of a managed object, but also is an integral part of SNMP.

Access Mode This is a specific level of access to a MIB. For example, this could be read-only, write-only, read-write, or none. The access mode is simply the amount of access that users have when accessing MIBs.

Figure 17-16
One virtual network
in two physical loca-
tions.

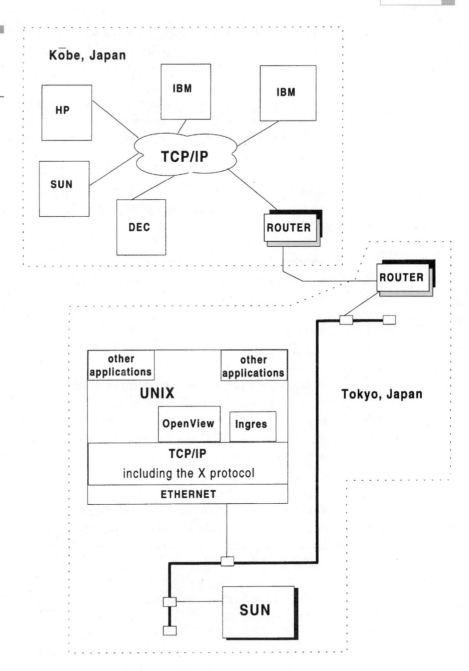

Agent This is a software application that is used by a network element to communicate with a manager. Network elements managed on a network contain an agent.

Community This is the identification, by association, between network elements and a manager application. The community is a method of identifying the agent and manager components functioning in a network.

Community Name A name is used to identify a community, as a password in messages confirming authorization to access information in a MIB.

Community Profile A community profile is a specific view of a MIB and the access mode that applies to an entire MIB view.

Enterprise-Specific MIB This refers to a management information base that is developed and supported by an individual vendor, usually for its own products.

Management Information Base (MIB) A MIB is a database of information about a network element. Technically, this information is used to comprise a logical database about the network; it includes information such as configuration, statistics, and status information about devices. Collectively, this information creates a logical representation of a network at any given period in time.

Manager This term refers to a software program that is used to manage a network. It communicates with network elements to obtain variables about devices that make up the network. This information is stored in a MIB.

Message Managers and agents use messages as the method by which they communicate.

Message Types The following are some message types used in SNMP:

- *GetRequest* A specific request from a manager to a network element, requesting a variable or list of variables from that network element's MIB.

- *GetNextRequest*   A method for a manager program to sequentially read a given network element's MIB.
- *GetResponse*   The response to a GetRequest, GetNextRequest, or SetRequest.
- *Trap*   A message that reports one of the following:
  - Functioning link
  - Incorrect authentication received with message
  - Initializing self
  - Link failure
  - Neighbor is not responding

Network Element This is a term used in conversations where SNMP or a variation thereof is the topic. It is nothing more than a reference term that refers to a device on a network. For example, a network element can be any of the following: bridge, gateway, host, router, or server. Another way of looking at this term is to consider it as a generic reference to devices.

Object An object is anything that has attributes and a name, and upon which a set of predefined operations can be operated. In SNMP, this object is considered a managed object. A MIB also contains managed objects. An object that is managed contains a name that is unique; this name is called an object identifier.

Object Definition This is a textual description of an object.

Object Descriptors These provide variable information about objects. A descriptor is the textual name for the object. An identifier is a numerical ID assigned to an object. This number is in the same form as used in the Domain Name System (DNS). Summary data about an object descriptor includes the access that explains the control rights to an object and the status that identifies a level of support for a given object.

Object Identifier This is a number assigned to an object.

Object Name An object name can also be referred to as an object identifier. It is a sequence of integers.

Structure of Management Information (SMI) This defines the structure regarding how network managers (people) describe network elements. SMI

SIM is a language. It is used to describe objects managed and as an identification method used against variables that read from, or are written to, a MIB.

Figure 17-17 is a simple illustration of SNMP components and concepts. Two components to focus upon are the agents and manager. Notice that each managed device (called a *network element*) is managed by the manager application program, a named SNMP manager operating on the HP host. The connections to the sites in Los Angeles and Miami can also be managed via SNMP in this scenario, assuming that agents are functioning on the network elements.

17.9   NNM Component Operations

NNM (Network Node Manager) can be explained in terms of its fundamental components and what they do. This section explains the following components, which work with OpenView's NNM:

- The SNMP management platform
- SNMP applications
- SNMP event configurator
- The MIB data collector
- MIB application builder

The SNMP management platform is a group of runtime applications. They are the essence of NNM. Customized applications can also work alongside these runtime applications, thus providing a user with a common user interface.

SNMP applications are nothing more than those software programs that make up NNM. Different applications provide different information. For example, one application acquires network traffic on the backbone and provides it in a real-time graphical fashion. Another application monitors the CPU load and disk space. Still another application monitors routing and configuration tables.

The SNMP event configurator is a program that governs how many events are displayed in a window in a given instant. The program enables users to customize the recognition of specific events they deem appropriate for their location.

The MIB data collector collects data from specific devices being monitored on a network. The MIB data collector works in conjunction with the SNMP event configurator. Between the two of them, a user can

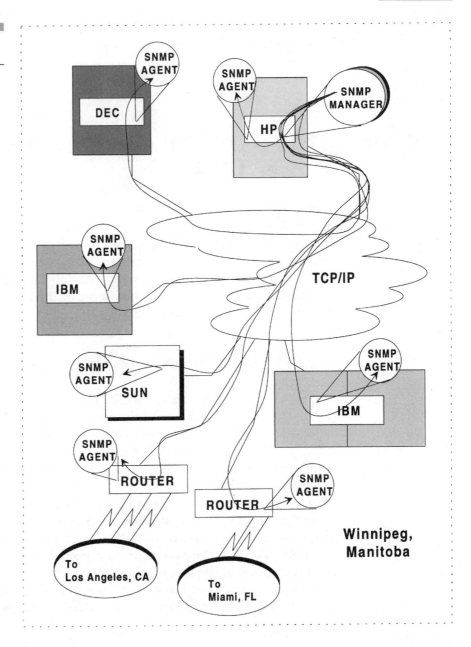

Figure 17-17
SNMP management.

define an event threshold that, when reached, causes the user to be notified via an alert.

The MIB application builder permits users to create customized screens. This program actually permits a user to display data in a text, graph, or tabular format. Information shown to the user comes from MIBs.

Other aspects of NNM exist in addition to the components discussed here. Hewlett Packard has many well-written manuals that explain NNM in greater detail. I suggest anyone working with it obtain the NNM library from Hewlett Packard.

17.10   TCP/IP and OpenView: A Closer Look

OpenView is a windowing interface. Consider Fig. 17-18, which shows OpenView using X protocol. X is not a transport-layer protocol; it is a protocol that can be divided into five layers itself. X supports a windowing interface and variations of programs that use its services.

OpenView, however, is more than a windowing interface. It is also the umbrella under which numerous network management components operate. The primary focus here is NNM.

TCP/IP and OpenView Interface

OpenView, to users, is seen as a working environment where "things" can be manipulated. To understand the operation of this interface, you must understand precisely the terms frequently used to describe and explain the OpenView environment. For example, the terms *object, map, submap,* and *symbol* are easily understood, but in OpenView, these terms convey particular meanings.

An *object* is best defined as that which has specific attributes maintained in the OpenView database. This means an object could be a router or a host computer. If objects were limited to only these two examples, life might be easier, but they are not. Since objects can be created, the notion of the intangible object emerges.

Considering that the broad discussion here is networks, it should not surprise you that objects can be any or all of the following:

- A network
- All computers in an office
- A building
- An interface
- A gateway

■ A country

■ A city

■ A software application

■ A region

■ A state (as in California or New York)

The key to understanding objects is realizing they can be virtually anything definable. This definition opens the door for a myriad of possible objects. In computer terms, this means an object can be logical or physical.

Figure 17-18
OpenView perceived as a windowing system.

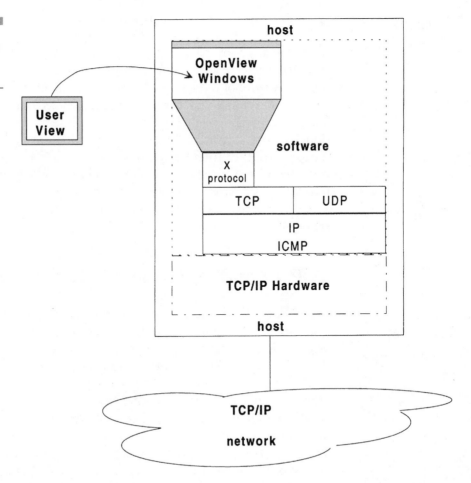

Closely related to an object is a *symbol*. Before exploring the definition of what a symbol is, one needs to understand the definition of a map.

A *map* is a set of objects, symbols, and submaps. Don't lose hope, this is not a circular definition! Consider just the first part here: a map is a set of objects. This makes sense; a collection of objects could be presented in such a way that one could call this presentation a map. A map is the presentation of data stored in the database that defines objects. Because this is true, intimate knowledge of a given site is required to understand the meaning portrayed by a map. Maps can be understood better by examining submaps.

A *submap* is a view of a specific part of a network. For example, consider Fig. 17-19, which conveys the relations between maps, submaps, and symbols in maps and submaps. Notice the map is at the top of the hierarchy. Within the map are lists of submaps. Each submap contains further submaps. Symbols are shown in some of the submaps.

Symbols deserve additional attention. In a sense, symbols are related to objects. A symbol represents an object by way of graphic portrayal on a submap. A symbol can have a type and status ascribed to it. The *type* refers to how the symbol appears on the screen. The *status* is its operational nature, displayed through colors. Different colors imply different statuses. Another interesting aspect of symbols is that they are said to have *behavior*. A symbol is either *executable* or *explodable*. The former implies a given function or event that happens once a symbol is selected. The latter suggests other symbols are beneath it in the structural arrangement of OpenView.

TCP/IP and OpenView SNA Node Manager

The SNA Node Manager is probably the most unusual (in historical perspective) of all OpenView products or functions. The philosophy, or original design intent, of Hewlett Packard was to be able to manage SNA the same way TCP/IP networks were managed, via SNMP.

The idea of managing SNA via the "eyes" of SNMP is difficult to conceive. One needs to understand SNA network management and how SNMP works in order to see how unusual this is. Until recently, few individuals had significant depth in both areas of network management (NetView and SNMP). It might seem strange, but a time existed when people generally focused on either SNA or TCP/IP management.

In fact, many professionals of the late 1970s to early 1990s tended to focus on just one area of networking technology. The intervention of

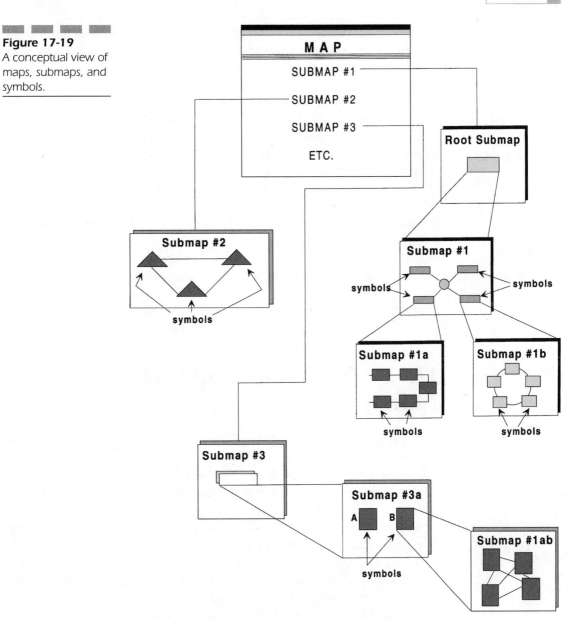

Figure 17-19
A conceptual view of maps, submaps, and symbols.

market forces changed this. In the 1980s, technology emerged providing the "glue" to bring together multiprotocol networks. This technology created the need for professionals to learn multiple network protocols.

In 1981, IBM introduced the first PC, which would, of course, evolve into a de facto standard. Prior to around 1986 or 1987, few devices existed to integrate multiprotocol networks. More scarce even than the devices to integrate networks were the individuals possessing the talent to perform the integration. At this time, NetWare was very much in its infancy, and Microsoft had merely announced Windows version 1.0. The point is that in the past, diverse network protocols were similar to human languages—few individuals attempted to learn more than one, and still fewer could learn one by the book.

SNA and TCP/IP are as different as day and night. SNA has traditionally used NetView, SystemView, or variations thereof to manage networks of any size. On the other hand, TCP/IP-based networks almost always have had SNMP at the heart of the network management.

17.11   Native SNA and TCP/IP Management

SNA management is well defined in the more than 1200 pages of the Network Management Vector Transport Protocol (NMVT) documentation. It explains, to the bit level, how "things" are managed in the world of SNA. NetView has been a popular user interface for humans through which the network can be visualized.

NetView (and other software used to manage SNA) provides everything from textual information to graphical illustrations about a SNA network. Histories can be saved and later evaluated against current events, real-time response can be monitored, and software distribution can be administered. There is another powerful aspect of SNA management: the ability to issue administrative commands via the management software and effect a part of the network that might be many time zones removed from the individual issuing the command.

Figure 17-20 illustrates this idea. In this illustration, only one copy of NetView is installed, but many sites can be managed. In this example, even the user is remote compared to where the copy of NetView is located.

Figure 17-20
Remote usage of
NetView.

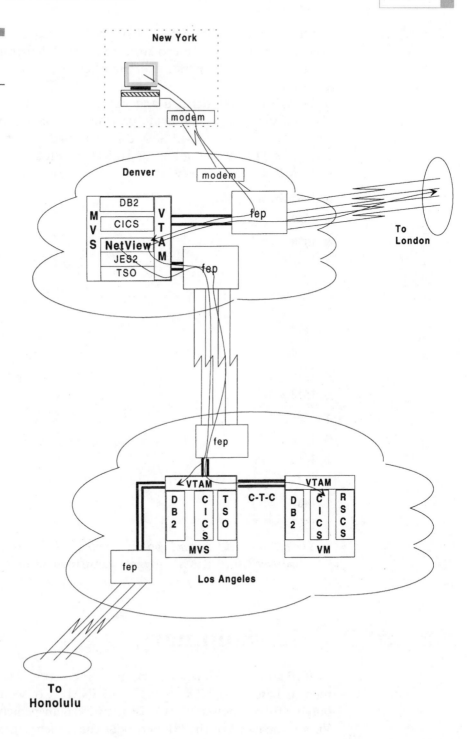

SNMP operation is different. Using SNMP implies operation of a windowing interface. SNMP can work in a remote environment, but many instances of SNMP operation occur where a user and SNMP are in the same place.

TCP/IP is a well-defined and well-documented network protocol. It also has significant worldwide presence. Its management software has been, and remains, SNMP. SNMP is explained in detail within numerous RFCs. The following is a partial RFC listing that covers various areas of SNMP or related topics:

- 1065 through 1067
- 1089
- 1090
- 1098
- 1155
- 1157
- 1158
- 1270
- 1351
- 1352
- 1381
- 1382
- 1414
- 1442 through 1451
- 1461
- 1503
- *The McGraw-Hill Internetworking Command Reference* has a more exhaustive list of RFCs. I suggest acquiring this book.

17.12   Summary

TCP/IP and OpenView are a natural combination. Much of HP's equipment is based in UNIX, TCP/IP, and SNMP. Consequently, the natural outgrowth is a powerful TCP/IP network management product. OpenView is the name of the HP network-management suite of products. Network Node Manager (NNM) is the most popular management software tool Hewlett Packard offers. Illustrations in this chapter have shown the power behind using NNM. To understand OpenView and NNM, you should become familiar with the terms associated with the product.

TCP/IP can be managed by vendor products other than HP; in fact, many vendors have TCP/IP network-management products. Most vendors put their own twist to products used to manage a network. Thus, just because different products use the same underlying protocol does not mean they will have similar user interfaces. In heterogeneous networks, complexity exists. Understanding the basic purpose and operation of major network-management programs is essential to manipulate data into meaningful information. TCP/IP and SNMP work together; typically, they are used by vendors as the core components of a network management tool.

This chapter presented an overview of some network management products used with TCP/IP management. Additional information was provided because of the presence TCP/IP has in installations throughout the world today. Some tools are more common than others, but each seems to have carved its own niche in the internetworking environment. Since most network managers or administrators must cope with multiple platforms, it is best to understand as much as possible about the different management tools in use.

18

Dynamic Host Configuration Protocol (DHCP)

The Dynamic Host Configuration Protocol (DHCP) is a way to pass configuration information to hosts on a TCP/IP network. It is based on the Bootstrap Protocol. DHCP adds the capability of automatic allocation of reusable network addresses and additional configuration options. DHCP functions with BOOTP relay agents.

18.1   Introduction

DHCP provides configuration parameters to Internet hosts. It consists of two components: a protocol for delivering host-specific configuration parameters from a DHCP server to a host, and a mechanism for allocation of network addresses to hosts.

DHCP is built on a client-server model, where designated DHCP server hosts allocate network addresses and deliver configuration parameters to dynamically configured hosts. Use of the term *server* here refers to a host providing initialization parameters through DHCP; the term *client* refers to a host requesting initialization parameters from a DHCP server.

A host should not act as a DHCP server unless explicitly configured to do so by a system administrator. The diversity of hardware and protocol implementations in the Internet would preclude reliable operation if random hosts were allowed to respond to DHCP requests.

For example, IP requires the setting of many parameters within the protocol implementation software. Because IP can be used on many dissimilar kinds of network hardware, values for those parameters cannot be guessed or assumed to have correct defaults. Also, distributed address-allocation schemes depend on a polling/defense mechanism to discover addresses that are already in use. IP hosts might not always be able to defend their network addresses, so that such a distributed address-allocation scheme cannot be guaranteed to avoid allocation of duplicate network addresses.

Address Allocation

DHCP supports three mechanisms for IP address allocation. In *automatic allocation,* DHCP assigns a permanent IP address to a client. *Dynamic allocation* means DHCP assigns an IP address to a client for a limited period of time (or until the client explicitly relinquishes the address). In *manual allocation,* a client's IP address is assigned by the network administrator, and DHCP is used simply to convey the assigned address to the client. A particular network will use one or more of these mechanisms, depending on the policies of the network administrator.

Dynamic allocation is the only one of the three mechanisms that allows automatic reuse of an address that is no longer needed by the client to which it was assigned. Thus, dynamic allocation is particularly useful for assigning an address to a client that will be connected to the network only temporarily, or for sharing a limited pool of IP addresses among a group of clients that do not need permanent IP addresses. Dynamic allocation might also be a good choice for assigning an IP address to a new client being permanently connected to a network where IP addresses are so scarce that it is important to reclaim them when old clients are retired.

Manual allocation allows DHCP to be used to eliminate the error-prone process of manually configuring hosts with IP addresses in environments where (for whatever reasons) it is desirable to manage IP address assignment outside of the DHCP mechanisms.

DHCP Message Format

The format of DHCP messages is based on the format of BOOTP messages, to capture the BOOTP relay-agent behavior described as part of the BOOTP specification and to allow interoperability of existing BOOTP clients with DHCP servers. Using BOOTP relay agents eliminates the necessity of having a DHCP server on each physical network segment.

Recent DHCP Additions

The DHCP message type DHCPINFORM is a recent addition to the protocol specification. Also, the classing mechanism for identifying DHCP clients to DHCP servers has been extended to include vendor classes. The minimum lease time restriction has been removed. Finally, many editorial changes have been made to clarify the text as a result of experience gained in DHCP interoperability tests.

DHCP Information

There are several Internet protocols and related mechanisms that address parts of the dynamic host configuration problem. Some of these include the following:

- The *Reverse Address Resolution Protocol (RARP)* explicitly addresses the problem of network address discovery and includes an automatic IP address assignment mechanism.

- *Trivial File Transfer Protocol (TFTP)* provides for the transport of a boot image from a boot server.

- *Internet Control Message Protocol (ICMP)* provides for informing hosts of additional routers via ICMP redirect messages. ICMP can also provide subnet-mask information by way of the ICMP mask request message. Additional information can be obtained by the ICMP information-request message. Network hosts can locate routers through the ICMP router discovery mechanism.

- *BOOTP* is a transport mechanism for a collection of configuration information.

- The *Network Information Protocol (NIP)*, used by Athena at MIT, is a distributed mechanism for dynamic IP address assignment.

- The *Resource Location Protocol (RLP)* provides for the location of higher-level services.

Sun Microsystems' diskless workstations use a boot procedure called *bootparams* that employs RARP, TFTP, and an RPC mechanism. bootparams deliver configuration information and operating system code to diskless network hosts. Some SUN networks also use DRARP and an autoinstallation mechanism to automate the configuration of new hosts in an existing, functional network.

In other related work, the *path minimum transmission unit (MTU)* discovery algorithm can determine the MTU of an arbitrary internet path. The *Address Resolution Protocol (ARP)* has been proposed as a transport protocol for resource location and selection.

DHCP Considerations

DHCP was designed to supply clients with the configuration parameters defined in the host requirement's RFCs. After obtaining parameters via DHCP, a client should be able to exchange packets with any other host in the Internet. Not all of these parameters are required for a newly initialized client. A client and server may negotiate transmission of those parameters required by the client or specific to a particular subnet.

DHCP allows but does not require the configuration of client parameters not directly related to the IP protocol. DHCP also does not address registration of newly configured clients with the Domain Name System. DHCP's original design was not intended to configure routers.

DHCP Terms

The following are some terms specific to DHCP:

DHCP client A DHCP client is an Internet host using DHCP to obtain configuration parameters, such as a network address.

DHCP server A DHCP server is an Internet or internet host that returns configuration parameters to DHCP clients.

BOOTP relay agent A BOOTP relay agent is an Internet host or router that passes DHCP messages between DHCP clients and DHCP servers. DHCP is designed to use the same relay-agent behavior as specified in the BOOTP protocol specification.

binding A binding is a collection of configuration parameters, including at least an IP address, associated with or bound to a DHCP client. Bindings are managed by DHCP servers.

DHCP Design Intent

The original design intent of DHCP includes the following principles:

- DHCP should be a mechanism, rather than a policy. DHCP must allow local system administrators control over configuration parameters where desired; local system administrators should be able to enforce local policies concerning allocation and access to local resources.
- Clients should require no manual configuration. Each client should be able to discover appropriate local configuration parameters without user intervention and incorporate those parameters into its own configuration.
- Networks should require no manual configuration for individual clients. Under normal circumstances, a network manager should not have to manually enter any per-client configuration parameters.

- DHCP should not require a server on each subnet. To allow for scale and economy, DHCP must work across routers or through the intervention of BOOTP relay agents.

- A DHCP client must be prepared to receive multiple responses to a request for configuration parameters. Some installations may include multiple, overlapping DHCP servers to enhance reliability and increase performance.

- DHCP must coexist with statically configured, nonparticipating hosts and with existing network protocol implementations.

- DHCP must work with the BOOTP relay-agent behavior as described by RFCs 951 and 1542.

- DHCP must provide service to existing BOOTP clients.

DHCP requirements include the following, specific to the transmission of network layer parameters:

- Guarantee that any network address will not be in use by more than one DHCP client at a time.

- Retain DHCP client configuration across a DHCP client reboot. A DHCP client should, whenever possible, be assigned the same configuration parameters (e.g., network address) in response to each request.

- Retain DHCP client configuration across server reboots. A DHCP client should be assigned the same configuration parameters despite restarts of the DHCP mechanism.

- Allow automated assignment of configuration parameters to new clients to avoid manual configuration for new clients.

- Support fixed or permanent allocation of configuration parameters to specific clients.

18.2   DHCP Protocol

From the client's point of view, DHCP is an extension of the BOOTP mechanism. This behavior allows existing BOOTP clients to operate with DHCP servers without requiring any change to the clients' initialization software. RFC 1542 details the interactions between BOOTP and DHCP clients and servers.

DHCP Message Format

The following DHCP message format describes each of the fields in the DHCP message. The numbers in parentheses indicate the size of each field, in octets. Two primary differences between DHCP and BOOTP exist. First, DHCP defines mechanisms through which clients can be assigned a network address for a finite lease, allowing for the serial reassignment of network addresses to different clients.

Second, DHCP provides the mechanism for a client to acquire all of the IP configuration parameters that it needs in order to operate. DHCP introduces a small change in terminology intended to clarify the meaning of one of the fields. What was at one time referred to as the *vendor extensions* field in BOOTP is know referred to as the *options* field in DHCP. *Tagged* data items that were used inside the BOOTP *vendor extensions* field, which were formerly referred to as *vendor extensions,* are now called *options.*

Consider the following representation of DHCP message format:

```
 0                   1                   2                   3
 0 1 2 3 4 5 6 7 8 9 0 1 2 3 4 5 6 7 8 9 0 1 2 3 4 5 6 7 8 9 0 1
+-+-+-+-+-+-+-+-+-+-+-+-+-+-+-+-+-+-+-+-+-+-+-+-+-+-+-+-+-+-+-+-+
|     op (1)    |   htype (1)   |   hlen (1)    |   hops (1)    |
+-+-+-+-+-+-+-+-+-+-+-+-+-+-+-+-+-+-+-+-+-+-+-+-+-+-+-+-+-+-+-+-+
|                            xid (4)                            |
+-+-+-+-+-+-+-+-+-+-+-+-+-+-+-+-+-+-+-+-+-+-+-+-+-+-+-+-+-+-+-+-+
|           secs (2)            |           flags (2)           |
+-+-+-+-+-+-+-+-+-+-+-+-+-+-+-+-+-+-+-+-+-+-+-+-+-+-+-+-+-+-+-+-+
|                          ciaddr  (4)                          |
+-+-+-+-+-+-+-+-+-+-+-+-+-+-+-+-+-+-+-+-+-+-+-+-+-+-+-+-+-+-+-+-+
|                          yiaddr  (4)                          |
+-+-+-+-+-+-+-+-+-+-+-+-+-+-+-+-+-+-+-+-+-+-+-+-+-+-+-+-+-+-+-+-+
|                          siaddr  (4)                          |
+-+-+-+-+-+-+-+-+-+-+-+-+-+-+-+-+-+-+-+-+-+-+-+-+-+-+-+-+-+-+-+-+
|                          giaddr  (4)                          |
+-+-+-+-+-+-+-+-+-+-+-+-+-+-+-+-+-+-+-+-+-+-+-+-+-+-+-+-+-+-+-+-+
|                                                               |
|                          chaddr  (16)                         |
|                                                               |
|                                                               |
+-+-+-+-+-+-+-+-+-+-+-+-+-+-+-+-+-+-+-+-+-+-+-+-+-+-+-+-+-+-+-+-+
|                                                               |
|                          sname   (64)                         |
+-+-+-+-+-+-+-+-+-+-+-+-+-+-+-+-+-+-+-+-+-+-+-+-+-+-+-+-+-+-+-+-+
|                                                               |
|                          file    (128)                        |
+-+-+-+-+-+-+-+-+-+-+-+-+-+-+-+-+-+-+-+-+-+-+-+-+-+-+-+-+-+-+-+-+
|                                                               |
|                          options (variable)                   |
+-+-+-+-+-+-+-+-+-+-+-+-+-+-+-+-+-+-+-+-+-+-+-+-+-+-+-+-+-+-+-+-+
```

DHCP defines a new *client identifier* option that is used to pass an explicit client identifier to a DHCP server. This change eliminates the overloading of the chaddr field in BOOTP messages, where chaddr is used both as a hardware address for transmission of BOOTP reply messages and as a client identifier.

The client identifier is an opaque key, not to be interpreted by the server. It may contain a hardware address, identical to the contents of the chaddr field, or it may contain another type of identifier, such as a DNS name. The client identifier chosen by a DHCP client *must* be unique to that client within the subnet to which the client is attached. If a client uses a client identifier in one message, it *must* use that same identifier in all subsequent messages, to ensure that all servers correctly identify the client.

DHCP clarifies the interpretation of the siaddr field as the address of the server to use in the next step of the client's bootstrap process. A DHCP server can return its own address in the siaddr field if the server is prepared to supply the next bootstrap service such as the delivery of an operating system executable image. A DHCP server always returns its own address in the server identifier option.

DHCP Message Field Explanation

Consider the following explanations for fields in a DHCP message:

Field	Octets	Description
op	1	Message op code/message type; 1= BOOTREQUEST, 2 = BOOTREPLY
htype	1	Hardware address type
hlen	1	Hardware address length.
hops	1	Client sets to zero; optionally used by relay agents when booting via a relay agent
xid	4	Transaction ID, a random number chosen by the client; used by the client and server to associate messages and responses between a client and a server
secs	2	Filled in by client; seconds elapsed since client began address acquisition or renewal process
flags	2	Flags

Field	Octets	Description
ciaddr	4	Client IP address; only filled in if client is in BOUND, RENEW, or REBINDING state and can respond to ARP requests
yiaddr	4	Your (client) IP address
siaddr	4	IP address of next server to use in bootstrap; returned in DHCPOFFER, DHCPACK by server
giaddr	4	Relay agent IP address, used in booting via a relay agent
chaddr	16	Client hardware address
sname	64	Optional server host name, null terminated string
file	128	Boot file name, null terminated string; "generic" name or null in DHCPDISCOVER, fully qualified directory path name in DHCPOFFER
options	var	Optional parameters field.

The options field is now variable length. A DHCP client must be prepared to receive DHCP messages with an options field of at least 312 octets. This requirement implies that a DHCP client must be prepared to receive a message of up to 576 octets, the minimum IP datagram size an IP host must accept.

It is possible, however, for DHCP clients to negotiate the use of larger DHCP messages through the *maximum DHCP message size* option. The options field may be further extended into the file and sname fields.

When DHCP is used in initial configuration (prior to the client's TCP/IP software's complete configuration), DHCP requires mental agility on behalf of the user. TCP/IP software installed should accept and forward any IP packets to the IP layer delivered to the client's hardware address before the IP address is configured. DHCP servers and BOOTP relay agents might not be able to deliver DHCP messages to clients that cannot accept hardware unicast datagrams before the TCP/IP software is configured. Be aware of this when you find yourself in this scenario.

In order to work around some clients that cannot accept IP unicast datagrams before the TCP/IP software is configured, DHCP uses the flags field. Here, the leftmost bit is defined as the BROADCAST (B) flag.

DHCP Flags Field Format

The flags field format is as follows:

```
                    1 1 1 1 1 1
0 1 2 3 4 5 6 7 8 9 0 1 2 3 4 5
+-+-+-+-+-+-+-+-+-+-+-+-+-+-+-+-+
|B|             MBZ              |
+-+-+-+-+-+-+-+-+-+-+-+-+-+-+-+-+

B:   BROADCAST flag
MBZ:  Must Be Zero
```

18.3   DHCP Configuration Parameters Repository

The first service provided by DHCP is to provide persistent storage of network parameters for network clients. The model of DHCP persistent storage is this: The DHCP service stores a key-value entry for each client (a unique identifier such as an IP subnet number and a unique identifier within the subnet), and the value contains the configuration parameters for the client.

For example, the key might be the IP-subnet-number and hardware-address. Alternately, the key might be the pair (IP-subnet-number, host-name), thus allowing the server to assign parameters intelligently to a DHCP client moved to a different subnet or one that has changed hardware addresses. The protocol defines that the key will be IP-subnet-number, hardware-address unless the client explicitly supplies an identifier using the *client identifier* option. A client can query the DHCP service to retrieve its configuration parameters. The client interface to the configuration parameters repository consists of protocol messages to request configuration parameters and responses from the server carrying the configuration parameters.

18.4   Network Address Dynamic Allocation

The second service provided by DHCP is the allocation of temporary or permanent network (IP) addresses to clients. The basic mechanism for the dynamic allocation of network addresses is simple: a client requests the use of an address for some period of time. The allocation mechanism (the collection of DHCP servers) guarantees not to reallocate that address within the requested time and attempts to return the same network address each time the client requests an address. The period over which a network address is allocated to a client is referred to here as a *lease*. The client may extend its lease with subsequent requests. It may issue a message to release the address back to the server when it no longer needs the address. The client may ask for a permanent assignment by asking for an infinite lease. Even when assigning *permanent* addresses, a server may choose to give out lengthy but non-infinite leases to allow detection of the fact that the client has been retired.

Some environments require reassignment of network addresses due to the exhaustion of available addresses. In such environments, the allocation mechanism will reuse an address whose lease has expired. The server should use whatever information is available in the configuration information repository to choose an address to reuse. For example, the server may choose the least recently assigned address. As a consistency check, the allocating server should probe the reused address before allocating the address, for example with an ICMP ECHO request, and the client should probe the newly received address with ARP.

18.5   Client-Server Protocol

DHCP uses the BOOTP message format shown on the next page.

The *op* field of each DHCP message sent from a client to a server contains a BOOTREQUEST. The BOOTREPLY is used in the op field of each DHCP message sent from a server to a client.

The first four octets of the options field of the DHCP message contain decimal values 99, 130, 83 and 99. The remainder of the options field consists of a list of tagged parameters called *options*. All vendor extensions are also DHCP options. For additional information, RFC 1533 provides a complete set of options defined for use with DHCP.

Some options included in the RFC are presented in this chapter. One particular option, the DHCP message type option, must be included in each DHCP message. This option defines the *type* of the DHCP message. Additional options may be allowed, required, or not allowed, depending on the DHCP message type.

```
 0                   1                   2                   3
 0 1 2 3 4 5 6 7 8 9 0 1 2 3 4 5 6 7 8 9 0 1 2 3 4 5 6 7 8 9 0 1
+-+-+-+-+-+-+-+-+-+-+-+-+-+-+-+-+-+-+-+-+-+-+-+-+-+-+-+-+-+-+-+-+
|     op (1)    |   htype (1)   |   hlen (1)    |   hops (1)    |
+-+-+-+-+-+-+-+-+-+-+-+-+-+-+-+-+-+-+-+-+-+-+-+-+-+-+-+-+-+-+-+-+
|                            xid (4)                           |
+-+-+-+-+-+-+-+-+-+-+-+-+-+-+-+-+-+-+-+-+-+-+-+-+-+-+-+-+-+-+-+-+
|           secs (2)            |           flags (2)          |
+-+-+-+-+-+-+-+-+-+-+-+-+-+-+-+-+-+-+-+-+-+-+-+-+-+-+-+-+-+-+-+-+
|                          ciaddr  (4)                         |
+-+-+-+-+-+-+-+-+-+-+-+-+-+-+-+-+-+-+-+-+-+-+-+-+-+-+-+-+-+-+-+-+
|                          yiaddr  (4)                         |
+-+-+-+-+-+-+-+-+-+-+-+-+-+-+-+-+-+-+-+-+-+-+-+-+-+-+-+-+-+-+-+-+
|                          siaddr  (4)                         |
+-+-+-+-+-+-+-+-+-+-+-+-+-+-+-+-+-+-+-+-+-+-+-+-+-+-+-+-+-+-+-+-+
|                          giaddr  (4)                         |
+-+-+-+-+-+-+-+-+-+-+-+-+-+-+-+-+-+-+-+-+-+-+-+-+-+-+-+-+-+-+-+-+
|                                                              |
|                          chaddr  (16)                        |
|                                                              |
|                                                              |
+-+-+-+-+-+-+-+-+-+-+-+-+-+-+-+-+-+-+-+-+-+-+-+-+-+-+-+-+-+-+-+-+
|                                                              |
|                          sname  (64)                         |
+-+-+-+-+-+-+-+-+-+-+-+-+-+-+-+-+-+-+-+-+-+-+-+-+-+-+-+-+-+-+-+-+
|                                                              |
|                          file   (128)                        |
+-+-+-+-+-+-+-+-+-+-+-+-+-+-+-+-+-+-+-+-+-+-+-+-+-+-+-+-+-+-+-+-+
|                                                              |
|                       options (variable)                     |
+-+-+-+-+-+-+-+-+-+-+-+-+-+-+-+-+-+-+-+-+-+-+-+-+-+-+-+-+-+-+-+-+
```

18.6   DHCP Messages and Meanings

Here are some common DHCP messages:

DHCPDISCOVER Client broadcasts to locate available servers.

DHCPOFFER In response to DHCPDISCOVER, server offers client configuration parameters.

DHCPREQUEST Client issues message to servers either requesting offered parameters from one server and implicitly declining offers from all others, or confirming the correctness of the previously allocated address. Examples include a system reboot or extending the lease on a particular network address.

DHCPACK Server responds to client with configuration parameters, including committed network address.

DHCPNAK Server responds to client, indicating client's notion of network address is incorrect (e.g., client has moved to a new subnet) or client's lease has expired.

DHCPDECLINE Client issues message to server indicating network address is already in use.

DHCPRELEASE Client asks server to relinquish network address and cancel remaining lease.

DHCPINFORM Client asks server only for local configuration parameters; client already has externally configured network address.

DHCP Message Timeline

On the top of page 406 is shown an example of the timeline concept between DHCP clients and a DHCP server when new address allocation is performed.

DHCP Client Usage

A client should use DHCP to reacquire or verify its IP address and network parameters whenever the local network parameters may have changed, such as at system boot time or after a disconnection from the local network. The local network configuration may change without the client's or user's knowledge.

If a client has knowledge of a previous network address and is unable to contact a local DHCP server, the client may continue to use the previous network address until the lease for that address expires. If the lease expires before the client can contact a DHCP server, the client must immediately discontinue use of the previous network address and might inform local users of the problem.

18.7 DHCP Client/Server Protocol Specification

It is important to understand the DHCP client/server specification. This section makes basic assumptions to provide helpful insights. The first assumption is that a DHCP server has a block of network addresses from

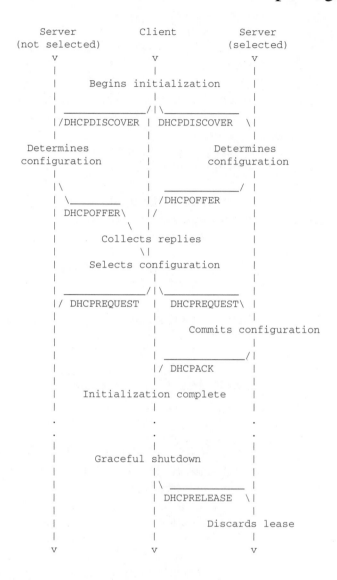

```
        Server            Client            Server
     (not selected)                       (selected)
          v                 v                 v
          |                 |                 |
          |        Begins initialization      |
          |                 |                 |
          | _____/|_____     |
          |/DHCPDISCOVER | DHCPDISCOVER  \|
          |                 |                 |
      Determines            |           Determines
     configuration          |          configuration
          |                 |                 |
          |\                | _____/  |
          | _____       | /DHCPOFFER      |
          | DHCPOFFER\    |/                  |
          |            \  |                    |
          |        Collects replies            |
          |               \|                   |
          |       Selects configuration        |
          |                 |                 |
          | _____/|_____     |
          |/ DHCPREQUEST  | DHCPREQUEST\ |
          |                 |                 |
          |                 |    Commits configuration
          |                 |                 |
          |                 | _____/|
          |               |/ DHCPACK          |
          |                 |                 |
          |      Initialization complete       |
          |                 |                 |
          .                 .                 .
          .                 .                 .
          |                 |                 |
          |        Graceful shutdown           |
          |                 |                 |
          |               |\ _____    |
          |               | DHCPRELEASE  \|
          |                 |                 |
          |                 |      Discards lease
          |                 |                 |
          v                 v                 v
```

which it can satisfy requests for new addresses. Another assumption is that each server maintains a database of allocated addresses and leases in its local permanent storage.

Constructing and Sending DHCP Messages

DHCP clients and servers construct DHCP messages by filling in fields in the fixed-format section of the message and appending tagged data items in the variable-length option area. The option area includes, first, a four-octet *magic cookie*, followed by the options. The last option must always be the *end* option.

DHCP uses UDP as its transport protocol. Its messages from client to server are sent to the DHCP server port 67. DHCP messages are sent from a server to a client on the DHCP client port 68. A server with multiple network address (a *multi-homed host*) can use any of its network addresses in outgoing DHCP messages.

The *server identifier* field is used both to identify a DHCP server in a DHCP message and as a destination address from clients to servers. A server with multiple network addresses must be prepared to accept any of its network addresses as identifying that server in a DHCP message. To accommodate potentially incomplete network connectivity, a server is required to choose an address as a server identifier. This address is reachable from the client. For example, if the DHCP server and the DHCP client are connected to the same subnet, the server should select the IP address it is using for communication on that subnet as the server identifier. If the server is using multiple IP addresses on that subnet, any such address may be used. If the server has received a message through a DHCP relay agent, the server should choose an address from the interface on which the message was received as the server identifier. DHCP clients are required to use the IP address provided in the server identifier option for any unicast requests to the DHCP server.

DHCP messages that are broadcast by a client prior to that client obtaining its IP address must have the source address field in the IP header set to zero.

If the *giaddr* field in a DHCP message from a client is nonzero, the server sends any return messages to the DHCP server port on the BOOTP relay agent whose address appears in giaddr. If the giaddr field is zero and the *ciaddr* field is nonzero, then the server unicasts DHCPOFFER and DHCPACK messages to the address in ciaddr. If both giaddr and ciaddr are zero, and the broadcast bit is set, then the server broadcasts DHCPOFFER and DHCPACK messages to 0xffffffff. If the broadcast bit is not set and both giaddr and ciaddr are zero, then the server unicasts DHCPOFFER and DHCPACK messages to the client's hardware address and *yiaddr* address. In all cases, when giaddr is zero, the server broadcasts any DHCPNAK messages to 0xffffffff.

If the options in a DHCP message extend into the sname and file fields, the *option overload* option is required to appear in the options field, with a value of one, two, or three. If the option overload option is present in the options field, the options in the options field must be terminated by an *end* option, and may contain one or more *pad* options to fill the options field. The options in the sname and file fields are required to begin with the first octet of the field, be terminated by an end option, and be followed by pad options to fill the remainder of the field. Any individual option in the options, sname, and file fields are required to be entirely contained in that field. The options in the options field must be interpreted first, so that any option overload options may be interpreted. The file field must be interpreted next, followed by the sname field.

The values to be passed in an option tag may be too long to fit in the 255 octets available to a single option. Options may appear only once, unless otherwise specified in the options document. The client concatenates the values of multiple instances of the same option into a single parameter list for configuration.

DHCP clients are responsible for all message retransmission. The client must adopt a retransmission strategy that incorporates a randomized exponential backoff algorithm to determine the delay between retransmissions. The delay between retransmissions should allow sufficient time for replies from the server to be delivered, based on the characteristics of the internetwork between the client and the server.

For example, in a 10Mb/sec ETHERNET network, the delay before the first retransmission should be 4 seconds, randomized by the value of a uniform random number chosen from the range -1 to +1. Clients with clocks that provide resolution granularity of less than one second may choose a noninteger randomization value. The delay before the next retransmission should be 8 seconds, randomized by the value of a uniform number chosen from the range -1 to +1. The retransmission delay should be doubled with subsequent retransmissions, up to a maximum of 64 seconds. The client can provide an indication of retransmission attempts to the user as an indication of the progress of the configuration process.

The *xid* field is used by the client to match incoming DHCP messages with pending requests. A DHCP client must choose xids in such a way as to minimize the chance of using an xid identical to one used by another client. For example, a client may choose a different, random initial xid each time the client is rebooted, and subsequently use sequential xids until the next reboot. Selecting a new xid for each retransmission is an implementation decision. A client may choose to reuse the same xid or select a new one for each retransmitted message.

Normally, DHCP servers and BOOTP relay agents attempt to deliver DHCPOFFER, DHCPACK, and DHCPNAK messages directly to the client using unicast delivery. The IP destination address is set to the DHCP yiaddr address and the link-layer destination address is set to the DHCP chaddr address. Unfortunately, some client implementations are unable to receive such unicast IP datagrams until the implementation has been configured with a valid IP address.

A client that cannot receive unicast IP datagrams until its protocol software has been configured with an IP address should set the BROADCAST bit in the flags field to one in any DHCPDISCOVER or DHCPREQUEST messages that client sends. The BROADCAST bit will provide a hint to the DHCP server and BOOTP relay agent to broadcast any messages to the client on the client's subnet. A client that can receive unicast IP datagrams before its protocol software has been configured should clear the BROADCAST bit to zero. The BOOTP clarifications document discusses the ramifications of the use of the BROADCAST bit.

A server or relay agent sending or relaying a DHCP message directly to a DHCP client should examine the BROADCAST bit in the flags field. If this bit is set to one, the DHCP message should be sent as an IP broadcast using an IP broadcast address (preferably 0xffffffff) as the IP destination address and the link-layer broadcast address as the link-layer destination address. If the BROADCAST bit is cleared to zero, the message should be sent as an IP unicast to the IP address specified in the yiaddr field and the link-layer address specified in the chaddr field. If unicasting is not possible, the message can be sent as an IP broadcast using an IP broadcast address (preferably 0xffffffff) as the IP destination address and the link-layer broadcast address as the link-layer destination address.

DHCP Server Administrative Controls

DHCP servers are not required to respond to every DHCPDISCOVER and DHCPREQUEST message they receive. For example, a network administrator, to retain control over the clients attached to the network, might configure DHCP servers to respond only to clients that have been previously registered through some external mechanism. The DHCP specification describes only the interactions between clients and servers when the clients and servers choose to interact; it is beyond the scope of the DHCP specification to describe all of the administrative controls that system administrators might want to use. Specific DHCP server implementations may incorporate any controls or policies desired by a network administrator.

In some environments, a DHCP server will also have to consider the values of the vendor class options included in DHCPDISCOVER or DHCPREQUEST messages when determining the correct parameters for a particular client. A DHCP server needs to use some unique identifier to associate a client with its lease. The client may choose to explicitly provide the identifier through the *client identifier* option. If the client supplies a client identifier, the client must use the same identifier in all subsequent messages, and the server must use that identifier to identify the client. If the client does not provide a client identifier option, the server must use the contents of the chaddr field to identify the client.

It is important for a DHCP client to use an identifier unique within the subnet to which the client is attached in the client identifier option. Use of chaddr as the client's unique identifier may cause unexpected results, as that identifier may be associated with a hardware interface that could be moved to a new client. Some sites may choose to use a manufacturer's serial number as the client identifier, to avoid unexpected changes in a client's network address due to transfer of hardware interfaces among computers. Sites may also choose to use a DNS name as the client identifier, causing address leases to be associated with the DNS name rather than a specific hardware box.

DHCP clients are free to use any strategy in selecting a DHCP server among those from which the client receives a DHCPOFFER message. The client implementation of DHCP should provide a mechanism for the user to select the vendor class identifier values directly.

18.8   DHCP Server Function

A DHCP server processes incoming DHCP messages from a client based on the current state of the binding for that client. The following messages can be received by a DHCP server from a client:

- DHCPDISCOVER
- DHCPREQUEST
- DHCPDECLINE
- DHCPRELEASE
- DHCPINFORM

An illustration later in this chapter shows the correlation between the use of the fields and options in a DHCP message by a server. After this correlation information is provided, a description of the DHCP server action for each possible incoming message is presented.

DHCPDISCOVER Message

When a server receives a DHCPDISCOVER message from a client, the server chooses a network address for the requesting client. If no address is available, the server may choose to report the problem to the system administrator. If an address is available, the new address should be selected as follows:

1. The client's current address as recorded in the client's current binding, ELSE

2. The client's previous address as recorded in the client's binding, if that address is in the server's pool of available addresses and not already allocated, ELSE

3. The address requested in the Requested IP Address option, if that address is valid and not already allocated, ELSE

4. A new address allocated from the server's pool of available addresses; the address is selected based on the subnet from which the message was received (if giaddr is zero) or on the address of the relay agent that forwarded the message (if giaddr is not zero).

A server can assign an address other than the one requested or refuse to allocate an address to a particular client even though free addresses are available. In some network architectures (internets with more than one IP subnet assigned to a physical network segment), it may be that the DHCP client should be assigned an address from a different subnet than the address recorded in giaddr. Hence, DHCP does not require that the client be assigned an address from the subnet in giaddr; a server is free to choose some other subnet.

While not required for correct operation of DHCP, the server should not reuse the selected network address before the client responds to the server's DHCPOFFER message. The server may choose to record the address as offered to the client. The server must also choose an expiration time for the lease, as follows:

1. IF the client has not requested a specific lease in the DHCPDISCOVER message and the client already has an assigned network address, the server returns the lease expiration time previously assigned to that address (note that the client must explicitly request a specific lease to extend the expiration time on a previously assigned address), ELSE

2. IF the client has not requested a specific lease in the DHCPDISCOVER message and the client does not have an assigned network address, the server assigns a locally configured default lease time, ELSE

3. IF the client has requested a specific lease in the DHCPDISCOVER message (regardless of whether the client has an assigned network address), the server may choose either to return the requested lease (if the lease is acceptable to local policy) or select another lease.

The rules for the fields and options used in DHCP are summarized in the following table.

Fields and Options Used by DHCP Servers

Field	DHCPOFFER	DHCPACK	DHCPNAK
op	BOOTREPLY	BOOTREPLY	BOOTREPLY
htype			
hlen	Hardware address length in octets		
hops	0	0	0
xid	xid from client DHCPDISCOVER message 0	xid from client DHCPREQUEST message 0	DHCPREQUEST message 'secs' 0
ciaddr	0	ciaddr from DHCPREQUEST or 0	0
yiaddr	IP address offered to client	IP address assigned to client	0
siaddr	IP address of next bootstrap server	IP address of next bootstrap server	0
flags	flags from client DHCPDISCOVER message	flags from client DHCPREQUEST message	flags from client DHCPREQUEST me
giaddr	giaddr from client DHCPDISCOVER message	giaddr from client DHCPREQUEST message	giaddr from client DHCPREQUEST me
chaddr	chaddr from client DHCPDISCOVER message	chaddr from client DHCPREQUEST message	chaddr from client DHCPREQUEST me
sname	Server host name or options	Server host name or options	(unused)
file	Client boot filename or options	Client boot filename or options	(unused)

Option	DHCPOFFER	DHCPACK	DHCPNAK
Requested IP address	Must not	Must not	Must not
IP address lease time	Must	Must (DHCPREQUEST), must not (DHCPINFORM)	Must not
File and sname fields	May	May	Must not (DHCP)

Option	DHCPOFFER	DHCPACK	DHCPNAK
Message type	DHCPOFFER	DHCPACK	DHCPNAK
Parameter request list	Must not	Must not	Must
NOT Message	Should	Should	Should
Client identifier	Must not	Must not	May
Vendor class identifier	May	May	May
Server identifier	Must	Must	Must
Maximum message size	Must not	Must not	Must not
All others	May	May	Must not

Once the network address and lease have been determined, the server constructs a DHCPOFFER message with the offered configuration parameters. It is important for all DHCP servers to return the same parameters (with the possible exception of a newly allocated network address) to ensure predictable client behavior regardless of which server the client selects. The configuration parameters must be selected by applying the rules in the order given. The network administrator is responsible for configuring multiple DHCP servers to ensure uniform responses from those servers.

The server must return the following to the client, in the order given:

1. The client's network address, as determined by the rules given earlier in this section

2. The expiration time for the client's lease, as determined by the rules given earlier in this section

3. Parameters requested by the client, according to the following rules:

   ■ IF the server has been explicitly configured with a default value for the parameter, the server must include that value in an appropriate option in the option field, ELSE

   ■ IF the server recognizes the parameter as a parameter defined in the Host Requirements Document, the server must include the default value for that parameter as given in the Host Requirements Document in an appropriate option in the option field, ELSE

   ■ The server must not return a value for that parameter, the server must supply as many of the requested parameters as possible and must omit any parameters it cannot provide.

4. Any parameters from the existing binding that differ from the Host Requirements Document defaults

5. Any parameters specific to this client (as identified by the contents of chaddr or client identifier in the DHCPDISCOVER or DHCPREQUEST message) as configured by the network administrator

6. Any parameters specific to this client's class (as identified by the contents of the vendor class identifier option in the DHCPDISCOVER or DHCPREQUEST message) as configured by the network administrator for which there is an exact match between the client's vendor class identifiers and the client's classes identified in the server

7. Parameters with non-default values on the client's subnet

The server *may* choose to return the vendor class identifier used to determine the parameters in the DHCPOFFER message to assist the client in selecting which DHCPOFFER to accept. The server inserts the xid field from the DHCPDISCOVER message into the xid field of the DHCPOFFER message and sends the DHCPOFFER message to the requesting client.

DHCPREQUEST Message

A DHCPREQUEST message may come from a client responding to a DHCPOFFER message from a server, from a client verifying a previously allocated IP address, or from a client extending the lease on a network address. If the DHCPREQUEST message contains a server identifier option, the message is in response to a DHCPOFFER message. Otherwise, the message is a request to verify or extend an existing lease. If the client uses a client identifier in a DHCPREQUEST message, it must use that same client identifier in all subsequent messages. If the client includes a list of requested parameters in a DHCPDISCOVER message, it must include that list in all subsequent messages.

Any configuration parameters in the DHCPACK message should not conflict with those in the earlier DHCPOFFER message to which the client is responding. The client should use the parameters in the DHCPACK message for configuration. The following sections provide details of the DHCPREQUEST messages that clients send in different states.

DHCPREQUEST Generated during SELECTING The client inserts the address of the selected server in the server identifier, ciaddr must be zero, and the requested IP address must be filled in with the yiaddr value from the chosen DHCPOFFER.

The client may choose to collect several DHCPOFFER messages and select the best. A client indicates its selection by identifying the offering server in the DHCPREQUEST message. If the client receives no acceptable offers, the client may choose to try another DHCPDISCOVER message. Therefore, the servers may not receive a specific DHCPREQUEST from which they can decide whether or not the client has accepted the offer. Because the servers have not committed any network address assignments on the basis of a DHCPOFFER, servers are free to reuse offered network addresses in response to subsequent requests. As an implementation detail, servers should not reuse offered addresses and may use an implementation-specific timeout mechanism to decide when to reuse an offered address.

DHCPREQUEST Generated during INIT-REBOOT   A server identifier must not be filled in, and the requested IP address option must be filled in with client's notion of its previously assigned address. The ciaddr must be zero. The client is seeking to verify a previously allocated, cached configuration. The server should send a DHCPNAK message to the client if the requested IP address is incorrect, or is on the wrong network.

Determining whether a client in the INIT-REBOOT state is on the correct network is done by examining the contents of giaddr and the requested IP address option, and performing a database lookup. If the DHCP server detects that the client is on the wrong network, the server should send a DHCPNAK message to the client.

If the network is correct, the DHCP server should check if the client's notion of its IP address is correct. If not, then the server should send a DHCPNAK message to the client. If the DHCP server has no record of this client, then it must remain silent, and can output a warning to the network administrator. This behavior is necessary for peaceful coexistence of noncommunicating DHCP servers on the same wire.

If giaddr is 0x0 in the DHCPREQUEST message, the client is on the same subnet as the server. The server must broadcast the DHCPNAK message to the 0xffffffff broadcast address because the client may not have a correct network address or subnet mask, and the client may not be answering ARP requests.

If giaddr is set in the DHCPREQUEST message, the client is on a different subnet. The server must set the broadcast bit in the DHCPNAK, so that the relay agent will broadcast the DHCPNAK to the client, because the client may not have a correct network address or subnet mask, and the client may not be answering ARP requests.

DHCPREQUEST Generated during RENEWING A server identifier must not be filled in, the requested IP address option must not be filled in, and ciaddr must be filled in with the client's IP address. In this situation, the client is completely configured, and is trying to extend its lease. This message will be unicast, so no relay agents will be involved in its transmission. Because giaddr is therefore not filled in, the DHCP server will trust the value in ciaddr, and use it when replying to the client.

A client may choose to renew or extend its lease prior to T1. The server may choose not to extend the lease, but should return a DHCPACK message regardless.

DHCPREQUEST Generated during REBINDING The server identifier must not be filled in, the requested IP address option must not be filled in, and ciaddr must be filled in with client's IP address. In this situation, the client is completely configured and is trying to extend its lease. This message must be broadcast to the 0xffffffff IP broadcast address. The DHCP server should check ciaddr for correctness before replying to the DHCPREQUEST.

The DHCPREQUEST from a REBINDING client is intended to accommodate sites that have multiple DHCP servers and a mechanism for maintaining consistency among leases managed by multiple servers. A DHCP server may extend a client's lease only if it has local administrative authority to do so.

DHCPDECLINE Message

If the server receives a DHCPDECLINE message, the client has discovered through some other means that the suggested network address is already in use. The server must mark the network address as not available and should notify the local system administrator of a possible configuration problem.

DHCPRELEASE Message

Upon receipt of a DHCPRELEASE message, the server marks the network address as not allocated. The server should retain a record of the client's initialization parameters for possible reuse in response to subsequent requests from the client.

DHCPINFORM Message

The server responds to a DHCPINFORM message by sending a DHCPACK message directly to the address given in the ciaddr field of the DHCPINFORM message. The server must not send a lease expiration time to the client and should not fill in yiaddr.

Client Messages

Consider the following example of differences between messages from clients in various states:

```
-+-+-+-+-+-+-+-+-+-+-+-+-+-+-+-+-+-+-+-+-+-+-+-+-+-+-+-+-+-+-
|             | INIT-REBOOT | SELECTING  |RENEWING    |REBINDING |
|-+-+-+-+-+-+-+-+-+-+-+-+-+-+-+-+-+-+-+-+-+-+-+-+-+-+-+-|
| broad/unicast | broadcast  | broadcast  | unicast    | broadcast  |
| server-ip     | MUST NOT   | MUST       | MUST NOT   | MUST NOT   |
| requested-ip  | MUST       | MUST       | MUST NOT   | MUST NOT   |
| ciaddr        | zero       | zero       | IP address| IP address |
-+-+-+-+-+-+-+-+-+-+-+-+-+-+-+-+-+-+-+-+-+-+-+-+-+-+-+-+-+-+-
```

18.9   DHCP Client Function

A DHCP client can receive the following messages from a server:

- DHCPOFFER
- DHCPACK
- DHCPNAK

The remainder of this section describes the action of the DHCP client for each possible incoming message.

The client begins in INIT state and forms a DHCPDISCOVER message. The client should wait a random time between one and ten seconds to desynchronize the use of DHCP at startup. The client sets ciaddr to 0x00000000. The client may request specific parameters by including the parameter request list option. The client may suggest a network address and/or lease time by including the requested IP address and IP address lease time options. The client must include its hardware address in the chaddr field, if that is necessary for the delivery of DHCP reply messages. The client can include a different unique identifier in the client identifier option, as discussed in the previous section. If the client includes a list of requested parameters in a DHCPDISCOVER message, it must include the list in all subsequent messages.

The client generates and records a random transaction identifier and inserts that identifier into the xid field. The client records its own local time for later use in computing the lease expiration. The client then broadcasts the DHCPDISCOVER message on the local hardware broadcast address to the 0xffffffff IP broadcast address and the DHCP server's UDP port.

If the xid of an arriving DHCPOFFER message does not match the xid of the most recent DHCPDISCOVER message, the DHCPOFFER message must be silently discarded. Any arriving DHCPACK messages must also be silently discarded.

The client collects DHCPOFFER messages over a period of time, selects one DHCPOFFER message from the incoming DHCPOFFER messages, and extracts the server address from the server identifier option in the DHCPOFFER message. The time over which the client collects messages and the mechanism used to select one DHCPOFFER are implementation-dependent.

If the parameters are acceptable, the client records the address of the server that supplied the parameters from the server identifier field and sends that address in the server identifier field of a DHCPREQUEST broadcast message. Once the DHCPACK message from the server arrives, the client is initialized and moves to BOUND state. The DHCPREQUEST message contains the same xid as the DHCPOFFER message. The client records the lease expiration time as the sum of the time at which the original request was sent and the duration of the lease from the DHCPACK message.

The client should perform a check on the suggested address to ensure that the address is not already in use. For example, if the client is on a network that supports ARP, the client may issue an ARP request for the suggested request. When broadcasting an ARP request for the suggested address, the client must fill in its own hardware address as the sender's hardware address, and zero as the sender's IP address, to avoid confusing ARP caches in other hosts on the same subnet. If the network address appears to be in use, the client must send a DHCPDECLINE message to the server. The client should broadcast an ARP reply to announce the client's new IP address and clear any outdated ARP cache entries in hosts on the client's subnet.

Initialization with a Known Network Address

The client begins in INIT-REBOOT state and sends a DHCPREQUEST message. The client must insert its known network address as a requested IP address option in the DHCPREQUEST message. The client may request specific configuration parameters by including the parameter request list option. The client generates and records a random transaction identifier and inserts that identifier into the xid field. The client records its own local time for later use in computing the lease expiration. The client must not include a server identifier in the DHCPREQUEST message. The client then broadcasts the DHCPREQUEST on the local hardware broadcast address to the DHCP server's UDP port.

Once a DHCPACK message with an xid field matching that in the client's DHCPREQUEST message arrives from any server, the client is initialized and moves to BOUND state. The client records the lease expiration time as the sum of the time at which the DHCPREQUEST message was sent and the duration of the lease from the DHCPACK message.

Initialization with External Assigned Network Addresses

The client sends a DHCPINFORM message. The client may request specific configuration parameters by including the parameter request list option. The client generates and records a random transaction identifier and inserts that identifier into the xid field. The client places its own network address in the ciaddr field. The client should not request lease time parameters.

The client then unicasts the DHCPINFORM to the DHCP server if it knows the server's address. Otherwise, it broadcasts the message to the limited (all ones) broadcast address. DHCPINFORM messages must be directed to the DHCP server's UDP port. Once a DHCPACK message with an xid field matching that in the client's DHCPINFORM message arrives from any server, the client is initialized.

Use of Broadcast and Unicast

The DHCP client broadcasts DHCPDISCOVER, DHCPREQUEST, and DHCPINFORM messages, unless the client knows the address of a DHCP server. The client unicasts DHCPRELEASE messages to the server. Because the client is declining the use of the IP address supplied by the server, the client broadcasts DHCPDECLINE messages.

When the DHCP client knows the address of a DHCP server, in either INIT or REBOOTING state, the client may use that address in the DHCPDISCOVER or DHCPREQUEST rather than the IP broadcast address. The client may also use unicast to send DHCPINFORM messages to a known DHCP server. If the client receives no response to DHCP messages sent to the IP address of a known DHCP server, the DHCP client reverts to using the IP broadcast address.

Reacquisition and Expiration

The client maintains two times, T1 and T2, that specify the times at which the client tries to extend its lease on its network address. T1 is the time at which the client enters the RENEWING state and attempts to contact the server that originally issued the client's network address. T2 is the time at which the client enters the REBINDING state and attempts to contact any server. T1 must be earlier than T2, which, in turn, must be earlier than the time at which the client's lease will expire.

To avoid the need for synchronized clocks, T1 and T2 are expressed in options as relative times. At time T1, the client moves to RENEWING state and sends (via unicast) a DHCPREQUEST message to the server to extend its lease. The client sets the ciaddr field in the DHCPREQUEST to its current network address. The client records the local time at which the DHCPREQUEST message is sent for computation of the lease expiration time. The client must not include a server identifier in the DHCPREQUEST message.

Any DHCPACK messages that arrive with an xid that does not match the xid of the client's DHCPREQUEST message are silently discarded. When the client receives a DHCPACK from the server, the client computes the lease expiration time as the sum of the time at which the client sent the DHCPREQUEST message and the duration of the lease in the DHCPACK message. The client has successfully reacquired its network address, so it returns to BOUND state and may continue network processing.

If no DHCPACK arrives before time T2, the client moves to REBINDING state and sends (via broadcast) a DHCPREQUEST message to extend its lease. The client sets the ciaddr field in the DHCPREQUEST to its current network address. The client must not include a server identifier in the DHCPREQUEST message.

Times T1 and T2 are configurable by the server through options. T1 defaults to

$$0.5 < x > \text{duration\_of\_lease}$$

T2 defaults to

$$0.875 < x > \text{duration\_of\_lease}$$

Times T1 and T2 should be chosen with some random "fuzz" around a fixed value, to avoid synchronization of client reacquisition.

A client can choose to renew or extend its lease prior to T1. The server can choose to extend the client's lease according to policy set by the network administrator. The server should return T1 and T2, and their values should be adjusted from their original values to take account of the time remaining on the lease.

In both RENEWING and REBINDING states, if the client receives no response to its DHCPREQUEST message, the client should wait one-half of the remaining time until T2 (in RENEWING state) and one-half of the remaining lease time (in REBINDING state), down to a minimum of 60 seconds, before retransmitting the DHCPREQUEST message.

If the lease expires before the client receives a DHCPACK, the client moves to INIT state, must immediately stop any other network processing, and requests network initialization parameters as if the client were uninitialized. If the client then receives a DHCPACK allocating that client its previous network address, the client should continue network processing. If the client is given a new network address, it must not continue using the previous network address and should notify the local users of the problem.

DHCPRELEASE

If the client no longer requires use of its assigned network address, the client sends a DHCPRELEASE message to the server. Note that the correct operation of DHCP does not depend on the transmission of DHCPRELEASE messages.

CHAPTER 19

The Domain Name System

This chapter introduces domain style names, their use for internet mail and host address support, and the protocols and servers used to implement domain name facilities.

19.1   A Historical Perspective on Domain Naming

The original motive for the development of the domain system was the growth of the Internet. The following were considerations:

■ Host-name-to-address mappings were maintained by the Network Information Center (NIC) in a single file (hosts.txt) which was FTPed by all hosts.

■ The bandwidth consumed in distributing a new version by this scheme is proportional to the square of the number of hosts in the network, and even when multiple levels of FTP were used, the outgoing FTP load on the NIC host was considerable. Explosive growth in the number of hosts didn't lend itself to ease of distribution.

■ The network population was also changing in character.

■ The timeshared hosts that made up the original ARPAnet were being replaced with local networks of workstations.

■ Local organizations were administering their own names and addresses, but had to wait for the NIC to change hosts.txt to make changes visible to the Internet at large. Organizations also wanted some local structure on the name space.

■ The applications on the Internet were getting more sophisticated and creating a need for a general-purpose name service.

The result was several ideas about name spaces and their management, proposed in IEN 116 and RFCs 799, 819, and 830. Proposals varied, but a common thread was the idea of a hierarchical name space, with the hierarchy roughly corresponding to organizational structure, and using a period as the character to mark the boundary between hierarchy levels. A design using a distributed database and generalized resources was described in RFCs 882 and 883. Based on experience with several implementations, the system evolved into the scheme described in this chapter.

Note that the term *domain* or *domain name* is used in many contexts beyond the DNS explained here. Often, the term refers to a name with a structure indicated by dots, but bears no other relation to the DNS.

19.2   DNS Design Goals

The design goals of the DNS influence its structure. The primary goal is a consistent name space that will be used for referring to resources. In order to avoid the problems caused by ad hoc encodings, names should not be required to contain network identifiers, addresses, routes, or similar information as part of the name. Also, the sheer size of the name database and the frequency of updates suggests that it must be maintained in a distributed manner, with local caching to improve performance. Approaches that attempt to collect a consistent copy of the entire database will become more and more expensive and difficult, and hence should be avoided. The same principle holds for the structure of the name space, and in particular mechanisms for creating and deleting names; these should also be distributed.

Other design considerations include the following:

■ Where there are tradeoffs between the cost of acquiring data, the speed of updates, and the accuracy of caches, the source of the data controls the tradeoff.

■ The costs of implementing such a facility dictate that it be generally useful, and not restricted to a single application. Users should be able to use names to retrieve host addresses, mailbox data, and other as-yet-undetermined information. All data associated with a name are tagged with a type, and queries can be limited to a single type.

■ Because the purpose is to use the naming wisely across dissimilar networks and applications, the ability to use the same name space with different protocol families or management is provided. For example, host address formats differ between protocols, although all protocols have the notion of addressing. The DNS tags all data with a class as well as the type, to allow parallel use of different formats.

■ Name server transactions need to be independent of the communications system that carries them. Some systems might want to use datagrams for queries and responses, and only establish virtual circuits for transactions that need reliability (such as database updates or long transactions); other systems will use virtual circuits exclusively.

■ The system should be useful across a wide spectrum of host capabilities. Both personal computers and large, timeshared hosts can use the system, although perhaps in different ways.

19.3 Assumptions about DNS Usage

The organization of the domain system derives from some assumptions about the needs and usage patterns of its user community. It is designed to avoid many of the complicated problems found in general-purpose database systems. Assumptions made include the following:

■ The size of the total database will initially be proportional to the number of hosts using the system, but will eventually grow to be proportional to the number of users on those hosts, as mailboxes and other information are added to the domain system.

■ Most of the data in the system will change very slowly (such as mailbox bindings and host addresses), but the system should be able to deal with subsets that change more rapidly (on the order of seconds or minutes).

■ The administrative boundaries used to distribute responsibility for the database will usually correspond to organizations that have one or more hosts. Each organization that has responsibility for a particular set of domains will provide redundant name servers, either on the organization's own hosts or on other hosts that the organization arranges to use.

■ Clients of the domain system should be able to identify a set of trusted name servers they prefer to use before accepting referrals to name servers outside of this set.

■ Access to information is more critical than instantaneous updates or guarantees of consistency. Hence, the update process allows updates to percolate out through the users of the domain system rather than guaranteeing that all copies are simultaneously updated. When updates are unavailable due to network or host failure, the usual course is to believe old information while continuing efforts to update it. The general model is that copies are distributed with timeouts for refreshing. The distributor sets the timeout value, and the recipient of the distribution is responsible for performing the refresh. In special situations, very short intervals can be specified, or the owner can prohibit copies. In any system that has a distributed database, a particular name server may be presented with a query that can only be answered by some other server. The two general approaches to dealing with this problem are *recursive*, in which the first server pursues the query for the client at another server, and *iterative*, in which the server refers the client to another server and lets the client pursue the query. Both approaches have advantages and disadvantages, but the iterative approach is preferred for the datagram style of access. The domain system requires implementation of the iterative approach, but allows the recursive approach as an option.

■ The domain system assumes that all data originates in master files scattered through the hosts that use the domain system. These master files are updated by local system administrators. Master files are text files that are read by a local name server, and hence become available through the name servers to users of the domain system. The user programs access name servers through standard programs called *resolvers.*

■ The standard format of master files allows them to be exchanged between hosts (via FTP, mail, or some other mechanism). This facility is useful when an organization wants a domain, but doesn't want to support a name server. The organization can maintain the master files locally using a text editor, transfer them to a foreign host that runs a name server, and then arrange with the system administrator of the name server to get the files loaded.

■ Each host's name servers and resolvers are configured by a local system administrator, as explained in RFC 1033. For a name server, this configuration data includes the identity of local master files and instructions about which nonlocal master files are to be loaded from foreign servers. The name server uses the master files (or copies) to load its zones. For resolvers, the configuration data identifies the name servers that should be the primary sources of information.

■ The domain system defines procedures for accessing the data and for referrals to other name servers. The domain system also defines procedures for caching retrieved data and for periodically refreshing data, as defined by the system administrator. A system administrator provides:

- The definition of zone boundaries
- Master files of data
- Updates to master files
- Statements of the refresh policies desired
- The domain system provides:
- Standard formats for resource data
- Standard methods for querying the database
- Standard methods for name servers to refresh local data from foreign name servers

19.4   Elements of DNS

The DNS has three major components: domain name space and resource records, name servers, and resolvers.

Domain name space and *resource records* are specifications for a tree-structured name space and data associated with the names. Conceptually, each node and leaf of the domain-name-space tree names a set of information, and query operations attempt to extract specific types of information from a particular set. A query names the domain name of interest and describes the type of resource information that is desired. For example, the Internet uses some of its domain names to identify hosts; queries for address resources return Internet host addresses.

Name Servers are server programs that hold information about the domain tree's structure and set information. A name server may cache structure or set information about any part of the domain tree, but in general a particular name server has complete information about a subset of the domain space, and pointers to other name servers that can be used to lead to information from any part of the domain tree. Name servers know the parts of the domain tree for which they have complete information; a name server is said to be an *authority* for these parts of the name space. Authoritative information is organized into units called *zones*, and these zones can be automatically distributed to the name servers that provide redundant service for the data in a zone.

Resolvers, as mentioned in the previous section, are programs that extract information from name servers in response to client requests.

Resolvers must be able to access at least one name server and use that name server's information to answer a query directly, or pursue the query using referrals to other name servers. A resolver is typically a system routine that is directly accessible to user programs; hence, no protocol is necessary between the resolver and the user program.

These three components roughly correspond to the three layers, or views, of the domain system. The following explanation correlates this:

- From the user's point of view, the domain system is accessed through a simple procedure or OS call to a local resolver.

- The domain space consists of a single tree, and the user can request information from any section of the tree.

- From the resolver's point of view, the domain system is composed of an unknown number of name servers. Each name server has one or more pieces of the whole domain tree's data, but the resolver views each of these databases as essentially static.

- From a name server's point of view, the domain system consists of separate sets of local information called zones. The name server has local copies of some of the zones. The name server must periodically refresh its zones from master copies in local files or foreign name servers. The name server must concurrently process queries that arrive from resolvers.

In the interest of performance, implementations may couple these functions. For example, a resolver on the same machine as a name server might share a database consisting of the zones managed by the name server and the cache managed by the resolver.

19.5   Domain Name Space and Resource Records

The domain name space is a tree structure. Each node and leaf on the tree corresponds to a resource set (which may be empty). The domain system makes no distinctions between the uses of the interior nodes and leaves, and this chapter uses the term *node* to refer to both. Each node has a label, which is zero to 63 octets in length. Sibling nodes may not have the same label, although the same label can be used for nodes that are not siblings. One label is reserved: the null (zero length) label is used for the root.

The domain name of a node is the list of the labels on the path from the node to the root of the tree. By convention, the labels that compose a domain name are printed or read left to right, from the most specific (lowest, farthest from the root) to the least specific (highest, closest to the root). Internally, programs that manipulate domain names should represent them as sequences of labels, where each label is a length octet followed by an octet string. Because all domain names end at the root, which has a null string for a label, these internal representations can use a length byte of zero to terminate a domain name.

By convention, domain names can be stored with arbitrary case, but domain name comparisons for all present domain functions are done in a case-insensitive manner, assuming an ASCII character set, and a high-order zero bit. This means that you are free to create a node with label *A* or a node with label *a*, but not both as siblings; you could refer to either using *a* or *A*. When you receive a domain name or label, you should preserve its case. The rationale for this choice is that full binary domain names might someday need to be added for new services; existing services would not be changed.

When a user needs to type a domain name, the length of each label is omitted and the labels are separated by periods (.). Since a complete domain name ends with the root label, this leads to a printed form that ends in a dot. This property distinguishes between

■ A character string that represents a complete domain name (often called *absolute*), such as *joejones.ISI.EDU*.

■ A character string that represents the starting labels of an incomplete domain name, which should be completed by local software using knowledge of the local domain (often called *relative*), such as *joejones* used in the ISI.EDU domain.

Relative names are considered relative either to a well-known origin or to a list of domains used as a search list. Relative names appear mostly at the user interface, where their interpretation varies from implementation to implementation, and in master files, where they are relative to a single-origin domain name. The most common interpretation uses the root period character as either the single origin or as one of the members of the search list, so a multilabel relative name is often one where the trailing period has been omitted to save typing.

To simplify implementations, the total number of octets that represent a domain name (the sum of all label octets and label lengths) is limited to 255. A domain is identified by a domain name, and consists of that part of the domain name space that is at or below the domain name which specifies the domain. A domain is a subdomain of another domain if it is contained within that domain. This relationship can be tested by seeing if the subdomain's name ends with the containing domain's name. For example, A.B.C.D is a subdomain of B.C.D, C.D, D, and the root.

DNS's technical specifications do not mandate a particular tree structure or rules for selecting labels; its goal is to be as general as possible, so that it can be used to build arbitrary applications. In particular, the system was designed so that the name space does not have to be organized along the lines of network boundaries, name servers, etc. The rationale for this is not that the name space should have no implied semantics, but rather that the choice of implied semantics should be left open to be used for the problem at hand, and that different parts of the tree can have different implied semantics. For example, the IN-ADDR.ARPA domain is organized and distributed by network and host address because its role is to translate from network or host numbers to names; NetBIOS domains (as discussed in RFCs 1001 and 1002) are flat because that is appropriate for that application.

However, there are some guidelines that apply to the "normal" parts of the name space used for such things as hosts and mailboxes, that will make the name space more uniform, provide for growth, and minimize problems as software is converted from the older host table. The political decisions about the top levels of the tree originated in RFC 920. Current policy for the top levels is discussed in RFC 1032. MILNET conversion issues are covered in RFC 1031.

Lower domains that will eventually be broken into multiple zones should provide branching at the top of the domain so that the eventual decomposition can be done without renaming. Node labels that use such things as special characters or leading digits are likely to break older software, which depends on more restrictive choices.

Before the DNS can be used to hold naming information for some kind of object, two needs must be met:

- A convention for mapping between object names and domain names must exist. This describes how information about an object is accessed.
- RR types and data formats for describing the object must exist.

These rules can be quite simple or fairly complex. Very often, the designer must take into account existing formats and plan for upward compatibility for existing usage. Multiple mappings or levels of mapping may be required. For hosts, the mapping depends on the existing syntax for host names, which is a subset of the usual text representation for domain names, together with RR formats for describing host addresses, etc. Because a reliable inverse mapping is needed from address to host name, a special mapping for addresses into the IN-ADDR.ARPA domain is also defined.

For mailboxes, the mapping is slightly more complex. The usual mail address *<local-part>@<mail-domain>* is mapped into a domain name by converting the local part into a single label (regardless of the period characters it contains), converting the mail domain into a domain name using the usual text format for domain names (where periods denote label breaks), and concatenating the two to form a single domain name. Thus, the mailbox HOSTMASTER@SRI-NIC.ARPA is represented as a domain name by HOSTMASTER.SRI-NIC.ARPA. An appreciation for the reasons behind this design also must take into account the scheme for mail exchanges (see RFC 974).

A typical user is not concerned with defining these rules, but should understand that they usually are the result of compromises between desires for upward compatibility with old usage, interactions between different object definitions, and the inevitable urge to add new features when defining the rules. The way the DNS is used to support objects is often more crucial than the restrictions inherent in the DNS.

The following shows a part of the current domain name space, and is used in many examples in this chapter. Note that the tree is a very small subset of the actual name space.

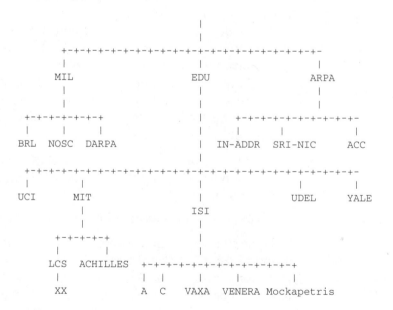

In this example, the root domain has three immediate subdomains: MIL, EDU, and ARPA. The LCS.MIT.EDU domain has one immediate subdomain, named XX.LCS.MIT.EDU. All of the leaves are also domains.

19.6   DNS Name Syntax

The DNS specifications attempt to be as general as possible in the rules for constructing domain names. The idea is that the name of any existing object can be expressed as a domain name with minimal changes. However, when assigning a domain name for an object, the prudent user will select a name that satisfies both the rules of the domain system and any existing rules for the object, whether these rules are published or implied by existing programs. For example, when naming a mail domain, the user should satisfy the rules explained in this chapter and those in RFC 822. When creating a new host name, the old rules for hosts.txt should be followed. This avoids problems when old software is converted to use domain names.

The following syntax will result in fewer problems with many applications that use domain names, such as mail and TELNET:

```
<domain> ::= <subdomain> | " "
<subdomain> ::= <label> | <subdomain> . <label>
<label> ::= <letter> [ [ <ldh-str> ] <let-dig> ]
<ldh-str> ::= <let-dig-hyp> | <let-dig-hyp> <ldh-str>
<let-dig-hyp> ::= <let-dig> | -
<let-dig> ::= <letter> | <digit>
<letter> ::= any one of the 52 alphabetic characters A through Z in
uppercase and a through z in lowercase
<digit> ::= any one of the ten digits 0 through 9
```

While upper- and lowercase letters are allowed in domain names, no significance is attached to the case. That is, two names with the same spelling but different case are treated as identical.

The labels must follow the rules for ARPAnet host names. They must start with a letter, end with a letter or digit, and have as interior characters only letters, digits, and hyphens. There are also some restrictions on the length. Labels must be 63 characters or less. For example, the following strings identify hosts in the Internet:

```
A.ISI.EDU
XX.LCS.MIT.EDU
SRI-NIC.ARPA
```

A domain name identifies a node. Each node has a set of resource information, which may be empty. The set of resource information associated with a particular name is composed of separate resource records (RRs). The order of RRs in a set is not significant, and need not be preserved by name servers, resolvers, or other parts of the DNS.

When talking about a specific RR, it is assumed that it has an owner, which is the domain name where the RR is found, and a type, which is an encoded 16-bit value that specifies the type of the resource in this resource record. Types refer to abstract resources.

19.7 DNS Queries

Queries are messages that may be sent to a name server to provoke a response. In the Internet, queries are carried in UDP datagrams or over TCP connections. The response by the name server answers the question posed in the query, refers the requester to another set of name servers, or signals some error condition.

In general, the user does not generate queries directly, but instead makes a request to a resolver, which in turn sends one or more queries to name servers and deals with the error conditions and referrals that may result. Of course, the possible questions that can be asked in a query shape the kind of service a resolver can provide.

DNS queries and responses are carried in a standard message format. The message format has a header containing a number of fixed fields that are always present, and four sections that carry query parameters and RRs. The most important field in the header is a four-bit field called an *opcode*, which separates different queries. Of the possible 16 values, one (standard query) is part of the official protocol, two (inverse query and status query) are options, one (completion) is obsolete, and the rest are unassigned. The four sections are as follows:

■ *Question* carries the query name and other query parameters.

■ *Answer* carries RRs that directly answer the query.

■ *Authority* carries RRs that describe other authoritative servers. It may, optionally, carry the SOA RR for the authoritative data in the answer section.

■ *Additional* carries RRs that may be helpful in using the RRs in the other sections.

19.8 Standard DNS Queries

A standard query specifies a target domain name (QNAME), query type (QTYPE), and query class (QCLASS), and asks for RRs which match. This type of query makes up such a vast majority of DNS queries that the term *query* is used to mean this standard query unless otherwise specified. The QTYPE and QCLASS fields are each 16 bits long, and are a superset of defined types and classes.

The QTYPE field may contain the following:

■ \<any type\>   matches just that type, for example, A or PTR.

■ AXFR   is a special zone-transfer QTYPE.

■ MAILB   matches all mailbox-related RRs (MB and MG).

■ *   matches all RR types.

The QCLASS field may contain the following:

■ \<any class\>   matches just that class, such as IN or CH.

■ *   matches all RR classes.

The query domain-name, QTYPE, and QCLASS use the name server to look for matching RRs. In addition to relevant records, the name server may return RRs that point toward a name server that has the desired information or RRs that are expected to be useful in interpreting the relevant RRs. For example, a name server that doesn't have the requested information might know a name server that does; a name server that returns a domain name in a relevant RR might also return the RR that binds that domain name to an address.

Suppose a mailer is trying to send mail to joejones@ISI.EDU. It might ask the resolver for mail information about ISI.EDU, resulting in a query for QNAME=ISI.EDU, QTYPE=MX, QCLASS=IN. The response's answer section would be similar to the following:

```
ISI.EDU.           MX    10     VENERA.ISI.EDU.
                   MX    10     VAXA.ISI.EDU.
```

An additional section might be

```
VAXA.ISI.EDU.      A     10.2.0.27
                   A     128.9.0.33
VENERA.ISI.EDU.    A     10.1.0.52
                   A     128.9.0.32
```

because the server assumes that if the requester wants mail exchange information, it will probably want the addresses of the mail exchanges soon afterward. The QCLASS=* construct requires special interpretation regarding authority. Since a particular name server might not know all of the classes available in the domain system, it can never know if it is authoritative for all classes. Hence, responses to QCLASS=* queries can never be authoritative.

19.9   DNS Name Servers

Name servers are the repositories of information that make up the domain database. The database is divided up into sections called zones, which are distributed among the name servers. While name servers can have several optional functions and sources of data, the essential task of a name server is to answer queries using data in its zones. By design, name servers can answer queries in a simple manner; the response can always be generated using only local data, and either contains the answer to the question or a referral to other name servers that are "closer" to the desired information.

A given zone will be available from several name servers to ensure its availability in spite of host failure or communication-link failure. By administrative fiat, every zone is required to be available on at least two servers, and many zones have more redundancy than that.

A given name server will typically support one or more zones, but this gives it authoritative information about only a small section of the domain tree. It might also have some cached nonauthoritative data about other parts of the tree. The name server marks its responses to queries so that the requester can tell whether the response comes from authoritative data or not.

DNS Database Zone Division

The domain database is partitioned in two ways: by class, and by "cuts" made in the name space between nodes. The class partition is simple. The database for any class is organized, delegated, and maintained separately from all other classes. Since, by convention, the name spaces are the same for all classes, the separate classes can be thought of as an array of parallel namespace trees. Note that the data attached to nodes will be different for these different parallel classes. The most common reason for creating a new class is the necessity for a new data format for existing types or a desire for a separately managed version of the existing name space.

Within a class, cuts in the name space can be made between any two adjacent nodes. After all cuts are made, each group of connected name space is a separate zone. The zone is said to be authoritative for all names in the connected region. Note that the cuts in the name space may be in different places for different classes, the name servers may be different, etc.

This means that every zone has at least one node, and hence domain name, for which it is authoritative, and all of the nodes in a particular zone are connected. Given the tree structure, every zone has a highest node that is closer to the root than any other node in the zone. The name of this node is often used to identify the zone.

It would be possible, though not particularly useful, to partition the name space so that each domain name was in a separate zone or so that all nodes were in a single zone. Instead, the database is partitioned at points where a particular organization wants to take over control of a subtree. Once an organization controls its own zone, it can unilaterally change the data in the zone, grow new tree sections connected to the zone, delete existing nodes, or delegate new subzones under its zone.

If the organization has a substructure, it might want to make further internal partitions to achieve nested delegations of name space control. In some cases, such divisions are made purely to make database maintenance more convenient.

The data that describes a zone has four major parts:

1. Authoritative data for all nodes within the zone

2. Data that defines the top node of the zone, which can be thought of as part of the authoritative data

3. Data that describes delegated subzones, i.e., cuts around the bottom of the zone

4. Data that allows access to name servers for subzones, sometimes called *glue* data.

All of this data is expressed in the form of RRs, so a zone can be completely described in terms of a set of RRs. Whole zones can be transferred between name servers by transferring the RRs, either carried in a series of messages or by FTPing a master file, which is a textual representation. The authoritative data for a zone is simply all of the RRs attached to all of the nodes from the top node of the zone down to leaf nodes or nodes above cuts around the bottom edge of the zone.

Though logically part of the authoritative data, the RRs that describe the top node of the zone are especially important to the zone's management. These RRs are of two types: name server RRs that list, one per RR, all of the servers for the zone, and a single SOA RR that describes zone-management parameters. The RRs that describe cuts around the bottom of the zone are NS RRs that name the servers for the subzones. Since the cuts are between nodes, these RRs are not part of the authoritative data of the zone, and should be exactly the same as the corresponding RRs in the top node of the subzone. Since name servers are always associated with zone boundaries, NS RRs are only found at nodes that are the top of some zone.

The data that makes up a zone, NS RRs are found at the top node of the zone (and are authoritative) and at cuts around the bottom of the zone (where they are not authoritative), but never in between. One of the goals of the zone structure is that any zone should have all the data required to set up communications with the name servers for any subzones. That is, parent zones have all the information needed to access servers for their children zones. The NS RRs that name the servers for subzones are often not enough for this task, since they name the servers, but do not give their addresses. In particular, if the name of the name server is itself in the subzone, you could be faced with the situation where the NS RR says that in order to learn a name server's address, you should contact the server using the address you wish to learn. To fix this problem, a zone contains glue RRs that are not part of the authoritative data, and are address RRs for the servers. These RRs are only necessary if the name server's name is "below" the cut, and are only used as part of a referral response.

DNS Administration Considerations

When some organization wants to control its own domain, the first step is to identify the proper parent zone and get the parent zone's owners to agree to the delegation of control. While there are no particular technical constraints dealing with where in the tree this can be done, there are some administrative groupings discussed in RFC 1032 that deal with top-level organization, and middle-level zones are free to create their own rules. For example, one university might choose to use a single zone, while another might choose to organize by subzones dedicated to individual departments or schools. RFC 1033 catalogs the available DNS software and discusses administration procedures.

Once the proper name for the new subzone is selected, the new owners should be required to demonstrate redundant name-server support. Note that there is no requirement that the servers for a zone reside in a host that has a name in that domain. In many cases, a zone will be more accessible to the Internet at large if its servers are widely distributed rather than being within the physical facilities controlled by the same organization that manages the zone. For example, in the current DNS, one of the name servers for the United Kingdom, or UK domain, is found in the U.S. This allows U.S. hosts to get UK data without using limited transatlantic bandwidth.

As the last installation step, the delegation NS RRs and glue RRs necessary to make the delegation effective should be added to the parent zone. The administrators of both zones should ensure that the NS and glue RRs that mark both sides of the cut are consistent, and that they remain so.

The principal activity of name servers is to answer standard queries. Both the query and its response are carried in a standard message format, which is described in RFC 1035. The query contains a QTYPE, QCLASS, and QNAME, which describe the types and classes of desired information and the name of interest.

The way that the name server answers the query depends on whether it is operating in recursive mode or not. The simplest mode for the server is nonrecursive, since it can answer queries using only local information: the response contains an error, the answer, or a referral to some other server "closer" to the answer. All name servers must implement nonrecursive queries. The simplest mode for the client is recursive, since in this mode the name server acts in the role of a resolver and returns either an error or the answer, but never referrals. This service is optional in a name server, and the name server may also choose to restrict the clients who can use recursive mode.

Recursive service is helpful in several situations:

- A relatively simple requester lacks the ability to use anything other than a direct answer to the question.

- A request needs to cross protocols or other boundaries and can be sent to a server, which can act as intermediary.

- The preference is for the network to concentrate the cache rather than having a separate cache for each client.

Nonrecursive service is appropriate if the requester is capable of pursuing referrals and interested in information that will aid future requests. The use of recursive mode is limited to cases where both the client and the name server agree to its use. The agreement is negotiated through the use of two bits in query and response messages.

The *recursion available*, or *RA*, bit is set or cleared by a name server in all responses. The bit is TRUE if the name server is willing to provide recursive service for the client, regardless of whether the client requested recursive service. That is, RA signals availability rather than use.

The other bit involved in the agreement process is called *recursion desired*, or *RD*. This bit specifies whether the requester wants recursive service for this query. Clients may request recursive service from any name server, though they should depend on receiving it only from servers that have previously sent an RA, or servers that have agreed to provide service through private agreement or some other means outside of the DNS protocol.

The recursive mode occurs when a query with RD set arrives at a server that is willing to provide recursive service. The client can verify that recursive mode was used by checking that both RA and RD are set in the reply. Note that the name server should never perform recursive service unless asked via RD, since this interferes with troubleshooting name servers and their databases.

If recursive service is requested and available, the recursive response to a query will be one of the following:

- The answer to the query, possibly prefaced by one or more CNAME RRs that specify aliases encountered on the way to an answer
- A name error indicating that the name does not exist, which may include CNAME RRs indicating that the original query name was an alias for a name that does not exist
- A temporary error indication

If recursive service is not requested or is not available, the nonrecursive response will be either an authoritative name error indicating that the name does not exist, a temporary error indication, or some combination of the following:

- RRs that answer the question, together with an indication of whether the data comes from a zone or is cached
- A referral to name servers that have zones which are closer ancestors to the name than the server sending the reply
- RRs that the name server thinks will prove useful to the requester

19.10   DNS Resolvers

Resolvers are programs that provide an interface between user programs and domain name servers. In the simplest case, a resolver receives a request from a user program (such as mail programs, TELNET, or FTP), typically in the form of a subroutine call or system call, and returns the desired information in a form compatible with the local host's data formats.

The resolver is located on the same machine as the program that requests the resolver's services, but it might need to consult name servers on other hosts. Because a resolver might need to consult several name servers or might have the requested information in a local cache, the amount of time that a resolver takes to complete a task can vary quite a bit, from milliseconds to several seconds.

One important goal of the resolver is to eliminate network delay and name-server load from most requests by answering them from its cache of prior results. It follows, then, that caches which are shared by multiple processes, users, machines, etc., are more efficient than nonshared caches.

Interface

The client-resolver interface to the resolver is influenced by the local host's conventions, but the typical resolver-client interface has three functions:

1. *Host-name-to-host-address translation*   This function is often defined to mimic a previous hosts.txt-based function. Given a character string, the caller wants one or more 32-bit IP addresses. Under the DNS, this translates into a request for type-A RRs. Since the DNS does not preserve the order of RRs, this function may choose to sort the returned addresses or select the "best" address if the service returns only one choice to the client. Note that a multiple-address return is recommended, but a single address might be the only way to emulate prior hosts.txt services.

2. *Host-address-to-host-name translation*   This function will often follow the form of previous functions. Given a 32-bit IP address, the caller wants a character string. The octets of the IP address are reversed, used as name components, and suffixed with IN-ADDR.ARPA. A type-PTR query is used to get the RR with the primary name of the host. For example, a request for the host name corresponding to IP address 1.2.3.4 looks for PTR RRs for domain name 4.3.2.1.IN-ADDR.ARPA.

3. *General lookup function*   This function retrieves arbitrary information from the DNS, and has no counterpart in previous systems. The caller supplies a QNAME, QTYPE, and QCLASS, and wants all of the matching RRs. This function often uses the DNS format for all RR data instead of the local host's, and returns all RR content (e.g., TTL) instead of a processed form with local quoting conventions.

When the resolver performs the indicated function, it usually has one of the following results to pass back to the client:

■ *One or more RRs giving the requested data*   In this case, the resolver returns the answer in the appropriate format.

■ *A name error (NE)*   This happens when the referenced name does not exist. For example, a user might have mistyped a host name.

■ *A data-not-found error*   This happens when the referenced name exists, but data of the appropriate type does not. For example, a host address function applied to a mailbox name would return this error, since the name exists, but no address RR is present.

The functions for translating between host names and addresses may combine the name-error and data-not-found error conditions into a single type of error return, but the general function should not. One reason for this is that applications may ask first for one type of information about a name, followed by a second request to the same name for some other type of information. If the two errors are combined, then useless queries might slow the application.

Resources

In addition to its own resources, the resolver might also have shared access to zones maintained by a local name server. This gives the resolver the advantage of more rapid access, but the resolver must be careful to never let cached information override zone data. In this discussion, the term *local information* means the union of the cache and such shared zones, with the understanding that authoritative data is always used in preference to cached data when both are present.

The resolver algorithm assumes that all functions have been converted to a general lookup function, and uses the following data structures to represent the state of a request in progress in the resolver:

- *SNAME*   The domain name being searched for.
- *STYPE*   The QTYPE of the search request.
- *SCLASS*   The QCLASS of the search request.
- *SLIST*   A structure that describes the name servers and the zone that the resolver is currently trying to query. This structure keeps track of the resolver's current best guess about which name servers hold the desired information; it is updated when arriving information changes the guess. This structure includes the equivalent of a zone name, the known name servers for the zone, the known addresses for the name servers, and historical information that can be used to suggest which server is likely to be the best one to try next. The zone-name equivalent is a match count of the number of labels, from the root down, that SNAME has in common with the zone being queried; this is used as a measure of how "close" the resolver is to SNAME.
- *SBELT*   A "safety belt" structure of the same form as SLIST, which is initialized from a configuration file, and lists servers that should be used when the resolver doesn't have any local information to guide name-server selection. The match count will be -1, to indicate that no labels are known to match.
- *CACHE*   A structure that stores the results from previous responses. Since resolvers are responsible for discarding old RRs whose TTLs have expired, most implementations convert the interval specified in arriving RRs to some sort of absolute time when the RR is stored in the cache. Instead of counting the TTLs down individually, the resolver just ignores or discards old RRs when it runs across them in the course of a search, or discards them during periodic sweeps to reclaim the memory consumed by old RRs.

19.11   Summary

The DNS is complex. Its design was intended to eliminate much of the work involved in updating domain-name databases.

This chapter has presented an overview of the DNS. More detailed information is available in the following works:

Dyer, S., and F. Hsu. "Hesiod, Project Athena Technical Plan—Name Service," version 1.9. April 1987. Describes the fundamentals of the Hesiod name service.

Postel, J. "Internet Name Server," IEN-116. USC/Information Sciences Institute, August 1979. Describes a name service obsoleted by the Domain Name System, but still in use.

Quarterman, J., and J. Hoskins. "Notable Computer Networks," Communications of the ACM, vol. 29, no. 10 (October 1986).

Solomon, M., L. Landweber, and D. Neuhengen. "The CSNET Name Server," Computer Networks, vol. 6, no. 3 (July 1982). Describes a name service for CSNET that is independent from the DNS and DNS used in the CSNET.

and in the following RFCs:

742

768

793

799   Suggests introduction of a hierarchy in place of a flat name space for the Internet.

805

819   Early thoughts on the design of the domain system. The current implementation is completely different.

821

830   Early thoughts on the design of the domain system. Current implementation is completely different.

882

883

920   Explains the naming scheme for top level domains.

952   Specifies the format of hosts.txt, the host/address table replaced by the DNS.

953   Contains the official specification of the hostname server protocol, which is obsoleted by the DNS. This TCP-based protocol accesses information stored in the RFC-952 format, and is used to obtain copies of the host table.

973   Describes changes to RFCs 882 and 883, and the reasons for them. Now obsolete.

974   Describes the transition from hosts.txt-based mail addressing to the more powerful MX system used with the domain system.

1001 and 1002   Provide a preliminary design for NetBIOS on top of TCP/IP, which proposes to base NetBIOS name service on top of the DNS.

1010   Contains socket numbers and mnemonics for host names, operating systems, etc.

1031   Describes a plan for converting the MILNET to the DNS.

1032   Describes the registration policies used by the NIC to administer the top-level domains and delegate subzones.

1033   A cookbook for domain administrators.

20

Remote Procedure Call (RPC)

Open Network Computing (ONC) acknowledges Remote Procedure Call version 2, as of this writing. RPC is explained in this chapter first in overview form, then in more detail, including some actual program examples and internals.

20.1   RPC and XDR: An Overview

RPC is a protocol. Technically, it can operate over TCP or UDP as a transport mechanism. This is important because sometimes illustrations show it on one or the other. Applications use RPC to call a routine, thus executing like a client, and make a call against a server on a remote host. This type of application programming is a high-level, peer relationship between an application and an RPC server. Consequently, these applications are portable to the extent that RPC is implemented.

External Data Representation Protocol (XDR) is within RPC. XDR's data description language can be used to define datatypes when heterogeneous hosts are integrated. Having the capability to overcome the inherent characteristics of different architectures lends RPC and XDR a robust solution for distributed application communication. This language permits parameter requests to be made against a file of an unlike type. In short, XDR permits datatype definition in the form of parameters, and the transmission of these encoded parameters.

XDR provides data transparency by way of encoding (or encapsulating) data at the application layer, so lower layers and hardware do not have to perform any conversions. A powerful aspect of XDR is automatic data conversion, performed via declaration statements and the XDR compiler. The XDR compiler generates required XDR calls, thus making the operation less manual by nature. Figure 20-1 is an example of this implementation.

RPC implements what is called a *port mapper*. It starts on RPC server initialization. When RPC services start, the operating system assigns a port number to each service. Each service informs the port mapper of its port number, program number, and other information required by the port mapper to match a service with a requester.

Client applications issue service requests to a port mapper. The port mapper in turn identifies a requested service and returns the appropriate parameters to the requesting client application. In other words, the port mapper is similar in function to a manager, knowing what services are available and their specific addressable locations.

Figure 20-1
A conceptual view of
RPC and XDR.

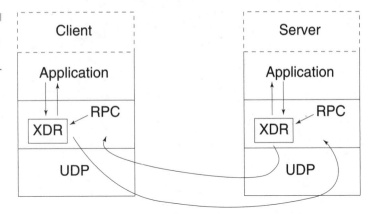

The port mapper can be used in a broadcast scenario. For example, a requesting RPC call can broadcast a call to all hosts on a network. Applicable port mappers report the information sought after by the client (hence the term *remote procedure call*).

20.2   A Perspective on RPC and NFS

NFS is a product of Sun Microsystems. It permits users to execute files without knowing the location of these files; they may be local or remote in respect to the user. Users can create, read, or remove directories. Files themselves can be written to or deleted. NFS provides a distributed file system that permits users to capitalize on access capabilities beyond their local file systems.

NFS uses RPC to make execution of a routine on a remote server possible. Conceptually, NFS, RPC, and UDP (which it typically uses) appear as in Fig. 20-2.

The idea behind NFS is to have one copy of it on a server that all users on a network can access. The consequence of this is that software (and updates) can be installed on one server in a networked environment, instead of on multiple hosts. NFS is based on a client/server model; a single NFS server can function to serve many client requests.

This chapter concentrates on the specification of version 2 of the message protocol used in RPC. The message protocol itself is specified with the XDR language. (The assumption here is that you are somewhat familiar with XDR.)

Figure 20-2

A conceptual view of
NFS, RPC, and UDP.

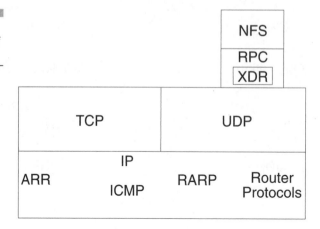

Each remote procedure call has two sides to it. One is an active client side that makes the call to a server, which sends back a reply. A *network service* is a collection of one or more remote programs. A *remote program* implements one or more remote procedures. These procedures, their parameters, and the results are documented in the specific program's protocol specification. A server may support more than one version of a remote program in order to be compatible with changing protocols.

For example, a network file service might be composed of two programs. One program might deal with high-level applications, such as file system access-control and locking. The other might deal with low-level file input and output and have procedures like READ and WRITE. A client of the network file service would call the procedures associated with the two programs of the service on behalf of the client.

The terms *client* and *server* in this framework apply to a particular transaction; that is, a particular host or piece of software (a process or program) that could operate in both roles at different times. For example, a program that supplies remote execution service could also be a client of a network file service.

20.3 The RPC Model

The RPC protocol is based on the *remote-procedure-call model*, which is similar to the *local-procedure-call model*. The latter is where the caller places arguments to a procedure in some well-specified location. It transfers control to the procedure, and eventually regains control. Next, the results of the procedure are extracted from the specified location and the caller continues execution.

The remote-procedure-call model is where one thread of control logically winds through two processes: the caller's process and a server's process. The caller process first sends a call message to the server process, and waits (blocks) for a reply message. The call message includes the procedure's parameters, and the reply message includes the procedure's operational results. Once the reply message is received, the results of the procedure are extracted, and the caller's execution is resumed.

On the server side, a process is dormant, awaiting the arrival of a call message. When one arrives, the server process extracts the procedure's parameters, computes the results, sends a reply message, and then awaits the next call message.

In this model, only one of the two processes is active at any given time. However, this model is only given as an example. The ONC RPC protocol makes no restrictions on the concurrency model implemented, and others are possible. For example, an implementation may have RPC calls operate as asynchronous, so that the client can do useful work while waiting for the reply from the server. Another option is to have the server create a separate task to process an incoming call, so that the original server can be free to receive other requests.

Remote procedure calls differ from local procedure calls in the following aspects:

- *Error handling* The failure of the remote server or network must be handled when using remote procedure calls.

- *Global variables and side effects* Since the server does not have access to the client's address space, hidden arguments cannot be passed as global variables or returned as side effects.

- *Performance* Remote procedures usually operate more slowly, by one or more orders of magnitude, than local procedure calls.

- *Authentication* Remote procedure calls can be transported over unsecured networks, so authentication may be necessary. Authentication prevents one entity from pretending to be another (different) entity.

Although there are tools to automatically generate client and server libraries for a given service, protocols must still be designed carefully. Careful attention to the details of each system can prevent major problems during the implementation phase.

20.4 RPC Transports and RPC Semantics

The RPC protocol can be implemented on several different transport protocols. The RPC protocol itself does not care how a message is passed from one process to another. It is only concerned with the specification and interpretation of messages. Applications might want to obtain information about the transport layer through an interface not specified here. For example, the transport protocol might impose a restriction on the maximum size of RPC messages, or it might be stream-oriented like TCP, with no size limit. The client and server must agree on their transport protocol choices.

RPC does not try to implement any kind of reliability. The application programmer must be aware of this, since an application might need to know the type of transport protocol underneath RPC. If it knows it is running on top of a reliable transport such as TCP, then most of the work is already done for it. On the other hand, if it is running on top of UDP, it must implement its own timeout, retransmission, and duplicate-detection policies because the RPC protocol does not provide these services.

Because of transport independence, the RPC protocol does not attach specific semantics to the remote procedures or their execution requirements. Semantics can be inferred from (but should be explicitly specified by) the underlying transport protocol. For example, consider RPC running on top of an unreliable transport such as UDP. If an application retransmits RPC call messages after timeouts and does not receive a reply, the application cannot infer anything about the number of times the procedure was executed. If the application does receive a reply, then it can infer that the procedure was executed at least once.

A server might want to retain previous information, such as granted requests from a client, and not re-grant them in order to ensure some degree of at-most-one-execution semantics. A server can do this by taking advantage of the transaction ID that is packaged with every RPC message. The main use of this transaction ID is by the client RPC entity in matching replies to calls. However, a client application may choose to reuse its previous transaction ID when retransmitting a call. The server may choose to remember this ID after executing a call, and not execute calls with the same ID in order to achieve some degree of execute at-most-one-time-type semantics. The server is not allowed to examine this ID in any other way except as a test for equality.

On the other hand, if RPC uses TCP, the application can infer from a reply message that the procedure was executed exactly once, but if it receives no reply message, it cannot assume that the remote procedure was not executed. Even if a connection-oriented transport protocol like TCP is used, an application still needs timeouts and reconnection to handle server crashes and other system glitches.

There are other possibilities for transports besides datagram- or connection-oriented protocols. For example, a request-reply protocol such as VMTransport Protocol (VMTP) is perhaps a natural transport for RPC. RPC uses both TCP and UDP transport protocols.

The notion of binding a particular client to a particular service and transport parameters is not part of this particular RPC protocol specification. Binding is important and required, but it is left up to a higher-level operating software. You could think of the RPC protocol as the old jump-subroutine instruction (JSR). The loader (binder) makes JSR useful, and the loader itself uses JSR to accomplish its task.

The RPC protocol provides the fields necessary for a client to identify itself to a service, and vice versa, in each call-and-reply message. Security and access control mechanisms can be built on top of this message authentication, if desired. Several different authentication protocols can be implemented and supported. A required field in the RPC header indicates which protocol is being used.

Authentication parameters are opaque and open-ended to the rest of the RPC protocol. The *flavor* of a credential or verifier refers to the value of the flavor field in the opaque_auth structure. Flavor numbers, like RPC program numbers, are administered centrally. It is possible, however, to request new flavor numbers by applying through electronic mail to rpc@sun.com, or you can perform more research into the nature of requesting new flavor numbers. Credentials and verifiers are represented as variable-length opaque data (reference the body field in the opaque_auth structure).

Null authentication is mandatory—it must be available in all implementations. System authentication is typically a program designer's choice. Many applications use this style of authentication, and the availability of this flavor in an implementation will enhance interoperability.

Many times, calls must be made where the client does not care about its identity or the server does not care who the client is. In such a case, the flavor of the RPC message's credential, verifier, and reply verifier is AUTH_NONE. Opaque data associated with AUTH_NONE is undefined. It is standard consideration for most applications that the length of the opaque data be zero.

20.5   RPC Protocol Requirements

RPC protocol must provide for the following:

- Unique specification of a procedure to be called
- Provisions for matching response messages to request messages
- Provisions for authenticating the caller to the service and vice versa

Besides these requirements, features that detect protocol roll-over errors, system-specific implementation bugs, user error, and network administration are worth supporting. To detect these things, RPC should provide the following:

- RPC protocol mismatches
- Remote program protocol version mismatches
- Protocol errors, such as a misspecification of a procedure's parameters
- Reasons why remote authentication failed
- Any other reasons why the desired procedure was not called

20.6   RPC Programs and Procedures

The RPC call message has three unsigned integer fields: the remote program number, the remote program version number, and the remote procedure number. These uniquely identify the procedure to be called. Program numbers are administered by a central authority via e-mail to rpc@sun.com. Once program designers have a program number, they can implement their remote programs. Generally, the first implementation has the version number 1. By having a version field identifying which version of the protocol the caller is using, it is easier to understand, troubleshoot, and interpret any problems. Version numbers enable support for new and old protocols through the same server process.

The procedure number identifies the procedure to be called. These numbers are documented in the specific program's protocol specification. For example, a file service's protocol specification might state that its procedure number 5 is READ and procedure number 12 is WRITE.

Remote program protocols often change over the timespan of several versions, and so can the actual RPC message protocol. Hence, the call message also has in it the RPC version number, which is always equal to two, reflecting version 2 of RPC as presented here.

The reply message to a request message has enough information to distinguish the following error conditions:

- The remote implementation of RPC does not support protocol version 2. The lowest and highest supported RPC version numbers are returned.

- The remote program is not available on the remote system.

- The remote program does not support the requested version number. The lowest and highest supported remote program version numbers are returned.

- The requested procedure number does not exist.

- The parameters to the remote procedure appear to be garbage from the server's point of view.

20.7   RPC Authentication

Provisions for authentication of caller-to-service and vice versa are part of the RPC protocol. The call message has two authentication fields: the credential and the verifier. The reply message has one authentication field, the response verifier. The RPC protocol specification defines all three fields to be an opaque type (in the XDR language). Consider the following:

```
enum auth_flavor {
AUTH_NONE  = 0,
AUTH_SYS   = 1,
AUTH_SHORT = 2
/* and more to be defined */
};
struct opaque_auth {
auth_flavor flavor;
opaque body<400>;};
```

Any opaque_auth structure is an auth_flavor enumeration followed by up to 400 bytes, which are opaque to (uninterpreted by) the RPC protocol implementation. Interpretation and semantics of the data contained within the authentication fields is specified by individual, independent, authentication protocol specifications. If authentication parameters are rejected, the reply message contains information stating why they were rejected.

20.8   RPC Program Number Assignment

RPC program numbers are given out in groups of hexadecimal 20000000 (decimal 536870912) according to the following chart:

0 - 1fffffff	Defined by rpc@sun.com
20000000 - 3fffffff	Defined by user
40000000 - 5fffffff	Transient
60000000 - 7fffffff	Reserved
80000000 - 9fffffff	Reserved
a0000000 - bfffffff	Reserved
c0000000 - dfffffff	Reserved
e0000000 - ffffffff	Reserved

The first group is a range of numbers administered by rpc@sun.com, and should be identical for all sites. The second range is for applications peculiar to a particular site. This range is intended primarily for debugging new programs. Program designers can request blocks of RPC program numbers in the first range. All correspondence about RPC numbers is between SUN and program developers or companies owning or developing the program. The third group is for applications that generate program numbers dynamically. The final groups are reserved for future use and should not be used.

20.9   Functionality of the RPC Protocol

The RPC protocol was originally designed for calling remote procedures. Typically, each call message is matched with a reply message. However, the protocol itself is a message-passing protocol with which other (non-procedure call) protocols can be implemented.

RPC Batch

RPC batch is useful to send a large sequence of call messages to a server. RPC batch typically uses reliable byte-stream protocols for its transport, such as the TCP transport mechanism. With RPC batch, the client never waits for a reply from the server, and the server does not send replies to RPC batch calls. A sequence of RPC batch calls is usually terminated by a legitimate RPC operation. The purpose for this is to clear the link and get positive acknowledgement.

Broadcast Remote Procedure Calls

In broadcast protocols, the client sends a broadcast call to the network and waits for numerous replies. This requires the use of packet-based protocols (UDP) as the transport protocol mechanism. Servers that support broadcast protocols generally respond only when the call is successfully processed and are silent in the face of errors; however, be aware that this varies by application. RPC broadcast principles also apply to multicasting; that is, an RPC request can be sent to a multicast address.

20.10   RPC Message Protocol

The focus in this section is to define RPC message protocol in the XDR data description language, as follows:

```
enum msg_type {
CALL  = 0,
REPLY = 1
};
```

A reply to a call message can take two forms: the message was either accepted or rejected. In XDR language, this is written as follows:

```
enum reply_stat {
MSG_ACCEPTED = 0,
MSG_DENIED   = 1
};
```

Given that a call message was accepted, the following is the status of an attempt to call a remote procedure:

```
enum accept_stat {
SUCCESS = 0, /* RPC executed successfully
*/ PROG_UNAVAIL  = 1, /* remote hasn't exported program
*/ PROG_MISMATCH = 2, /* remote can't support version #
*/ PROC_UNAVAIL  = 3, /* program can't support procedure
*/ GARBAGE_ARGS  = 4, /* procedure can't decode param
*/ SYSTEM_ERR    = 5  /* errors like memory allocation
failure
*/
};
```

Reasons why a call message was rejected are indicated like this:

```
enum reject_stat {
RPC_MISMATCH = 0, /* RPC version number != 2
*/ AUTH_ERROR = 1 /* remote can't authenticate caller
*/
};
```

The following code is used to indicate why authentication failed:

```
enum auth_stat {
AUTH_OK = 0,  /* success
*/
/*
* failed at remote end
*/ AUTH_BADCRED = 1, /* bad credential (seal broken)
*/ AUTH_REJECTEDCRED = 2, /* client must begin new session
*/ AUTH_BADVERF = 3,  /* bad verifier (seal broken)
*/ AUTH_REJECTEDVERF = 4,  /* verifier expired or replayed
*/ AUTH_TOOWEAK = 5,  /* rejected for security reasons
*/
/*
* failed locally
*/ AUTH_INVALIDRESP  = 6,  /* bogus response verifier
*/ AUTH_FAILED  = 7   /* reason unknown
*/
};
```

The RPC Message

All messages start with a transaction identifier, *xid,* followed by a two-armed *discriminated union.* The union's discriminant is a msg_type that switches to one of the two types of the message. The xid of a REPLY message always matches that of the initiating CALL message. The xid field is only used for clients matching reply messages with call messages or for servers detecting retransmissions; the service side cannot treat xid as any type of sequence number.

The RPC message looks like this:

```
struct rpc_msg {
unsigned int xid;
union switch (msg_type mtype) {
case CALL:
      call_body cbody;
      case REPLY:
      reply_body rbody;        } body;
};
```

The Body of an RPC Call

In version 2 of the RPC protocol specification, rpcvers must be equal to two. The fields prog, vers, and proc specify the remote program, its version number, and the procedure within the remote program to be called. After these fields have two authentication parameters: cred (authentication credential) and verf (authentication verifier). The two authentication parameters are followed by the parameters to the remote procedure, which are specified by the specific program protocol.

The purpose of the authentication verifier is to validate the authentication credential. Note that these two items are historically separate, but are always used together as one logical entity, as shown:

```
struct call_body {
unsigned int rpcvers;
/* must be equal to two (2)
*/ unsigned int prog;
unsigned int vers;
unsigned int proc;
opaque_auth   cred;
opaque_auth   verf;
/* procedure specific parameters start here */
};
Body of a reply to an RPC call:
union reply_body switch (reply_stat stat) {
case MSG_ACCEPTED:
              accepted_reply areply;
case MSG_DENIED:
              rejected_reply rreply;
} reply;
```

The Reply to an RPC Call Accepted by the Server

There could be an error in an RPC call even though it was accepted by the server. The first field is an authentication verifier that the server generates in order to validate itself to the client. It is followed by a union whose discriminant is an enum, accept_stat. The SUCCESS arm of the union is protocol specific. The PROG_UNAVAIL, PROC_UNAVAIL, GARBAGE_ARGS, and SYSTEM_ERR arms of the union are void. The PROG_MISMATCH arm specifies the lowest and highest version numbers of the remote program supported by the server. This is shown in XDR language as follows:

```
struct accepted_reply {
opaque_auth verf;
        union switch (accept_stat stat) {
        case SUCCESS:
        opaque results[0];      /*
        * procedure-specific results start here
        */
        case PROG_MISMATCH:
                struct {
                unsigned int low;
                unsigned int high;
                } mismatch_info;
        default:
                /*
                * Void.  Cases include PROG_UNAVAIL, PROC_UNAVAIL,
                * GARBAGE_ARGS, and SYSTEM_ERR.                    */
                        void;
                } reply_data;   };
```

The Reply to an RPC Call Rejected by the Server

An RPC call can be rejected by the server for two reasons: either the server is not running a compatible version of the RPC protocol (RPC_MISMATCH), or the server rejects the identity of the caller (AUTH_ERROR). In the case of an RPC version mismatch, the server returns the lowest and highest supported RPC version numbers. In the case of invalid authentication, failure status is returned, as shown:

```
union rejected_reply switch (reject_stat stat) {
case RPC_MISMATCH:
        struct {
```

```
            unsigned int low;
            unsigned int high;
            } mismatch_info;
    case AUTH_ERROR:
            auth_stat stat;
    };
```

20.11   RPC Record-Marking Standard

When RPC messages are passed on top of a byte-stream transport protocol such as TCP, it is necessary to delimit one message from another in order to detect and possibly recover from protocol errors. This is called *record marking* (RM). One RPC message fits into one RM record.

A record is composed of one or more record fragments. A record fragment is a four-byte header followed by zero to 2^{31-1} bytes of fragment data. The bytes encode an unsigned binary number; as with XDR integers, the byte order is from highest to lowest. The number encodes two values: a boolean that indicates whether the fragment is the last fragment of the record (bit value 1 implies the fragment is the last fragment), and a 31-bit unsigned binary value that is the length in bytes of the fragment's data. The boolean value is the highest-order bit of the header; the length is the 31 low-order bits. This record specification is *not* in XDR standard form.

20.12   RPC Language

Just as there is a need to describe the XDR data types in a formal language, there is also a need to describe the procedures that operate on these XDR data types in a formal language. The RPC language is an extension of the XDR language, with the addition of program, procedure, and version declarations. The following example describes the essence of RPC language.

RPC Language Example Service

Here is an example of the specification of a simple ping program:

```
program PING_PROG {
/*
* Latest version
*/
version PING_VERS_PINGBACK {
void
        PINGPROC_NULL(void) = 0;
/*
* Ping the client, return the round-trip time
* (in microseconds). Return -1 if the operation timed out.
*/
int
PINGPROC_PINGBACK(void) = 1;
} = 2;
/*
* Original version
*/
        version PING_VERS_ORIG {
        void    PINGPROC_NULL(void) = 0;
} = 1;
} = 1;
const PING_VERS = 2; /* Latest version */
```

The first version that is described is PING_VERS_PINGBACK with two procedures, PINGPROC_NULL and PINGPROC_PINGBACK. PING-PROC_NULL takes no arguments and returns no results, but it is useful for computing round-trip times from the client to the server and back again. By convention, procedure zero of any RPC protocol should have the same semantics, and never require any kind of authentication. The second procedure is used for the client to have the server do a reverse ping operation back to the client, and it returns the amount of time (in microseconds) that the operation used. The next version, PING_VERS_ORIG, is the original version of the protocol; it does not contain the PINGPROC_PINGBACK procedure. It is useful for compatibility with old client programs, but as this program matures, it might be dropped from the protocol entirely.

RPC Language Specification

The RPC language is identical to the XDR language defined in RFC 1014, except for the added definition of the *program-def* described here:

```
program-def:
"program" identifier "{"
        version-def
        version-def *  "}" "=" constant ";"
        version-def:
"version" identifier "{"
        procedure-def
```

```
        procedure-def *
"}" "=" constant ";"
procedure-def:
type-specifier identifier "(" type-specifier
("," type-specifier )* ")" "=" constant ";"
Here are some notes on RPC language syntax:
```

- The keywords "program" and "version" are added and cannot be used as identifiers.

- A version name and version number cannot occur more than once within the scope of a program definition.

- A procedure name and procedure number cannot occur more than once within the scope of a version definition.

- Program identifiers are in the same name space as constant and type identifiers.

- Only unsigned constants can be assigned to programs, versions, and procedures.

The Method of RPC System Authentication

The client might wish to identify itself, for example, as it is identified on a UNIX system. The flavor of the client credential is AUTH_SYS. The opaque data constituting the credential encodes the following structure:

```
struct authsys_parms {
unsigned int stamp;
string machinename<255>;
unsigned int uid;
unsigned int gid;
unsigned int gids<16>;
};
```

The stamp is an arbitrary ID that the caller machine may generate. The machinename is the name of the caller's machine. The UID is the caller's effective user ID. The GID is the caller's effective group ID. GIDs are a counted array of groups that contain the caller as a member. The verifier accompanying the credential should have a flavor value of AUTH_NONE (defined in Secs. 20.4 through 20.7 of this chapter). Note that this credential is only unique within a particular domain of machine names, UIDs, and GIDs.

The flavor value of the verifier received in the reply message from the server may be AUTH_NONE or AUTH_SHORT. In the case of

AUTH_SHORT, the bytes of the reply verifier's string encode an opaque structure. This new opaque structure may now be passed to the server instead of the original AUTH_SYS flavor credential. The server may keep a cache that maps shorthand opaque structures (passed back by way of an AUTH_SHORT-style reply verifier) to the original credentials of the caller. The caller can save network bandwidth and server CPU cycles by using the shorthand credential.

The server may remove the shorthand opaque structure at any time. If this happens, the RPC message will be rejected due to an authentication error. The reason for the failure will be AUTH_REJECTEDCRED. At this point, the client might want to try the original AUTH_SYS style of credential.

Use of this particular way of authentication does not guarantee any security for the users or providers of a service. Authentication provided by this scheme can be considered legitimate only when both applications using this scheme and the network can be secured externally, and privileged transport addresses are used for the communicating endpoints. An example of this is the use of privileged TCP/UDP ports in UNIX systems—note that not all systems enforce privileged transport address mechanisms.

APPENDIX A

Acronyms and Abbreviations

Term	Meaning
10Base-T	Technical name for ETHERNET implemented on twisted wire
3270	Reference to a 3270 data-stream supporting entity
3770	Reference to Remote Job Entry
370/XA	370 Extended Architecture
5250	Reference to a 5250 data-stream supporting entity
576	The minimum datagram size that all hosts, including routers, must accommodate
AAA	Autonomous Administrative Area
AAI	Administration Authority Identifier
AAL	ATM Adaptation Layer
AARP	AppleTalk Address Resolution Protocol
ABOM	A-bis Operations and Maintenance
ACB	Application Control Block
ACB	Access Method Control Block
ACCS	Automated Calling Card Service
ACD	Automatic Call Distribution
ACDF	Access Control Decision Function
ACE	Access Control List Entry
ACF	Access Control Field
ACF	Advanced Communications Function
ACIA	Access Control Inner Areas
ACID	Atomicity, Consistency, Isolation, and Durability
ACK	Positive Acknowledgement
ACL	Access Control List

Term	Meaning
ACP	Ancillary Control Process
ACS	Access Control Store
ACSA	Access Control Specific Area
ACSE	Association Control Service Element
ACSP	Access Control Specific Point
ACTLU	Activate Logical Unit
ACTPU	Activate Physical Unit
ACU	Auto Calling Unit
AD	Addendum Document to an OSI standard
ADF	Adapter Description File
ADMD	Administrative Management Domain
ADP	Adapter Control Block
ADP	AppleTalk Data Stream Protocol
ADPCM	Adaptive Differential Pulse Code Modulation
ADSP	AppleTalk Data Stream Protocol
AE	Application Entity
AEI	Application Entity Invocation
AEP	AppleTalk Echo Protocol
AET	Application Entity Title
AF	Auxiliary Facility
AFI	Authority and Format Identifier
AFP	AppleTalk Filing Protocol
AI	Artificial Intelligence
AIFF	Audio Interchange File Format
AIX	Advanced Interactive Executive
ALS	Application Layer Structure
ALU	Application Layer User
AMI	Alternating Mark Inversion

Term	Meaning
ANI	Automatic Number Identification
ANS	American National Standard
ANSI	American National Standards Institute
AP	Application Process
AP	Argument Pointer
APB	Alpha Primary Bootstrap
APD	Avalanche Photodiode
APDU	Application Protocol Data Unit
API	Application Program Interface; Application Programming Interface
APIC	Advanced Programming Interrupt Controller
APLI	ACSE/Presentation Library Interface
APP	Applications Portability Profile
APPC	Advanced Peer-to-Peer Communications
APPC	Advanced Program-to-Program Communications
APPL	Application Program
APPN	Advanced Peer-to-Peer Networking
APT	Application Process Title
ARPA	Advanced Research Projects Agency
ARP	Address Resolution Protocol
ARQ	Automatic Repeat Request
ARS	Automatic Route Selection
AS/400	Application System/400
ASC	Accredited Standard Committee
ASCII	American Standard Code for Information Interchange
ASDC	Abstract Service Definition Convention
ASE	Application Service Element
ASN	Abstract Syntax Notation
ASN.1	Abstract Syntax Notation One

Term	Meaning
ASO	Application Service Object
ASP	Abstract Service Primitive; AppleTalk Session Protocol
AST	Asynchronous System Trap
ASTLVL	Asynchronous System Trap Level
ASTSR	Asynchronous System Trap Summary Register
ATM	Asynchronous Transfer Mode; Abstract Text Method
ATP	AppleTalk Transaction Protocol
ATS	Abstract Test Suite
AU	Access Unit
AVA	Attribute Value Assertion
B-ISDN	Broadband ISDN
B8ZS	Bipolar 8-Zeros Substitution
BACM	Basic Access Control Model
BAR	Base Address Register
BAS	Basic Activity Subset
BASIC	Beginners All-purpose Instruction Code
BB	Begin Bracket
BBS	Bulletin Board System
BC	Begin Chain
BCC	Block Check Character
BCS	Basic Combined Subset
BCVT	Basic Class Virtual Terminal
Bellcore	Bell Communications Research, Inc.
BER	Box Event Records
BER	Bit Error Rate
GBP	Border Gateway Protocol
BIB	Backward Indicator Bit
BIS	Bracket Initiation Stopped

Term	Meaning
BISUP	Broadband ISUP
BITS	Building Integrated Timing Systems
BITNET	Because Its Time Network
BIU	Basic Information Unit
BMS	Basic Mapping Support
BMU	Basic Measurement Unit
BNN	Boundary Network Node
BOC	Bell Operating Company
BOM	Beginning of Message
bps	Bits Per Second
BRI	Basic Rate Interface
BSC	Binary Synchronous Communications
BSS	Basic Synchronization Subset; Base Station Subsystem
BSSMAP	Base Station Subsystem Mobile Application Part
BTAM	Basic Telecommunications Access Method
BTU	Basic Transmission Unit
CA	Channel Adapter
CA	Certification Authority
CAD	Computer-Aided Design
CAE	Common Applications Environment
CAF	Channel Auxiliary Facility
CAI	Computer-Assisted Instruction
CASE	Common Application Service Elements
CATV	Community Antenna Television
CBD	Changeback Declaration
CBEMA	Computer and Business Equipment Manufacturers Association
CCA	Conceptual Communication Area
CCAF	Call Control Agent Function

Term	Meaning
CCAF+	Call Control Agent Function Plus
CCB	Connection Control Block
CCB	Channel Control Block
CCIRN	Coordinating Committee for Intercontinental Research Networking
CCIS	Common Channel Interoffice Signaling
CCITT	Consultative Committee for International Telephone and Telegraph
CCO	Context Control Object
CCR	Commitment, Concurrency, and Recovery
CCS	Common Communications Support
CCS	Common Channel Signaling
CCU	Central Control Unit
CCU	Communications Control Unit
CCW	Channel Command Word
CD	Countdown Counter
CD	Committee Draft
CDF	Configuration Dataflow
CDI	Change Direction Indicator
CDRM	Cross-Domain Resource Manager
CDRSC	Cross-Domain Resource
CDS	Conceptual Data Storage
CDS	Conceptual Data Store
CEBI	Conditional End-Bracket Indicator
CEI	Connection Endpoint Identifier
CEN/ELEC	Committee European de Normalization Electrotechnique
CEP	Connection Endpoint
CEPT	Conference of European Postal and Telecommunications Administrations
CESID	Caller Emergency Service Identification
CF	Control Function

Term	Meaning
CGM	Computer Graphics Metafile
CHILL	CCITT High-Level Language
CICS	Customer Information Control System
CIDR	Classless Inter-Domain Routing
CIGOS	Canadian Interest Group on Open Systems
CIM	Computer Integrated Manufacturing
CIS	Card Information Structure
CLI	Connectionless Internetworking
CLIST	Command List
CLNP	Connectionless Network Protocol
CLNS	Connectionless Network Service
CLSDST	Close Destination
CLTP	Connectionless Transport Protocol
CLTS	Connectionless Transport Service
CMC	Communications Management Configurations
CMIP	Common Management Information Protocol
CMIS	Common Management Information Service
CMISE	Common Management Information Service Element
CMOL	CMIP Over Logical Link Control
CMOT	CMIP Over TCP/IP
CMS	Conversational Monitor System
CMT	Connection Management
CN	Composite Node
CNM	Communication Network Management
CNMA	Communication Network for Manufacturing Applications
CNMI	Communication Network Management Interface
CNOS	Change Number of Sessions
CNT	Communications Name Table

Term	Meaning
CO	Central Office
COCF	Connection-Oriented Convergence Function
CODEC	Coder/Decoder
COI	Connection-Oriented Internetworking
COM	Continuation-of-Message DMPDU
CONF	Confirm
CONS	Connection-Oriented Network Service
CORBA	Common Object-Oriented Request Broker Architecture
COS	Class of Service
COS	Corporation for Open Systems
COTP	Connection-Oriented Transport Protocol
COTS	Connection-Oriented Transport Service
CP	Control Point
CP	Control Program
CPE	Customer Premises Equipment
CPH	Call Party Handling
CPF	Control Program Facility
CPI	Common Programming Interface
CPI-C	Common Programming Interface with C Language
CPMS	Control Point Management Services
CRACF	Call-Related Radio Access Control Function
CRC	Cyclical Redundancy Check
CRST	Cluster-Route-Set-Test
CRT	Cathode Ray Tube
CRV	Call Reference Value
CSm	Call Segment Model
CS-MUX	Circuit-Switching Multiplexer
CSMA/CA	Carrier-Sense Multiple Access with Collision Avoidance

Term	Meaning
CSMA/CD	Carrier-Sense Multiple Access with Collision Detection
CSP	Communications Scanner Processor
CSN	Card Select Number Register
CSNET	Computer Science Network
CSS	Control, Signalling, and Status Store
CSU	Channel Service Unit
CTC	Channel to Channel
CTCA	Channel-to-Channel Adaptor
CTCP	Communication and Transport Control Program
CTS	Clear to Send
CUA	Channel Unit Address; Common User Access
CURACF	Call Unrelated Service Function
CUT	Control Unit Terminal
CVS	Connection View State
CVT	Communications Vector Table
DACD	Directory Access Control Domain
DAD	Draft Addendum
DAF	Distributed Architecture Framework; Distributed Applications Framework; Destination Address Field
DAP	Directory Access Protocol
DAS	Dual Attachment Station; Dynamically Assigned Sockets
DAT	Dynamic Address Translation
dB	Decibels
DCA	Document Content Architecture
DCC	Data Country Code
DCE	Data Communications Equipment; Distributed Computing Environment; Data Circuit-Terminating Equipment
DCS	Defined Context Set
DDCMP	Digital Data Communication Message Protocol

Term	Meaning
DDIM	Device Driver Initialization Model
DDM	Distributed Data Management
DDN	Data Defense Network
DDP	Datagram Delivery Protocol
DDS	Digital Data Service
DES	Data Encryption Standard
DFC	Data Flow Control
DECNET	Digital Equipment Corporation's Network Architecture
DFI	DSP Format Identifier
DFT	Distributed Function Terminal
DHCP	Dynamic Host Configuration Protocol
DH	DMPDU Header
DIA	Document Interchange Architecture
DIB	Directory Information Base
DIS	Draft International Standard
DISP	Draft International Standardized Profile
DISP	Directory Information Shadowing Protocol
DIT	Directory Information Tree
DIU	Distribution Interchange Unit
DL	Distribution List
DLC	Data Link Control; Data Link Connection
DLCEP	Data Link Connection End Point
DLCI	Data Link Connection Identifier
DLPDU	Data Link Protocol Data Unit
DLS	Data Link Service
DLSAP	Data Link Service Access Point
DLSDU	Data Link Service Data Unit
DLU	Dependent Logical Unit

Term	Meaning
DMA	Direct Memory Access
DMD	Directory Management Domain
DMI	Digital Multiplexed Interface; Definition of Management Information; Desktop Management Interface
DMO	Domain Management Organization
DMPDU	Derived MAC Protocol Data Unit
DMTF	Desktop Management Task Force
DMUX	Double Multiplexer
DN	Distinguished Name
DNS	Domain Name System
DNHR	Dynamic Non-hierarchical Routing
DoD	U.S. Department of Defense
DOP	Directory Operational Binding Management Protocol
DOS	Disk Operating System
DP	Draft Proposal
DPG	Dedicated Packet Group
DPI	Dots Per Inch
DQDB	Distributed Queue Dual Bus
DR	Definite Response
DS	Directory Services
DS3	Telephony classification of leased line speed
DS-n	Digital Signaling Level n
DSA	Directory Service Agent
DSAP	Destination Service Access Point
DSD	Data Structure Definition
DSE	DSA Specific Entries
DSL	Digital Subscriber Line
DSP	Directory Service Protocol

Term	Meaning
DSP	Domain-Specific Part
DSS 1	Digital Subscriber Signaling System, Number 1
DSTINIT	Data Services Task Initialization
DSU	Digital Services Unit
DSUN	Distribution Services Unit Name
DT	DMPDU Trailer
DTE	Data Terminal Equipment
DTMF	Dual-Tone Multifrequency
DTR	Data Terminal Ready
DU	Data Unit
DUP	Data User Port
DUA	Directory User Agent
DVMRP	Distance Vector Multicast Routing Protocol
E.164	An ATM address format specified by the ITU-TS
E-Mail	Electronic Mail
EAS	Extended Area Service
EB	End Bracket
EBCDIC	Extended Binary-Coded Decimal Interchange Code
EACK	Extended Acknowledgement
EARN	European Academic Research
ECA	Event Detection Point
ECC	Enhanced Error Checking and Correction
ECH	Echo Canceller with Hybrid
ECMA	European Computer Manufacturers' Association
ECO	Echo Control Object
ECSA	Exchange Carriers Standards Association
EDI	Electronic Data Interchange
EDIFACT	EDI For Administration, Commerce, and Transport

Term	Meaning
EDIM	EDI Message
EDIME	EDI Messaging Environment
EDIMS	EDI Messaging System
EDI-MS	EDI Message Store
EDIN	EDI Notification
EDI-UA	EDI User Agent
EEI	External Environment Interface
EGP	Exterior Gateway Protocol
EIA	Electronic Industries Association
EISA	Extended Industry Standard Architecture
EIT	Encoded Information Type
EN	End Node
ENA	Extended Network Addressing
ENV	European Pre-standards
EOM	End-of-Message DMPDU
EOT	End of Transmission
EP	Emulation Program
ER	Explicit Route
ER	Exception Response
EREP	Environmental Recording Editing and Printing
ES	End System
ESA	Enterprise Systems Architecture
ESA	Enhanced Subarea Addressing
ESCON	Enterprise System Connectivity
ESF	Extended Superframe Format
ESH	End System Hello
ES-IS	End System Intermediate System
ESS	Electronic Switching System

Term	Meaning
ESTELLE	Extended-State Transition Language
ETB	End-of-Text Block
ETR	Early Token Release
ETX	End of Text
EUnet	European UNIX network
EUUG	European UNIX User's Group
EWOS	European Workshop on Open Systems
FA	Framework Advisory
FADU	File Access Data Unit
FARNET	Federation of American Research Networks
FAS	Frame Alignment Sequence
FAT	File Allocation Table
FC	Frame Control Field
FCC	Federal Communications Commission
FCS	Frame Check Sequence
FDCO	Field Definition Control Object
FDDI	Fiber Distributed Data Interface
FDDI-FO	FDDI Follow-on (FDDI)
FDM	Frequency Division Multiplexing
FDR	Field Definition Record
FDT	Formal Description Technique
FDX	Full Duplex
FEC	Field Entry Condition
FEE	Field Entry Event
FEI	Field Entry Instruction
FEICO	Field Entry Instruction Control Object
FEIR	Field Entry Instruction Record
FEP	Front End Processor

Term	Meaning
FEPCO	Field Entry Pilot Control Object
FEPR	Field Entry Pilot Record
FER	Field Entry Reaction
FFOL	FDDI Follow-on LAN
FID	Format Identification
FIPS	Federal Information Processing Standard
FISU	Fill-in Signal Unit
FM	Function Management
FMH	Function Management Header
FOD	Office Document Format
FOR	Forward Transfer
FNC	Federal Networking Council
FRICC	Federal Research Internet Coordinating Committee
FR	Family of Requirement
FRMR	Frame Reject
FS	Frame Status Field
FSG	SGML Interchange Format
FSM	Finite-State Machine
FTAM	File Transfer and Access Management
FTP	File Transfer Protocol in TCP/IP
FYI	For Your Information
FX	Foreign Exchange Service
Gb	Gigabits
Gbps	Gigabits per Second
GDMO	Guidelines for the Definition of Managed Objects
GDS	Generalized Data Stream
GFI	General Format Indicator
GFP	Global Functional Plane

Term	Meaning
GGP	Gateway-to-Gateway Protocol
GMT	Greenwich Mean Time
GPS	Global Positioning System
GOSIP	Government OSI Protocol
GSA	General Services Administration
GTF	Generalized Trace Facility
GWNCP	Gateway NCP
GWSSCP	Gateway SSCP
HAL	Hardware Abstraction Layer
H-MUX	Hybrid Multiplexer
HCL	Hardware Compatibility List
HCS	Header Check Sequence
HDB3	High-Density Bipolar—3 zeros
HDLC	High-Level Data Link Control
HDX	Half Duplex
HI-SAP	Hybrid Isochronous MAC Service Access Point
HLR	Home Location Register
HMP	Host Monitoring Protocol
HOB	Head of Bus
HP-SAP	Hybrid Packet MAC Service Access Point
HRC	Hybrid Ring Control
HSLN	High-Speed Local Network
HTML	Hypertext Markup Language
HTTP	Hypertext Transfer Protocol
Hz	Hertz (cycles per second)
IAB	Internet Architecture Board
IANA	Internet Assigned Number Authority

Term	Meaning
IADCS	Interactivity Defined Context Set
IAN	Integrated Analog Network
IAP	Inner Administrative Point
IBM	International Business Machines Corporations
IC	Interexchange Carrier
ICD	International Code Designator
ICF	Isochronous Convergence Function
ICI	Interface Control Information
ICMP	Internet Control Message Protocol
ICV	Integrity Check Value
IDI	Initial Domain Identifier
IDN	Integrated Digital Network; Interface Definition Notation
IDP	Initial Domain Part; Internetwork Datagram Packet
IDU	Interface Data Unit
IEC	Interexchange Carrier
IEC	International Electrotechnical Commission
IEEE	Institute of Electrical and Electronic Engineers
IEN	Internet Engineering Notes
IF	Information Flow
IETF	Internet Engineering Task Force
IESG	Internet Engineering Steering Group
IGP	Interior Gateway Protocol
IGMP	Internet Group Management Protocol
IGRP	Internet Gateway Routing Protocol
ILD	Injection Laser Diode
ILU	Independent Logical Unit
IMAC	Isochronous Media Access Control

Term	Meaning
IMIL	International Managed Information Library
IML	Initial Microcode Load
IMPDU	Initial MAC Protocol Data Unit
IMS	Information Management System
IN	Intelligent Network
IND	Indication
INN	Intermediate Network Node
INTAP	Interoperability Technology Association for Information Processing
IOC	Input/Output Control
IOCP	Input/Output Control Program
IONL	Internal Organization of Network Layer
IP	Internet Protocol
IPng	IP Next Generation
IPv4	IP version 4
IPv6	IP version 6
IPC	Interprocess Communication
IPDS	Intelligent Printer Data Stream
IPI	Initial Protocol Identifier
IPICS	ISP Implementation Conformance Statement
IPL	Initial Program Load
IPM	Interpersonal Message
IPM-UA	Interpersonal Messaging User Agent
IPMS	Interpersonal Messaging System
IPN	Interpersonal Notification
IPR	Isolated Pacing Response
IPX	Internetwork Packet Exchange
IR	Internet Router
IRN	Intermediate Routing Node

Term	Meaning
IRQ	Interrupt Request Lines
IRTF	Internet Research Task Force
IS	International Standard
ISA	Industry Standard Architecture
ISAM	Index-Sequential Access Method
ISC	Intersystem Communications in CICS
ISCF	Intersystems Control Facility
ISDN	Integrated Services Digital Network
ISH	Intermediate System Hello
IS-IS	Intermediate System to Intermediate System
ISO	International Standards Organization
ISODE	ISO Development Environment
ISP	International Standard Profile
ISPBX	Integrated Services Private Branch Exchange
ISPSN	Initial Synchronization Point Serial Number
ISSI	Interswitching System Interface
ISUP	ISDN User Part
IT	Information Technology
ITC	Independent Telephone Company
ITU	International Telecommunication Union
ITU-TS	International Telecommunication Union, Telecommunication Section
IUT	Implementation Under Test
IVDT	Integrated Voice/Data Terminal
IWU	Interworking Unit
IXC	Interexchange Carrier
JCL	Job Control Language
JES	Job Entry Subsystem
JTC	Joint Technical Committee

Term	Meaning
JTM	Job Transfer and Manipulation
KA9Q	TCP/IP implementation for amateur radio
kb	Kilobits
kbps	Kilobits per Second
kHz	Kilohertz
km	Kilometers
LAB	Latency Adjustment Buffer
LAB	Line Attachment Base
LAN	Local Area Network
LANSUP	LAN Adapter NDIS Support
LAP	Link Access Procedure
LAPB	Link Access Procedure Balanced
LAPD	Link Access Procedures on the D-channel
LAPS	LAN Adapter and Protocol Support
LATA	Local Access and Transport Area
LCF	Log Control Function
LCN	Logical Channel Number
LE	Local Exchange
LEC	Local Exchange Carrier
LED	Light-Emitting Diode
LEN	Low-Entry Networking
LI	Length Indicator
LIB	Line Interface Base
LIC	Line Interface Coupler
LIDB	Line Information Database
LIS	Logical IP Subnet
LLAP	LocalTalk Link Access Protocol
LLC	Logical Link Control

Term	Meaning
LME	Layer Management Entity
LMI	Layer Management Interface
LOCKD	Lock Manager Daemon
LOTOS	Language of Temporal Ordering Specifications
LPD	Line Printer Daemon
LPDA	Link Problem Determination Application
LPR	Line Printer
LRC	Longitudinal Redundancy Check
LSE	Local System Environment
LSL	Link Support Layer
LSS	Low-Speed Scanner
LSSU	Link Status Signal Unit
LT	Local Termination
LU	Logical Unit
m	Meters
MAC	Media Access Control; Medium Access Control
MACE	Macintosh Audio Compression and Expansion
MACF	Multiple Association Control Function
MAN	Metropolitan Area Network
MAP	Manufacturing Automation Protocol
MAU	Media Access Unit; Multistation Access Unit
Mb	Megabits
MBA	MASSBUS Adapter
MBONE	Multicast Backbone
Mbps	Megabits per Second
MBZ	Must Be Zero
MCA	Microchannel Architecture
MCF	MAC Convergence Function

Term	Meaning
MCI	Microwave Communications, Inc.
MCP	MAC Convergence Protocol
MCR	Monitor Console Routine
MD	Management Domain
MFA	Management Functional Areas
MFJ	Modified Final Judgment
MFS	Message Formatting Services in IMS
MH	Message-Handling Package
MHS	Message-Handling Service; Message-Handling System
MHz	Megahertz
MIB	Management Information Base
MIC	Media Interface Connector
MID	Message Identifier
MILNET	Military Network
MIM	Management Information Model
MIME	Multipurpose Internet Mail Extension
MIN	Mobile Identification Number; Multiple Interaction Negotiation
MIPS	Million Instructions per Second
MIS	Management Information Systems
MIT	Managed Information Tree
MLID	Multiple Link Interface Driver
MMF	Multimode Fiber
MMI	Man-Machine Interface
MMS	Manufacturing Message Specification
MOSS	Maintenance and Operator Subsystem
MOTIS	Message-Oriented Text Interchange System
MOT	Means of Testing

Term	Meaning
MPAF	Mid-Page Allocation Field
MRO	Multiregion Operation in CICS
ms	Millisecond
MS	Management Services; Message Store
MSC	Mobile Switching Center
MSCP	Mass Storage Control Protocol
MSN	Multiple Systems Networking
MSNF	Multiple Systems Networking Facility
MSS	MAN Switching System; Maximum Segment Size
MST	Multiplexed Slotted and Token Ring
MSU	Management Services Unit
MTA	Message Transfer Agent
MTACP	Magnetic Tape Ancillary Control Process
MTBF	Mean Time between Failures
MTTD	Mean Time of Diagnosis
MTOR	Mean Time of Repair
MTP	Message Transfer Part
MTS	Message Transfer System
MTSE	Message Transfer Service Element
MTU	Maximum Transfer Unit
MVS/XA	Multiple Virtual Storage/Extended Architecture
MVS/370	Multiple Virtual Storage/370
MVS	Multiple Virtual Systems
NAK	Negative Acknowledgement in BSC
NAP	Network Access Provider
NAU	Network Addressable Unit
NBP	Name-Binding Protocol

Term	Meaning
NC	Network Connection; Numerical Controller
NCB	Node Control Block
NCCF	Network Communications Control Facility
NCEP	Network Connection End-Point
NCP	Network Control Program; Network Core Protocol
NCS	Network Computing System
NCTE	Network Channel-Terminating Equipment
NDIS	Network Driver Interface Specification
NFS	Network File System
NIB	Node Identification Block
NIC	Network Interface Card
NIF	Network Information File
NISDN	Narrow Band ISDN
NIS	Names Information Socket
NIST	National Institute of Standards and Technology
NIUF	North American ISDN Users' Forum
NJE	Network Job Entry
NLM	NetWare Loadable Module
NLDM	Network Logical Data Manager
nm	Nanometer
NM	Network Management
NMP	Network Management Process
NMVT	Network Management Vector Transport
NMS	Network Management Station
NN	Network Node
NOC	Network Operations Center
NPA	Numbering Plan Area

Term	Meaning
NPAI	Network Protocol-Control Information
NPDA	Network Problem-Determination Application
NPDU	Network Protocol Data Unit
NPM	NetView Performance Monitor
NPSI	NCP Packet-Switching Interface
NRN	Non-receipt Notification
NRZ	Non-Return to Zero
NRZI	Non-Return-to-Zero Inverted
ns	Nanosecond
NS	Network Service
NSAP	Network Service Access Points
NSDU	Network Service Data Unit
NSF	National Science Foundation
NTFS	Windows NT File System
NTO	Network Terminal Option
NVLAP	National Voluntary Accreditation Program
OAF	Origination Address Field
OAM	Operations, Administration, and Maintenance
OAM&P	Operations Administration, Maintenance, and Provisioning
OC-n	Optical Carrier level n
OC3	155 million bits per second over fiber
OCA	Open Communication Architectures
OCC	Other Common Carrier
ODA	Office Document Architecture
ODI	Open Data-Link Interface
ODIF	Office Document Interchange Format
ODINSUP	ODI NSIS Support

Term	Meaning
ODP	Open Distributed Processing
OIT	Object Identifier Tree
OIW	OSI Implementation Workshop
OLRT	Online Realtime
OLU	Originating Logical Unit
OM	Object Management
ONA	Open Network Architecture
ONC	Open Network Computing
OPNDST	Open Destination
O/R	Originator/Recipient
OS/400	Operating System/400 for the AS/400 Computer
OS	Operating System
OSE	Open Systems Environment
OSF	Open Software Foundation
OSI	Open Systems Interconnection
OSI/CS	OSI Communications Subsystem
OSIE	Open Systems Interconnection Environment
OSILL	Open Systems Interconnection, Lower Layers
OSIUL	Open Systems Interconnection, Upper Layers
OSPF	Open Shortest Path First
OSNS	Open Systems Network Services
P-MAC	Packet-Switched Media Access Control
PA	Pre-Arbitrated
PABX	Private Automatic Branch Exchange
PAD	Packet Assembler/Disassembler
PAF	Pre-Arbitrated Function
PAI	Protocol Address Information

Term	Meaning
PAM	Pass Along Message
PANS	Pretty Amazing New Stuff
PAP	Printer Access Protocol
PBX	Private Branch Exchange
PC	Path Control; Personal Computer
PCCU	Physical Communications Control Unit
PCEP	Presentation Connection End-Point
PCI	Protocol Control Information; Presentation Context Identifier; Peripheral Component Interconnect bus
PCM	Pulse Code Modulation
PCO	Points of Control and Observation
PCTR	Protocol Conformance Test Report
PDAD	Proposed Draft Addendum
PDAU	Physical Delivery Access Unit
PDC	Packet Data Channel
pDISP	Proposed Draft International Standard Profile
PDN	Public Data Network
PDU	Protocol Data Unit
PDV	Presentation Data Value
PELS	Picture Elements
PEM	Privacy-Enhanced Mail
PEP	Partition Emulation Program
PETS	Parameterized Executable Test Suite
PH	Packet Handler; Packet Handling
PhC	Physical Layer Connection
PhCEP	Physical Connection End-Point
PhL	Physical Layer

Term	Meaning
PhPDU	Physical Layer Protocol Data Unit
PhS	Physical Layer Service
PhSAP	Physical Layer Service Access Point
PhSDU	Physical Layer Service Data Unit
PHY	Physical Layer
PICS	Protocol Information Conformance Statement
PIN	Positive-Intrinsic Negative Photodiode
PING	Packet Internet Groper
PIP	Program Initialization Parameters
PIU	Path Information Unit
PIXIT	Protocol Implementation, Extra Information for Testing
PKCS	Public Key Cryptosystems
PLC	Programmable Logic Controller
PLCP	Physical Layer Convergence Protocol
PLMN	Public Land Mobile Network
PLP	Packet Layer Protocol
PLS	Primary Link Station; Physical Signalling
PLU	Primary Logical Unit
PM	Protocol Machine
PMD	Physical Layer Medium Dependent
POI	Point of Initiation; Program Operator Interface
POP	Point of Presence
POSI	Promoting Conference for OSI
POSIX	Portable Operating System Interface
POTS	Plain Old Telephone Service
PPDU	Presentation Protocol Data Unit
PPO	Primary Program Operator
PPP	Point-to-Point Protocol

Term	Meaning
PPSDN	Public Packet-Switched Data Network
PRI	Primary Rate Interface
PRMD	Private Management Domain
PS	Presentation Services
PSAP	Public Safety Answering Point
PSC	Public Service Commission
PSDN	Packet Switched Data Network
PSN	Packet Switched Network
PSPDN	Packet Switched Public Data Network
PSTN	Public Switched Telephone Network
PTF	Program Temporary Fix
PTLXAU	Public Telex Access Unit
PTN	Public Telephone Network
PTT	Post, Telegraph, and Telephone
PU	Physical Unit
PUC	Public Utility Commission
PUCP	Physical Unit Control Point
PUMS	Physical Unit Management Services
PUP	Parc Universal Packet
PUT	Program Update Tape
PVC	Private Virtual Circuit
PVN	Private Virtual Network
P 1	Protocol 1 (Message Transfer Protocol/MHS/X.400)
P 2	Protocol 2 (Interpersonal Messaging MHS/X.400)
P 3	Protocol 3 (Submission and Delivery Protocol/MHS/X.400)
P 5	Protocol 5 (Teletext Access Protocol)
P 7	Protocol 7 (Message Store Access Protocol in X.400)
QA	Queued Arbitrated

Term	Meaning
QAF	Queued Arbitrated Function
QC	Quiesce Complete
QEC	Quiesce at End of Chain
QMF	Query Management Facility
QOS	Quality of Service
QPSX	Queued Packet and Synchronous Switch
QUIPU	X.500 Conformant Directory Services in ISODE
RAM	Random Access Memory
RARE	Reseaux Associes poir la Recherche Europeenne; European Association of Research Networks
RARP	Reverse Address Resolution Protocol
RAS	Remote Access Service
RBOC	Regional Bell Operating Company
RD	Routing Domain
RD	Route Redirection
RDA	Relative Distinguished Names
RDA	Remote Database Access
RDI	Restricted Digital Information
RDN	Relative Distinguished Name
RDP	Reliable Datagram Protocol
RDT	Resource Definition Table
RECFMS	Record Formatted Maintenance Statistics
REJ	Reject
REQ	Request
RESP	Response
RESYNC	Resynchronization
RFC	Request for Change
RFP	Request for Proposal

Term	Meaning
RFQ	Request for Price Quotation
RH	Response Header
RIB	Routing Information Base
RIF	Routing Information Field
RIM	Request Initialization Mode
RIP	Router Information Protocol
RIPE	Reseaux IP Europeens; European continental TCP/IP network operated by EUnet
RISC	Reduced Instruction Set Computer
RJE	Remote Job Entry
RM	Reference Model
RMT	Ring Management
RN	Receipt Notification
RNAA	Request Network Address Assignment
RNR	Receiver Not Ready
ROSE	Remote Operations Service Element
RPC	Remote Procedure Call in OSF/DCE; Remote Procedure Call
RPL	Request Parameter List; Remote Program Load
RPOA	Recognized Private Operating Agency
RQ	Request Counter
RR	Receiver Ready
RS	Relay System
RSF	Remote Support Facility
RSP	Response
RTM	Response-Time Monitor
RTMP	Routing Table Maintenance Protocol
RTO	Round Trip Time-Out
RTR	Ready to Receive

Term	Meaning
RTS	Request to Send
RTSE	Reliable-Transfer Service Element
RTT	Round-Trip Time
RU	Request Unit
RU	Response Unit
S/390	IBM's System/390 Hardware Architecture
s	Second
SA	Source Address field
SA	Subarea; Sequenced Application
SAA	System Applications Architecture; Specific Administrative Areas
SABM	Set Asynchronous Mode Balanced
SACF	Single Association Control Function
SACK	Selective Acknowledgement
SAF	SACF Auxiliary Facility
SALI	Source Address Length Indicator
SAM	Security Accounts Manager
SAMBE	Set Asynchronous Mode Balanced Extended
SAO	Single Association Object
SAP	Service Access Point; Service Advertising Protocol
SAPI	Service Access Point Identifier
SAS	Single-Attachment Station; Statically Assigned Sockets
SASE	Specific Application Service Element
SATS	Selected Abstract Test Suite
SAW	Session Awareness Data
SBA	Set Buffer Address
SBI	Stop Bracket Initiation
SC	Session Connection

Term	Meaning
SC	Subcommittee
SCC	Specialized Common Carrier
SCCP	Signaling Connection Control Part
SCEP	Session Connection End-Point
SCP	Service Control Point
SCS	System Conformance Statement
SCSI	Small Computer System Interface
SCTR	System Conformance Test Report
SDH	Synchronous Digital Hierarchy
SDIF	Standard Document Interchange Format
SDL	System Description Language
SDLC	Synchronous Data Link Control
SDN	Software-Defined Network
SDSE	Shadowed DSA Entries
SDU	Service Data Unit
SE	Session Entity
SG	Study Group
SGFS	Special Group on Functional Standardization
SGML	Standard Generalized Markup Language
SIA	Stable Implementation Agreement
SID	Security ID
SIM	Set Initialization Mode
SIO	Start I/O
SIP	SMDS Interface Protocol
SLU	Secondary Logical Unit
SMAE	System Management Application Entity
SMASE	Systems Management Application Service Element

Term	Meaning
SMB	Server Message Block
SMDR	Station Message Detail Recording
SMDS	Switched Multi-Megabit Data Service
SMF	Single-Mode Fiber; System Management Facility
SMFA	Systems Management Functional Area
SMI	Structure of the OSI Management Information Service
SMIB	Stored Message Information Base
SMP	System Modification Program
SMS	Service Management System
SMT	Station Management Standard
SMTP	Simple Mail Transfer Protocol
SNA	System Network Architecture
SNAP	Subnetwork Attachment Point
SNAcF	Subnetwork Access Function
SNAcP	Subnetwork Access Protocol
SNADS	SNA Distribution Services
SNARE	Subnetwork Address-Routing Entity
SNCP	Single Node Control Point
SNDCP	Subnetwork-Dependent Convergence Protocol
SNI	Subscriber-Network Interface; SNA Network Interconnection SNA Network Interface
SNICP	Subnetwork Independent Convergence Protocol
SNMP	Simple Network Management Protocol
SNPA	Subnetwork Point of Attachment
SNRM	Set Normal Response Mode
SOA	Start Of Authority
SONET	Synchronous Optical Network
SP	Signaling Point

Term	Meaning
SPAG	Standards Promotion and Applications Group
SPC	Signaling Point Code
SPDU	Session Protocol Data Unit
SPE	Synchronous Payload Envelope
SPF	Shortest Path First
SPI	Subsequent Protocol Identifier
SPM	FDDI to SONET Physical Layer Mapping Standard
SPSN	Synchronization Point Serial Number
SQL	Structured Query Language
SRH	SNARE Request Hello
SS	Session Service; Switching System
SS6	Signaling System, number 6
SS7	Signaling System, number 7
SSA	Subschema-Specific Area
SSAP	Source Service Access Point
SSCP	System Services Control Point
SSDU	Session Service Data Unit
SSM	Single Segment Message DMPDU
STA	Spanning Tree Algorithms
ST	Sequenced Terminal
STD	Standard
STM	Synchronous Transfer Mode; Station Management
STM-n	Synchronous Transport Module, level n
STP	Service Transaction Program in LU 6.2; Shielded Twisted Pair; Signal Transfer Point
STS-n	Synchronous Transport Signal, level n
SUERM	Signal Unit Error Rate Monitor
SUT	System Under Test

Term	Meaning
SVA	Shared Virtual Area
SVC	Switched Virtual Circuit
SWS	Silly Window Syndrome
SYN	Synchronizing Segment; Synchronous character in IBM's Bisync Protocol
SYNC	Synchronization
T3	A designation of telephony used over DS3 speed lines
T	Transport
TA	Terminal Adaptor
TC	Technical Committee
TCP	Transmission Control Protocol
TAG	Technology Advisory Group
TAP	Trace Analysis Program
TC	Transport Connection; Technical Committee
TCAM	Telecommunications Access Method
TCB	Task Control Block
TCEP	Transport Connection End-Point
TCM	Time Compression Multiplexing
TCP	Transmission Control Protocol
TCP/IP	Transmission Control Protocol/Internet Protocol
TCT	Terminal Control Table in CICS
TDM	Time-Division Multiplexing
TDMA	Time-Division Multiple AccessTE Terminal Equipment
TELNET	Remote Logon in TCP/IP
TEP	Transport End-Point
TFTP	Trivial File Transfer Protocol
TG	Transmission Group
TH	Transmission Header

Term	Meaning
THT	Token Holding Timer
TIC	Token Ring Interface Coupler
TINA	Telecommunications Information Network Architecture
TINA-C	Telecommunications Information Network Architecture Consortium
TI RPC	Transport-Independent RPC
TLI	Transport Layer Interface
TLMAU	Telematic Access Unit
TLV	Type, Length, and Value
TLXAU	Telex Access Unit
TMP	Text Management Protocol
TMS	Time-Multiplexed Switching
TOP	Technical and Office Protocol
TOS	Type of Service
TN3270	A version of TELNET that implements the IBM 3270 data stream
TP	Transaction Program; Transaction Processing; Transport Protocol
TPDU	Transport Protocol Data Unit
TP-PMD	Twisted-Pair PMD
TPS	Two-Processor Switch
TPSP	Transaction Processing Service Provider
TPSU	Transaction Processing Service User
TPSUI	TPSU Invocation
TP 0	TP class 0; simple
TP 1	TP class 1; basic error recovery
TP 2	TP class 2; multiplexing
TP 3	TP class 3; error recovery and multiplexing
TP 4	TP class 4; error detection and recovery
TR	Technical Report; Token Ring
TRA	Token Ring Adapter

Term	Meaning
TRPB	Truncated Reverse-Path Broadcast
TRSS	Token Ring Subsystem
TRT	Token Rotation Timer
TS	Transaction Service(s)
TSAP	Transport Service Access Point
TSC	Transmission Subsystem Controller
TSDU	Transport Service Data Unit
TSCF	Target System Control Facility
TSI	Time Slot Interchange
TSO	Time Sharing Option
TSR	Terminate and Stay Resident Program
TSS	Transmission Subsystem
TTCN	Tree and Tabular Combined Notation
TTL	Time to Live
TTP	Timed Token Protocol; Transport Test Platform
TTRT	Target Token Rotation Time
TTY	Teletype
TUP	Telephone User Part
TVX	Valid Transmission Timer (FDDI)
TWX	Teletypewriter Exchange Service
UA	Unnumbered Acknowledgement; User Agent; Unsequenced Application
UART	Universal Asynchronous Receiver and Transmitter
UDI	Unrestricted Digital Information
UDP	User Datagram Protocol
UOW	Unit of Work
UPS	Uninterruptable Power Supply
URL	Universal Resource Locator
User-ASE	User Application Service Element

Term	Meaning
USS	Unformatted System Services
UT	Unsequenced Terminal
UTC	Coordinated Universal Time
UTP	Unshielded Twisted Pair
UUCP	UNIX-to-UNIX Copy Program
VAC	Value-Added Carrier
VAN	Value-Added Network
VAS	Value-Added Service
vBNS	A reference to the 155 Mps deployment of an Internet backbone, to have been implemented in 1995
VCI	Virtual Channel Identifier (DQDB)
VDT	Video Display Terminal
VESA	Video Electronics Standards Association
VLR	Visitor Location Register
VLSI	Very Large Scale Integration
VPI/VCI	Virtual Path Identifier and a Virtual Call Identifier
VM	Virtual Machine
VMD	Virtual Manufacturing Device
VM/SP	Virtual Machine System Product
VPN	Virtual Private Network
VR	Virtual Route
VRPWS	Virtual Route Pacing Window Size
VS	Virtual Storage
VSAM	Virtual Storage Access Method
VSE	Virtual Storage Extended
VT	Virtual Terminal
VTAM	Virtual Telecommunications Access Method
VTE	Virtual Terminal Environment

Term	Meaning
VTP	Virtual Terminal Protocol
VTPM	Virtual Terminal Protocol Machine
VTSE	Virtual Terminal Service Element
WACA	Write-Access Connection Acceptor
WACI	Write-Access Connection Initiator
WAN	Wide Area Network
WAVAR	Write Access Variable
WBC	Wideband Channel
WD	Working Document
WG	Working Group
WNM	Workgroup Node Manger
WP	Working Party
WWW	Worldwide Web
X	The X Window System
X.25	An ITU-TX standard; a transport-layer service
X.400	The ITU-TS protocol for electronic mail
XAPIA	X.400 API Association
XDR	External Data Representation
XDS	X/Open Directory Services API
XI	SNA X.25 Interface
XID	Exchange Identification
XNS	Xerox Network Standard
XTI	X/Open Transport Interface
XUDTS	Extended Unitdata Service
ZIP	Zone Information Protocol
ZIS	Zone Information Socket

APPENDIX B

TCP/IP RFC References

Where *NOL* is used in this appendix, it means "not online."

RFC	Title
2200	Internet Official Protocol Standards. 06/12/1997
2177	IMAP4 IDLE command. 07/02/1997
2176	IPv4 over MAPOS Version 1. 06/25/1997
2175	MAPOS 16—Multiple Access Protocol over SONET/SDH with 16 Bit Addressing. 06/25/1997
2174	A MAPOS version 1 Extension—Switch-Switch Protocol. 06/25/1997
2173	MAPOS version 1 Extension—Node Switch Protocol. 06/25/1997
2172	MAPOS Version 1 Assigned Numbers. 06/25/1997
2171	MAPOS—Multiple Access Protocol over SONET/SDH Version 1. 06/25/1997
2170	Application Requested IP over ATM (AREQUIPA). 07/02/1997
2169	A Trivial Convention for using HTTP in URN Resolution . 06/23/1997
2168	Resolution of Uniform Resource Identifiers using the Domain Name System. 06/23/1997
2167	Referral Whois (RWhois) Protocol V1.5. 06/20/1997
2166	APPN Implementer's Workshop Closed Pages Document DLSw v2.0 Enhancements. 06/19/1997
2165	Service Location Protocol. 06/20/1997
2155	Definitions of Managed Objects for APPN Using SMIv2. 06/16/1997
2154	OSPF with Digital Signatures. 06/16/1997
2153	PPP Vendor Extensions. 06/03/1997
2152	A Mail-Safe Transformation Format of Unicode. 06/03/1997
2151	A Primer On Internet and TCP/IP Tools and Utilities 6/10/1997
2149	Multicast Server Architectures for MARS-based ATM multicasting. 05/23/1997
2147	TCP and UDP over IPv6 Jumbograms. 05/23/1997

RFC	Title
2146	U.S. Government Internet Domain Names. 05/23/1997
2145	Use and interpretation of HTTP version numbers. 05/23/1997
2144	The CAST-128 Encryption Algorithm. 05/21/1997
2143	Encapsulating IP with the Small Computer System Interface. 05/14/1997
2142	Mailbox Names for Common Services and Functions. 05/06/1997
2141	URN Syntax. 05/05/1997
2140	TCP Control Block Interdependence. 04/29/1997
2139	RADIUS Accounting. 04/18/1997
2138	Remote Authentication Dial In User Service (RADIUS). 04/18/1997
2137	Secure Domain Name System Dynamic Update. 04/21/1997
2136	Dynamic Updates in the Domain Name System. 04/21/1997
2135	Internet Society "Internet Society By-Laws". 04/30/1997
2134	Internet Society "Articles of Incorporation of the Internet Society". 05/01/1997
2133	Basic Socket Interface Extensions for IPv6. 04/21/1997
2132	DHCP Options and BOOTP Vendor Extensions. 04/07/1997
2131	Dynamic Host Configuration Protocol. 04/07/1997
2130	The Report of the IAB Character Set Workshop held 29 February—1 March, 1996. 04/21/1997
2129	Toshiba's Flow Attribute Notification Protocol (FANP) Specification. 04/21/1997
2128	Dial Control Management Information Base using SMIv2. 03/31/1997
2127	ISDN Management Information Base. 03/31/1997
2126	ISO Transport Service on top of TCP (ITOT). 03/28/1997
2125	The PPP Bandwidth Allocation Protocol (BAP). 03/31/1997
2124	Light-weight Flow Admission Protocol Specification Version 1.0. 03/28/1997
2123	Traffic Flow Measurement: Experiences with NeTraMet. 03/28/1997
2122	VEMMI URL Specification. 03/28/1997
2121	Issues affecting MARS Cluster Size. 03/28/199
2120	Managing the X.500 Root Naming Context. 03/28/1997
2119	Key words for use in RFCs to Indicate Requirement Levels. 03/26/1997

RFC	Title
2118	Microsoft Point-To-Point Compression (MPPC) Protocol. 03/20/1997
2117	Protocol Independent Multicast-Sparse Mode (PIM-SM) Protocol Specification. 06/16/1997
2116	X.500 Implementations Catalog-96. 04/24/1997
2114	Data Link Switching Client Access Protocol. 03/03/1997
2113	IP Router Alert Option. 02/28/1997
2112	The MIME Multipart/Related Content-type. 03/12/1997
2111	Content-ID and Message-ID Uniform Resource Locators. 03/12/1997
2110	MIME E-mail Encapsulation of Aggregate Documents, such as HTML (MHTML). 03/12/1997
2109	HTTP State Management Mechanism. 02/18/1997
2108	Definitions of Managed Objects for IEEE 802.3 Repeater Devices using SMIv2. 02/12/1997
2107	Ascend Tunnel Management Protocol—ATMP. 02/06/1997
2106	Data Link Switching Remote Access Protocol. 03/03/1997
2105	Cisco Systems' Tag Switching Architecture Overview. 02/06/1997
2104	HMAC: Keyed-Hashing for Message Authentication. 02/05/1997
2103	Mobility Support for Nimrod: Challenges and Solution Approaches. 02/06/1997
2102	Multicast Support for Nimrod: Requirements and Solution Approaches. 02/20/1997
2101	IPv4 Address Behavior Today. 02/04/1997
2100	The Naming of Hosts. 04/01/1997
2099	Request for Comments Summary RFC Numbers 2000-2099 3/13/1997
2098	Toshiba's Router Architecture Extensions for ATM: Overview. 02/04/1997
2097	The PPP NetBIOS Frames Control Protocol. (NBFCP) 01/30/1997
2096	IP Forwarding Table MIB. 01/30/1997
2095	IMAP/POP Authorize Extension for Simple Challenge/Response. 01/30/1997
2094	Group Key Management Protocol (GKMP) Architecture. 07/02/1997
2093	Group Key Management Protocol (GKMP) Specification. 07/02/1997
2092	Protocol Analysis for Triggered RIP. 01/24/1997
2091	Triggered Extensions to RIP to Support Demand Circuits. 01/24/1997

RFC	Title
2090	TFTP Multicast Option. 02/04/1997
2089	V2ToV1 Mapping SNMPv2 onto SNMPv1 within a Bilingual SNMP agent. 01/28/1997
2088	IMAP4 non-synchronizing literals. 01/22/1997
2087	IMAP4 QUOTA extension. 01/22/1997
2086	IMAP4 ACL extension. 01/22/1997
2085	HMAC-MD5 IP Authentication with Reply Prevention. 02/05/1997
2084	Considerations for Web Transaction Security. 01/22/1997
2083	PNG Portable Network Graphics Specification Version 1.0. 01/16/1997
2082	RIP-2 MD5 Authentication. 01/10/1997
2081	RIPng Protocol Applicability Statement. 01/10/1997
2080	RIPng for IPv6. 01/10/1997
2079	Definition of X.500 Attribute Types and an Object Class to Hold Uniform Resource Identifiers (URIs). 01/10/1997
2078	Generic Security Service Application Program Interface, version 2. 01/10/1997
2077	The Model Primary Content Type for Multipurpose Internet Mail Extensions. 01/10/1997
2076	Common Internet Message Headers. 02/24/1997
2075	IP Echo Host Service. 01/08/1997
2074	Remote Network Monitoring MIB Protocol Identifiers 1/16/1997
2073	An IPv6 Provider-Based Unicast Address Format. 01/08/1997
2072	Router Renumbering Guide. 01/08/1997
2071	Network Renumbering Overview: Why would I want it and what is it anyway? 01/08/1997
2070	Internationalization of the Hypertext Markup Language. 01/06/1997
2069	An Extension to HTTP: Digest Access Authentication. 01/03/1997
2068	Hypertext Transfer Protocol — HTTP/1.1. 01/03/1997
2067	IP over HIPPI. 01/03/1997
2066	TELNET CHARSET Option. 01/03/1997
2065	Domain Name System Security Extensions. 01/03/1997

RFC	Title
2064	Traffic Flow Measurement: Meter MIB. 01/03/1997
2063	Traffic Flow Measurement: Architecture. 01/03/1997
2062	Internet Message Access Protocol—Obsolete Syntax. 12/04/1996
2061	IMAP4 COMPATIBILITY WITH IMAP2BIS. 12/05/1996
2060	Internet Message Access Protocol, V4 Rev1. 12/04/1996
2059	RADIUS Accounting. 01/03/1997
2058	Remote Authentication Dial In User Service (RADIUS). 01/03/1997
2057	Source directed access control on the Internet. 11/11/1996
2056	Uniform Resource Locators for Z39.50. 11/05/1996
2055	WebNFS Server Specification. 10/31/1996
2054	WebNFS Client Specification. 10/31/1996
2053	The AM (Armenia) Domain. 10/31/1996
2052	A DNS RR for specifying the location of services (DNS SRV). 10/31/1996
2051	Definitions of Managed Objects for APPC. 10/30/1996
2050	Internet Registry IP Allocation Guidelines. 11/05/1996
2049	Multipurpose Internet Mail Extensions (MIME) Part Five: Conformance Criteria and Examples 2/02/1996
2048	Multipurpose Internet Mail Extensions (MIME) Part Four: Registration Procedures. 01/28/1997
2047	MIME (Multipurpose Internet Mail Extensions) Part Three: Message Header Extensions for Non-ASCII Text. 12/02/1996
2046	Multipurpose Internet Mail Extensions (MIME) Part Two: Media Types. 12/02/1996
2045	Multipurpose Internet Mail Extensions (MIME) Part One: Format of Internet Message Bodies. 12/02/1996
2044	UTF-8, a transformation format of Unicode and ISO 10646. 10/30/1996
2043	The PPP SNA Control Protocol (SNACP). 10/30/96
2042	Registering New BGP Attribute Types. 01/03/1997
2041	Mobile Network Tracing. 10/30/1996
2040	The RC5, RC5-CBC, RC5-CBC-Pad, and RC5-CTS Algorithms. 10/30/1996

RFC	Title
2039	Applicability of Standards Track MIBs to Management of World Wide Web Servers. 11/06/1996
2038	RTP Payload Format for MPEG1/MPEG2 Video. 10/30/1996
2037	Entity MIB. 10/30/1996
2036	Observations on the use of Components of the Class A Address Space within the Internet. 10/30/1996
2035	RTP Payload Format for JPEG-compressed Video. 10/30/1996
2034	SMTP Service Extension for Returning Enhanced Error Codes. 10/30/1996
2033	Local Mail Transfer Protocol. 10/30/1996
2032	RTP payload format for H.261 video streams. 10/30/1996
2031	IETF-ISOC relationship. 10/29/1996
2030	Simple Network Time Protocol (SNTP) Version 4 for IPv4, IPv6, and OSI. 10/30/1996
2029	RTP Payload Format of Sun's CellB Video Encoding. 10/30/1996
2028	The Organizations Involved in the IETF Standards Process. 10/29/1996
2027	IAB and IESG Selection, Confirmation, and Recall Process: Operation of the Nominating and Recall Committees. 10/29/1996
2026	The Internet Standards Process-Rev.3. 10/29/1996
2025	The Simple Public-Key GSS-API Mechanism (SPKM). 10/22/1996
2024	Definitions of Managed Objects for Data Link switching using SNMPv2. 10/22/1996
2023	IP Version 6 over PPP. 10/22/1996
2022	Support for Multicast over UNI 3.0/3.1 based ATM Networks. 11/05/1996
2021	Remote Network Monitoring Management Information Base Version 2 using SMIv2. 01/16/1997
2020	Definitions of Managed Objects for IEEE 802.12 Interfaces. 10/17/1996
2019	Transmission of IPv6 Packets Over FDDI. 10/17/1996
2018	TCP Selective Acknowledgment Options. 10/17/1996
2017	Definition of the URL MIME External-Body Access-Type. 10/14/1996
2016	Uniform Resource Agents (URAs). 10/31/1996
2015	MIME Security with Pretty Good Privacy (PGP). 10/14/1996

RFC	Title
2014	IRTF Research Group Guidelines and Procedures. 10/17/1996
2013	SNMPv2 Management Information Base for the User Datagram Protocol using SMIv2. 11/12/1996
2012	SNMPv2 Management Information Base for the Transmission Control Protocol. 11/12/1996
2011	SNMPv2 Management Information Base for the Internet Protocol using SMIv2. 11/12/1996
2010	Operational Criteria for Root Name Servers. 10/14/1996
2009	GPS-Based Addressing and Routing. 11/08/1996
2008	Implications of Various Address Allocation Policies for Internet Routing. 10/14/1996
2007	Catalogue of Network Training Materials. 10/14/1996
2006	The Definitions of Managed Objects for IP Mobility Support using SMIv2. 10/22/1996
2005	Applicability Statement for IP Mobility Support. 10/22/1996
2004	Minimal Encapsulation within IP. 10/22/1996
2003	IP Encapsulation within IP. 10/22/1996
2002	IP Mobility Support. 10/22/1996
2001	TCP Slow Start, Congestion Avoidance, Fast Retransmit, and Fast Recovery Algorithms. 01/24/1997
2000	Internet Official Protocol Standards. 02/24/1997
1999	Request for Comments Summary RFC Numbers 1900-1999. 01/06/1997
1998	An Application of the BGP Community Attribute in Multi-home Routing. 08/30/1996
1997	BGP Communities Attribute. 08/30/1996
1996	A Mechanism for Prompt Notification of Zone Changes. 08/28/1996
1995	Incremental Zone Transfer in DNS. 08/28/1996
1994	PPP Challenge Handshake Authentication Protocol (CHAP). 08/30/1996
1993	PPP Gandalf FZA Compression Protocol. 08/30/1996
1992	The Nimrod Routing Architecture. 08/30/1996
1991	PGP Message Exchange Formats. 08/16/1996

RFC	Title
1990	The PPP Multilink Protocol (MP). 08/16/1996
1989	PPP Link Quality Monitoring. 08/16/1996
1988	Conditional Grant of Rights to Specific Hewlett-Packard Patents In Conjunction With the Internet Engineering Task Force's Internet-Standard Network Management Framework. 08/16/1996
1987	Ipsilon's General Switch Management Protocol Specification Version 1.1. 08/16/1996
1986	Experiments with a Simple File Transfer Protocol for Radio Links using Enhanced Trivial File Transfer Protocol (ETFTP). 08/16/1996
1985	SMTP Service Extension for Remote Message Queue Starting. 08/14/1996
1984	IAB and IESG Statement on Cryptographic Technology and the Internet. 08/20/1996
1983	Internet Users' Glossary. 08/16/1996
1982	Serial Number Arithmetic. 09/03/1996
1981	Path MTU Discovery for IP version 6. 08/14/1996
1980	A Proposed Extension to HTML: Client-Side Image Maps. 08/14/1996
1979	PPP Deflate Protocol. 08/09/1996
1978	PPP Predictor Compression Protocol. 08/28/1996
1977	PPP BSD Compression Protocol. 08/09/1996
1976	PPP for Data Compression in Data Circuit-Terminating Equipment (DCE). 08/14/1996
1975	PPP Magnalink Variable Resource Compression. 08/09/1996
1974	PPP Stac LZS Compression Protocol. 08/13/1996
1973	PPP in Frame Relay. 06/19/1996
1972	A Method for the Transmission of IPv6 Packets over Ethernet Networks. 08/16/1996
1971	IPv6 Stateless Address Autoconfiguration. 08/16/1996
1970	Neighbor Discovery for IP Version 6 (IPv6). 08/16/1996
1969	The PPP DES Encryption Protocol (DESE). 06/19/1996
1968	The PPP Encryption Control Protocol (ECP). 06/19/1996
1967	PPP LZS-DCP Compression Protocol (LZS-DCP). 08/13/1996
1966	BGP Route Reflection An alternative to full mesh IBGP. 06/19/1996

RFC	Title
1965	Autonomous System Confederations for BGP. 06/19/1996
1964	The Kerberos Version 5 GSS-API Mechanism. 06/19/1996
1963	PPP Serial Data Transport Protocol (SDTP). 08/14/1996
1962	The PPP Compression Control Protocol (CCP). 06/19/1996
1961	GSS-API Authentication Method for SOCKS Version 5 6/19/1996
1960	A String Representation of LDAP Search Filters 6/19/1996
1959	An LDAP URL Format. 06/19/1996
1958	Architectural Principles of the Internet. 06/06/1996
1957	Some Observations on Implementations of the Post Office Protocol (POP3). 06/06/1996
1956	Registration in the MIL Domain. 06/06/1996
1955	New Scheme for Internet Routing and Addressing (ENCAPS) for IPN. 06/06/1996
1954	Transmission of Flow Labelled IPv4 on ATM Data Links Ipsilon Version 1.0. 05/22/1996
1953	Ipsilon Flow Management Protocol Specification for IPv4 Version 1.0. 05/23/1996
1952	GZIP file format specification version 4.3. 05/23/1996
1951	DEFLATE Compressed Data Format Specification V1.3. 05/23/1996
1950	ZLIB Compressed Data Format Specification V3.3. 05/23/1996
1949	Scalable Multicast Key Distribution. 05/17/1996
1948	Defending Against Sequence Number Attacks. 05/17/1996
1947	Greek Character Encoding for Electronic Mail Messages. 05/17/1996
1946	Native ATM Support for ST2+. 05/17/1996
1945	Hypertext Transfer Protocol- HTTP V1.0. 05/17/1996
1944	Benchmarking Methodology for Network Interconnect Devices. 05/17/1996
1943	Building an X.500 Directory Service in the U.S. 05/15/1996
1942	HTML Tables. 05/15/1996
1941	Frequently Asked Questions for Schools 5/15/1996
1940	Source Demand Routing: Packet Format and Forwarding Specification (V1). 05/14/1996
1939	Post Office Protocol V3. 05/14/1996

RFC	Title
1938	A One-Time Password System. 05/14/1996
1937	Local/Remote Forwarding Decision in Switched Data Link Subnetworks. 05/08/1996
1936	Implementing the Internet Checksum in Hardware. 04/10/1996
1935	What is the Internet, Anyway? 04/10/1996
1934	Ascend's Multilink Protocol Plus (MP+). 04/08/1996
1933	Transition Mechanisms for IPv6 Hosts and Routers 4/08/1996
1932	IP over ATM: A Framework Document. 04/08/1996
1931	Dynamic RARP Extensions and Administrative Support for Automatic Network Address Allocation. 04/03/1996
1930	Guidelines for creation, selection, and registration of an Autonomous System (AS). 04/03/1996
1929	Username/Password Authentication for SOCKS V5. 04/03/1996
1928	SOCKS Protocol V5. 04/03/1996
1927	Suggested Additional MIME Types for Associating Documents. 04/01/1996
1926	An Experimental Encapsulation of IP Datagrams on Top of ATM. 04/01/1996
1925	The Twelve Networking Truths. 04/01/1996
1924	A Compact Representation of IPv6 Addresses. 04/01/1996
1923	RIPv1 Applicability Statement for Historic Status. 03/25/1996
1922	Chinese Character Encoding for Internet Messages. 03/26/1996
1921	TNVIP protocol. 03/25/1996
1920	Internet Official Protocol Standards 3/22/1996
1919	Classical versus Transparent IP Proxies. 03/28/1996
1918	Address Allocation for Private Internets. 02/29/1996
1917	An Appeal to the Internet Community to Return Unused IP Networks (Prefixes) to the IANA. 02/29/1996
1916	Enterprise Renumbering: Experience and Information Solicitation. 02/28/1996
1915	Variance for The PPP Connection Control Protocol and The PPP Encryption Control Protocol. 02/28/1996
1914	How to interact with a Whois++ mesh. 02/28/1996

RFC	Title
1913	Architecture of the Whois++ Index Service. 02/28/1996
1912	Common DNS Operational and Configuration Errors 2/28/1996
1911	Voice Profile for Internet Mail. 02/19/1996
1910	User-based Security Model for SNMPv2. 02/28/1996
1909	An Administrative Infrastructure for SNMPv2. 02/28/1996
1908	Coexistence between V1 & V2 of the Internet-standard Network Management Framework. 01/22/1996
1907	Management Information Base for V2 of the Simple Network Management Protocol (SNMPv2). 01/22/1996
1906	Transport Mappings for V2 of the Simple Network Management Protocol (SNMPv2). 01/22/1996
1905	Protocol Operations for V2 of the Simple Network Management Protocol (SNMPv2). 01/22/1996
1904	Conformance Statements for V2 of the Simple Network Management Protocol (SNMPv2). 01/22/1996
1903	Textual Conventions for V2 of the Simple Network Management Protocol (SNMPv2). 01/22/1996
1902	Structure of Management Information for Version 2 of the Simple Network Management Protocol (SNMPv2). 01/22/1996
1901	Introduction to Community-based SNMPv2. 01/22/1996
1900	Renumbering Needs Work. 02/28/1996
1899	RFC Summary Numbers 1800-1899 1/06/1997
1898	CyberCash Credit Card Protocol Version 0.8. 02/19/1996
1897	IPv6 Testing Address Allocation. 01/25/1996
1896	The text/enriched MIME Content-type. 02/19/1996
1895	The Application/CALS-1840 Content-type. 02/15/1996
1894	An Extensible Message Format for Delivery Status Notifications. 01/15/1996
1893	Enhanced Mail System Status Codes. 01/15/1996
1892	The Multipart/Report Content Type for the Reporting of Mail System Administrative Messages. 01/15/1996
1891	SMTP Service Extension for Delivery Status Notifications. 01/15/1996

RFC	Title
1890	RTP Profile for Audio and Video Conferences with Minimal Control. 01/25/1996
1889	RTP: A Transport Protocol for Real-Time Applications. 01/25/1996
1888	OSI NSAPs and IPv6. 08/16/1996
1887	An Architecture for IPv6 Unicast Address Allocation. 01/04/1996
1886	DNS Extensions to support IP V6. 01/04/1996
1885	Internet Control Message Protocol (ICMPv6) for the Internet Protocol Version 6 (IPv6). 01/04/1996
1884	IP V6 Addressing Architecture. 01/04/1996
1883	Internet Protocol, Version 6 (IPv6) Specification. 01/04/1996
1882	The 12 Days of Technology Before Christmas. 12/26/1995
1881	IPv6 Address Allocation Management. 12/26/1995
1880	Internet Official Protocol Standards. 11/29/1995
1879	Class A Subnet Experiment Results and Recommendations. 01/15/1996
1878	Variable Length Subnet Table For IPv4. 12/26/1995
1877	PPP Internet Protocol Control Protocol Extensions for Name Server Addresses. 12/26/1995
1876	A Means for Expressing Location Information in the Domain Name System. 01/15/1996
1875	UNINETT PCA Policy Statements. 12/26/1995
1874	SGML Media Types. 12/26/1995
1873	Message/External-Body Content-ID Access Type. 12/26/1995
1872	The MIME Multipart/Related Content-type. 12/26/1995
1871	Addendum to RFC 1602—Variance Procedure. 11/29/1995
1870	SMTP Service Extension for Message Size Declaration. 11/06/1995
1869	SMTP Service Extensions. 11/06/1995
1868	ARP Extension—UNARP. 11/06/1995
1867	Form-based File Upload in HTML. 11/07/1995
1866	Hypertext Markup Language—2.0. 11/03/1995
1865	EDI Meets the Internet: Frequently Asked Questions about Electronic Data Interchange (EDI) on the Internet 1/4/1996

RFC	Title
1864	The Content-MD5 Header Field. 10/24/1995
1863	A BGP/IDRP Route Server alternative to a full mesh routing. 10/20/1995
1862	Report of the IAB Workshop on Internet Information infrastructure 1994. 11/03/1995
1861	Simple Network Paging Protocol—V3—Two-Way-Enhanced. 10/19/1995
1860	Variable Length Subnet Table For IPv4. 10/20/1995
1859	ISO Transport Class 2 Non-use of Explicit Flow Control over TCP RFC1006 extension. 10/20/1995
1858	Security Considerations for IP Fragment Filtering. 10/25/1995
1857	A Model for Common Operational Statistics. 10/20/1995
1856	The Opstat Client-Server Model for Statistics Retrieval. 10/20/1995
1855	Netiquette Guidelines. 10/20/1995
1854	SMTP Service Extension for Command Pipelining. 10/04/1995
1853	IP in IP Tunneling. 10/04/1995
1852	IP Authentication using Keyed SHA. 10/02/1995
1851	The ESP Triple DES-CBC Transform. 10/02/1995
1850	OSPF V2 Management Information Base. 11/03/1995
1848	MIME Object Security Services. 10/03/1995
1847	Security Multiparts for MIME: Multipart/Signed and Multipart/Encrypted. 10/03/1995
1846	SMTP 521 reply code. 10/02/1995
1845	SMTP Service Extension for Checkpoint/Restart. 10/02/1995
1844	Multimedia E-mail (MIME) User Agent checklist 8/24/1995
1843	HZ—A Data Format for Exchanging Files of Arbitrarily Mixed Chinese and ASCII characters. 08/24/1995
1842	ASCII Printable Characters-Based Chinese Character Encoding for Internet Messages. 08/24/1995
1841	PPP Network Control Protocol for LAN Extension 9/29/1995
1838	Use of the X.500 Directory to support mapping between X.400 and RFC 822 Addresses. 08/22/1995
1837	Representing Tables and Subtrees in the X.500 Directory. 08/22/1995

RFC	Title
1836	Representing the O/R Address hierarchy in the X.500 Directory Information Tree. 08/22/1995
1835	Architecture of the WHOIS++ service. 08/16/1995
1834	Whois and Network Information Lookup Service Whois++. 08/16/1995
1833	Binding Protocols for ONC RPC V2. 08/09/1995
1832	XDR: External Data Representation Standard. 08/09/1995
1831	RPC: Remote Procedure Call Protocol Specification V2. 08/09/1995
1830	SMTP Service Extensions for Transmission of Large and Binary MIME Messages. 08/16/1995
1829	The ESP DES-CBC Transform. 08/09/1995
1828	IP Authentication using Keyed MD5. 08/09/1995
1827	IP Encapsulating Security Payload (ESP). 08/09/1995
1826	IP Authentication Header. 08/09/1995
1825	Security Architecture for the Internet Protocol 8/9/1995
1824	The Exponential Security System TESS: An Identity-Based Cryptographic Protocol for Authenticated Key-Exchange (E.I.S.S.-Report 1995/4). 08/11/1995
1823	The LDAP Application Program Interface. 08/09/1995
1822	A Grant of Rights to Use a Specific IBM patent with Photuris. 08/14/1995
1821	Integration of Real-time Services in an IP-ATM Network Architecture. 08/11/1995
1820	Multimedia E-mail (MIME) User Agent Checklist. 08/22/1995
1819	Internet Stream Protocol V2 (ST2) Protocol Specification—Version ST2+. 08/11/1995
1818	Best Current Practices. 08/04/1995
1817	CIDR and Classful Routing. 08/04/1995
1816	U.S. Government Internet Domain Names. 08/03/1995
1815	Character Sets ISO-10646 and ISO-10646-J-1. 08/01/1995
1814	Unique Addresses are Good. 06/22/1995
1813	NFS V3 Protocol Specification. 06/21/1995
1812	Requirements for IP V4 Routers. 06/22/1995
1811	U.S. Government Internet Domain Names. 06/21/1995

RFC	Title
1810	Report on MD5 Performance. 06/21/1995
1809	Using the Flow Label Field in IPv6. 06/14/1995
1808	Relative Uniform Resource Locators. 06/14/1995
1807	A Format for Bibliographic Records. 06/21/1995
1806	Communicating Presentation Information in Internet Messages: The Content-Disposition Header. 06/07/1995
1805	Location-Independent Data/Software Integrity Protocol. 06/07/1995
1804	Schema Publishing in X.500 Directory. 06/09/1995
1803	Recommendations for an X.500 Production Directory Service. 06/07/1995
1802	Introducing Project Long Bud: Internet Pilot Project for the Deployment of X.500 Directory Information in Support of X.400 Routing 6/12/1995
1801	MHS use of the X.500 Directory to support MHS Routing. 06/09/1995
1800	Internet Official Protocol Standards. 07/11/1995
1799	Request for Comments Summary RFC Numbers 1700-1799. 01/06/1997
1798	Connection-less Lightweight Directory Access Protocol. 06/07/1995
1797	Class A Subnet Experiment. 04/25/1995
1796	Not All RFCs are Standards. 04/25/1995
1795	Data Link Switching: Switch-to-Switch Protocol AIW DLSw RIG: DLSw Closed Pages, DLSw Standard V1. 04/25/1995
1794	DNS Support for Load Balancing. 04/20/1995
1793	Extending OSPF to Support Demand Circuits. 04/19/1995
1792	TCP/IPX Connection MIB Specification. 04/18/1995
1791	TCP and UDP over IPX Networks with Fixed Path MTU 4/18/1995
1790	An Agreement between the Internet Society and Sun Microsystems, Inc. in the Matter of ONC RPC and XDR Protocols 4/17/1995
1789	INETPhone: Telephone Services and Servers on Internet. 04/17/1995
1788	ICMP Domain Name Messages. 04/14/1995
1787	Routing in a Multi-provider Internet. 04/14/1995
1786	Representation of IP Routing Policies in a Routing Registry (ripe-81++). 03/28/1995
1785	TFTP Option Negotiation Analysis 3/28/1995

RFC	Title
1784	TFTP Timeout Interval and Transfer Size Options 3/28/1995
1783	TFTP Blocksize Option. 03/28/1995
1782	TFTP Option Extension. 03/28/1995
1781	Using the OSI Directory to Achieve User Friendly Naming. 03/28/1995
1780	Internet Official Protocol Standards. 03/28/1995
1779	A String Representation of Distinguished Names. 03/28/1995
1778	The String Representation of Standard Attribute Syntaxes. 03/28/1995
1777	Lightweight Directory Access Protocol. 03/28/1995
1776	The Address is the Message. 04/01/1995
1775	To Be "On" the Internet. 03/17/1995
1774	BGP-4 Protocol Analysis. 03/21/1995
1773	Experience with the BGP-4 protocol. 03/21/1995
1772	Application of the Border Gateway Protocol in the Internet. 03/21/1995
1771	A Border Gateway Protocol 4 (BGP-4). 03/21/1995
1770	IPv4 Option for Sender Directed Multi-Destination Delivery. 03/28/1995
1769	Simple Network Time Protocol (SNTP). 03/17/1995
1768	Host Group Extensions for CLNP Multicasting. 03/03/1995
1767	MIME Encapsulation of EDI Objects. 03/02/1995
1766	Tags for the Identification of Languages. 03/02/1995
1765	OSPF Database Overflow. 03/02/1995
1764	The PPP XNS IDP Control Protocol (XNSCP). 03/01/1995
1763	The PPP Banyan Vines Control Protocol (BVCP). 03/01/1995
1762	The PPP DECnet Phase IV Control Protocol (DNCP). 03/1/1995
1761	Snoop V2 Packet Capture File Format. 02/09/1995
1760	The S/KEY One-Time Password System. 02/15/1995
1759	Printer MIB. 03/28/1995
1758	NADF Standing Documents: A Brief Overview. 02/09/1995
1757	Remote Network Monitoring Management Information Base. 02/10/1995

Appendix B

RFC	Title
1756	REMOTE WRITE PROTOCOL—V1. 01/19/1995
1755	ATM Signaling Support for IP over ATM. 02/17/1995
1754	IP over ATM Working Group's Recommendations for the ATM Forum's Multiprotocol BOF V1. 01/19/1995
1753	IPng Technical Requirements of the Nimrod Routing and Addressing Architecture. 01/05/1995
1752	The Recommendation for the IP Next Generation Protocol. 01/18/1995
1751	A Convention for Human-Readable 128-bit Keys. 12/29/1994
1750	Randomness Recommendations for Security. 12/29/1994
1749	IEEE 802.5 Station Source Routing MIB using SMIv2. 12/29/1994
1748	IEEE 802.5 MIB using SMIv2. 12/29/1994
1747	Definitions of Managed Objects for SNA Data Link Control: SDLC. 01/11/1995
1746	Ways to Define User Expectations. 12/30/1994
1745	BGP4/IDRP for IP-OSPF Interaction. 12/27/1994
1744	Observations on the Management of the Internet Address Space. 12/23/1994
1743	IEEE 802.5 MIB using SMIv2. 12/27/1994
1742	AppleTalk Management Information BaseII. 01/05/1995
1741	MIME Content Type for BinHex Encoded Files. 12/22/1994
1740	MIME Encapsulation of Macintosh files—MacMIME. 12/22/1994
1739	A Primer On Internet and TCP/IP Tools. 12/22/1994
1738	Uniform Resource Locators (URL). 12/20/1994
1737	Functional Requirements for Uniform Resource Names. 12/20/1994
1736	Functional Requirements for Internet Resource Locators. 02/09/1995
1735	NBMA Address Resolution Protocol (NARP). 12/15/1994
1734	POP3 Authentication command. 12/20/1994
1733	Distributed Electronic Mail Models In IMAP4. 12/20/1994
1732	IMAP4 Compatibility with IMAP2 and IMAP2BIS. 12/20/1994
1731	IMAP4 Authentication mechanisms. 12/20/1994
1730	Internet Message Access Protocol—V4. 12/20/1994

RFC	Title
1729	Using the Z39.50 Information Retrieval Protocol in the Internet Environment. 12/16/1994
1728	Resource Transponders. 12/16/1994
1727	A Vision of an Integrated Internet Information Service. 12/16/1994
1726	Technical Criteria for Choosing IP:The Next Generation (IPng). 12/20/1994
1725	Post Office Protocol—V3. 11/23/1994
1724	RIP V2 MIB Extension. 11/15/1994
1723	RIP V2 Carrying Additional Information. 11/15/1994
1722	RIP V2 Protocol Applicability Statement. 11/15/1994
1721	RIP V2 Protocol Analysis. 11/15/1994
1720	Internet Official Protocol Standards. 11/23/1994
1719	A Direction for IPng. 12/16/1994
1718	The Tao of IETF—A Guide for New Attendees of the Internet Engineering Task Force. 11/23/1994
1717	The PPP Multilink Protocol (MP). 11/21/1994
1716	Towards Requirements for IP Routers. 11/04/1994
1715	The H Ratio for Address Assignment Efficiency. 11/03/1994
1714	Referral Whois Protocol (RWhois). 12/15/1994
1713	Tools for DNS debugging. 11/03/1994
1712	DNS Encoding of Geographical Location. 11/01/1994
1711	Classifications in E-mail Routing. 10/26/1994
1710	Simple Internet Protocol Plus White Paper. 10/26/1994
1709	K-12 Internetworking Guidelines. 12/23/1994
1708	NTP PICS PROFORMA For the Network Time Protocol V3. 10/26/1994
1707	CATNIP: Common Architecture for the Internet. 11/02/1994
1706	DNS NSAP Resource Records. 10/26/1994
1705	Six Virtual Inches to the Left: The Problem with IPng. 10/26/1994
1704	On Internet Authentication. 10/26/1994
1703	Principles of Operation for the TPC.INT Subdomain: Radio Paging—Technical Procedures. 10/26/1994

RFC	Title
1702	Generic Routing Encapsulation over IPv4 networks. 10/21/1994
1701	Generic Routing Encapsulation (GRE). 10/21/1994
1700	ASSIGNED NUMBERS. 10/20/1994
1699	Request for Comments Summary RFC Numbers 1600-1699. 01/06/1997
1698	Octet Sequences for Upper-Layer OSI to Support Basic Communications Applications. 10/26/1994
1697	Relational Database Management System (RDBMS) Management Information Base (MIB) using SMIv2. 08/23/1994
1696	Modem Management Information Base (MIB) using SMIv2. 08/25/1994
1695	Definitions of Managed Objects for ATM Management Version 8.0 using SMIv2. 08/25/1994
1694	Definitions of Managed Objects for SMDS Interfaces using SMIv2. 08/23/1994
1693	An Extension to TCP: Partial Order Service. 11/01/1994
1692	Transport Multiplexing Protocol (TMux). 08/17/1994
1691	The Document Architecture for the Cornell Digital Library. 08/17/1994
1690	Introducing the Internet Engineering and Planning Group (IEPG). 08/17/1994
1689	A Status Report on Networked Information Retrieval: Tools and Groups. 08/17/1994
1688	IPng Mobility Considerations. 08/11/1994
1687	A Large Corporate User's View of IPng. 08/11/1994
1686	IPng Requirements: A Cable Television Industry Viewpoint. 08/11/1994
1685	Writing X.400 O/R Names. 08/11/1994
1684	Introduction to White Pages services based on X.500. 08/11/1994
1683	Multiprotocol Interoperability In IPng. 08/11/1994
1682	IPng BSD Host Implementation Analysis. 08/11/1994
1681	On Many Addresses per Host. 08/08/1994
1680	IPng Support for ATM Services. 08/08/1994
1679	HPN Working Group Input to the IPng Requirements Solicitation. 08/08/1994
1678	IPng Requirements of Large Corporate Networks. 08/08/1994
1677	Tactical Radio Frequency Communication Requirements for IPng. 08/08/1994

RFC	Title
1676	INFN Requirements for an IPng. 08/11/1994
1675	Security Concerns for IPng. 08/08/1994
1674	A Cellular Industry View of IPng. 08/08/1994
1673	Electric Power Research Institute Comments on IPng. 08/08/1994
1672	Accounting Requirements for IPng. 08/08/1994
1671	IPng White Paper on Transition and Other Considerations. 08/08/1994
1670	Input to IPng Engineering Considerations. 08/08/1994
1669	Market Viability as a IPng Criteria. 08/08/1994
1668	Unified Routing Requirements for IPng. 08/08/1994
1667	Modeling and Simulation Requirements for IPng. 08/08/1994
1666	Definitions of Managed Objects for SNA NAUs using SMIv2. 08/11/1994
1665	Definitions of Managed Objects for SNA NAUs using SMIv2. 07/22/1994
1664	Using the Internet DNS to Distribute RFC1327 Mail Address Mapping Tables. 08/11/1994
1663	PPP Reliable Transmission. 07/21/1994
1662	PPP in HDLC-like Framing. 07/21/1994
1661	The Point-to-Point Protocol (PPP). 07/21/1994
1660	Definitions of Managed Objects for Parallel-printer-like Hardware Devices using SMIv2. 07/20/1994
1659	Definitions of Managed Objects for RS-232-like Hardware Devices using SMIv2. 07/20/1994
1658	Definitions of Managed Objects for Character Stream Devices using SMIv2. 07/20/1994
1657	Definitions of Managed Objects for the Fourth Version of the Border Gateway Protocol (BGP-4) using SMIv2 7/21/1994
1656	BGP-4 Protocol Document Roadmap and Implementation Experience. 07/21/1994
1655	Application of the Border Gateway Protocol in the Internet. 07/21/1994
1654	A Border Gateway Protocol 4 (BGP-4). 07/21/1994
1653	SMTP Service Extension for Message Size Declaration. 07/18/1994
1652	SMTP Service Extension for 8-bit MIME transport. 07/18/1994
1651	SMTP Service Extensions. 07/18/1994

RFC	Title
1650	Definitions of Managed Objects for the Ethernet-like Interface Types using SMIv2. 08/23/1994
1649	Operational Requirements for X.400 Management Domains in the GO-MHS Community. 07/18/1994
1648	Postmaster Convention for X.400 Operations. 07/18/1994
1647	TN3270 Enhancements. 07/15/1994
1646	TN3270 Extensions for LUname and Printer Selection. 07/14/1994
1645	Simple Network Paging Protocol—V2. 07/14/1994
1644	T/TCP—TCP Extensions for Transactions Functional Specification. 07/13/1994
1643	Definitions of Managed Objects for the Ethernet-like Interface Types. 07/13/1994
1642	UTF-7: A Mail-Safe Transformation Format of Unicode. 07/13/1994
1641	Using Unicode with MIME. 07/13/1994
1640	The Process for Organization of Internet Standards Working Group (POISED). 06/09/1994
1639	FTP Operation Over Big Address Records (FOOBAR) 6/09/1994
1638	PPP Bridging Control Protocol (BCP). 06/09/1994
1637	DNS NSAP Resource Records. 06/09/1994
1636	Report of IAB Workshop on Security in the Internet Architecture, February 8-10, 1994. 06/09/1994
1635	How to Use Anonymous FTP. 05/25/1994
1634	Novell IPX over Various WAN Media (IPXWAN). 05/24/1994
1633	Integrated Services in the Internet Architecture: An Overview. 06/09/1994
1632	A Revised Catalog of Available X.500 Implementations. 05/20/1994
1631	The IP Network Address Translator (Nat). 05/20/1994
1630	Universal Resource Identifiers in WWW: A Unifying Syntax for the Expression of Names and Addresses of Objects on the Network as used in the World-Wide Web. 06/09/1994
1629	Guidelines for OSI NSAP Allocation in the Internet. 05/19/1994
1628	UPS Management Information Base. 05/19/1994
1627	Network 10 Considered Harmful (Some Practices Shouldn't be Codified). 07/01/1994
1626	Default IP MTU for use over ATM AAL5. 05/19/1994
1625	WAIS over Z39.50-1988. 06/09/1994

RFC	Title
1624	Computation of the Internet Checksum via Incremental Update. 05/20/1994
1623	Definitions of Managed Objects for the Ethernet-like Interface Types. 05/24/1994
1622	Pip Header Processing. 05/20/1994
1621	Pip Near-term Architecture. 05/20/1994
1620	Internet Architecture Extensions for Shared Media. 05/20/1994
1619	PPP over SONET/SDH. 05/13/1994
1618	PPP over ISDN. 05/13/1994
1617	Naming and Structuring Guidelines for X.500 Directory Pilots. 05/20/1994
1616	X.400(1988) for the Academic and Research Community in Europe. 05/19/1994
1615	Migrating from X.400(84) to X.400(88). 05/19/1994
1614	Network Access to Multimedia Information. 05/20/1994
1613	Cisco Systems X.25 over TCP (XOT). 05/13/1994
1612	DNS Resolver MIB Extensions. 05/17/1994
1611	DNS Server MIB Extensions. 05/17/1994
1610	Internet Official Protocol Standards. 07/08/1994
1609	Charting Networks in the X.500 Directory. 03/25/1994
1608	Representing IP Information in the X.500 Directory. 03/25/1994
1607	A View From the 21st Century. 04/01/1994
1606	A Historical Perspective On The Usage Of IP Version 9. 04/01/1994
1605	SONET to Sonnet Translation. 04/01/1994
1604	Definitions of Managed Objects for Frame Relay Service. 03/25/1994
1603	IETF Working Group Guidelines and procedures. 03/24/1994
1602	The Internet Standards Process—R2. 03/24/1994
1601	Charter of the Internet Architecture Board (IAB). 03/22/1994
1600	Internet Official Protocol Standards. 03/14/1994
1599	Request for Comments Summary RFC Numbers 1500—1599. 01/06/1997
1598	PPP in X.25. 03/17/1994
1597	Address Allocation for Private Internets. 03/17/1994

RFC	Title
1596	Definitions of Managed Objects for Frame Relay Service. 03/17/1994
1595	Definitions of Managed Objects for the SONET/SDH Interface Type. 03/11/1994
1594	FYI on Questions and Answers to Commonly asked New Internet User Questions. 03/11/1994
1593	SNA APPN Node MIB. 03/10/1994
1592	Simple Network Management Protocol Distributed Protocol Interface V2. 03/03/1994
1591	Domain Name System Structure and Delegation. 03/03/1994
1590	Media Type Registration Procedure. 03/02/1994
1589	A Kernel Model for Precision Timekeeping. 03/03/1994
1588	White Pages Meeting Reports. 02/25/1994
1587	The OSPF NSSA Option. 03/24/1994
1586	Guidelines for Running OSPF Over Frame Relay Networks. 03/24/1994
1585	MOSPF: Analysis and Experience. 03/24/1994
1584	Multicast Extensions to OSPF. 03/24/1994
1583	OSPF Version 2. 03/23/1994
1582	Extensions to RIP to Support Demand Circuits. 02/18/1994
1581	Protocol Analysis for Extensions to RIP to Support Demand Circuits. 02/18/1994
1580	Guide to Network Resource Tools. 03/22/1994
1579	Firewall-Friendly FTP. 02/18/1994
1578	FYI on Questions and Answers: Answers to Commonly Asked Primary and Secondary School Internet User Questions. 02/18/1994
1577	Classical IP and ARP over ATM. 01/20/1994
1576	TN3270 Current Practices. 01/20/1994
1575	An Echo Function for CLNP (ISO 8473). 02/18/1994
1574	Essential Tools for the OSI Internet. 02/18/1994
1573	Evolution of the Interfaces Group of MIB-II. 01/20/1994
1572	Telnet Environment Option. 01/14/1994
1571	Telnet Environment Option Interoperability Issues 01/14/1994

RFC	Title
1570	PPP LCP Extensions. 01/11/1994
1569	Principles of Operation for the TPC.INT Subdomain: Radio Paging—Technical Procedures. 01/07/1994
1568	Simple Network Paging Protocol—V1(b). 01/07/1994
1567	X.500 Directory Monitoring MIB. 01/11/1994
1566	Mail Monitoring MIB. 01/11/1994
1565	Network Services Monitoring MIB. 01/11/1994
1564	DSA Metrics (OSI-DS 34 (v3)). 01/14/1994
1563	The text/enriched MIME Content-type. 01/10/1994
1562	Naming Guidelines for the AARNet X.500 Directory Service. December 1993
1561	Use Of ISO CLNP in TUBA Environments. December 1993
1560	The MultiProtocol Internet. December 1993
1559	DECnet Phase IV MIB Extensions. (Obsoletes RFC 1289) December 1993
1558	A String Representation of LDAP Search Filters. December 1993
1557	Korean Character Encoding for Internet Messages. December 1993
1556	Handling of Bidirectional Texts in MIME. December 1993
1555	Hebrew Character Encoding for Internet Messages. December 1993
1554	ISO-2022-JP-2: Multilingual Extensions of ISO-2022-JP. December 1993
1553	Compressing IPX Headers Over WAN Media (CIPX). December 1993
1552	The PPP Internetwork Packet Exchange Control Protocol. December 1993
1551	Novell IPX Over Various WAN Media (IPXWAN). (Obsoletes RFC 1362) December 1993
1550	IP: Next Generation (IPng) White Paper Solicitation. December 1993
1549	PPP in HDLC Framing. December 1993
1548	The Point-to-Point Protocol (PPP). (Obsoletes RFC 1331). December 1993
1547	Requirements for an Internet Standard Point-to-Point Protocol. December 1993
1546	Host Anycasting Service. November 1993
1545	FTP Operation Over Big Address Records (FOOBAR). November 1993
1544	The Content-MD5 Header Field. November 1993

RFC	Title
1543	Instructions to RFC Authors. (Obsoletes RFC 1111) October 1993
1542	Clarifications and Extensions for the Bootstrap Protocol. (Obsoletes RFC 1532) October 1993
1541	Dynamic Host Configuration Protocol. (Obsoletes RFC 1531) October 1993
1540	Internet Official Protocol Standards. (Obsoletes RFC 1500) October 1993
1539	The Tao of IETF: A Guide for New Attendees of the Internet Engineering Task Force. (Obsoletes RFC 1391) October 1993
1538	Advanced SNA/IP: A Simple SNA Transport Protocol. October 1993
1537	Common DNS Data File Configuration Errors. October 1993
1536	Common DNS Implementation Errors and Suggested Fixes. October 1993
1535	A Security Problem and Proposed Correction With Widely Deployed DNS Software. October 1993
1534	Interoperation Between DHCP and BOOTP. October 1993
1533	DHCP Options and BootP Vendor Extensions. (Obsoletes RFC 1497) October 1993
1532	Clarifications and Extensions for the Bootstrap Protocol. (Obsoleted RFC 1542) October 1993
1531	Dynamic Host Configuration Protocol. (Obsoleted RFC 1541) October 1993
1530	Principles of Operation for the TPC.INT Subdomain: General Principles and Policy. October 1993
1529	Principles of Operation for the TPC.INT Subdomain: Remote Printing—Administrative Policies. (Obsoletes RFC 1486) October 1993
1528	Principles of Operation for the TPC.INT Subdomain: Remote Printing—Technical Procedures. (Obsoletes RFC 1486) October 1993
1527	What Should We Plan Given the Dilemma of the Network? September 1993
1526	Assignment of System Identifiers for TUBA/CLNP Hosts. September 1993
1525	Definitions of a Managed Objects for Source Routing Bridges. (Obsoletes RFC 1286) September 1993
1524	A User Agent Configuration Mechanism for Multimedia Mail Format Information. September 1993
1523	The text/enriched MIME Content-type. September 1993
1522	MIME (Multipurpose Internet Mail Extensions) Part Two: Message Header Extensions for Non-ASCII Text. (Obsoletes 1342) September 1993

RFC	Title
1521	MIME (Multipurpose Internet Mail Extensions) Part One: Mechanisms for Specifying and Describing the Format of Internet Message Bodies. (Obsoletes RFC 1341) September 1993
1520	Exchanging Routing Information Across Provider Boundaries in the CIDR Environment. September 1993
1519	Classless Inter-Domain Routing (CIDR): an Address Assignment and Aggregation Strategy. (Obsoletes RFC 1338) September 1993
1518	An Architecture for IP Address Allocation with CIDR. September 1993
1517	Applicability Statement for the Implementation of Classless Inter-Domain Routing (CIDR). September 1993
1516	Definitions of Managed Objects for IEEE 802.3 Repeater Devices. (Obsoletes RFC 1368) September 1993
1515	Definitions of Managed Objects for IEEE 802.3 Medium Attachment Units (MAUs). September 1993
1514	Host Resources MIB. September 1993
1513	Token Ring Extensions to the Remote Network Monitoring MIB. (Obsoletes RFC 1271) September 1993
1512	FDDI Management Information Base. (Updates RFC 1285) September 1993
1511	Common Authentication Technology Overview. September 1993
1510	The Kerberos Network Authentication Service (V5). September 1993
1509	Generic Security Service API: C-bindings. September 1993
1508	Generic Security Service Application Program Interface. September 1993
1507	DASS Distributed Authentication Security Service. September 1993
1506	A Tutorial on Gatewaying between X.400 and Internet Mail. August 1993
1505	Encoding Header Field for Internet Messages. (Obsoletes RFC 1154) August 1993
1504	AppleTalk Update-Based Routing Protocol: Enhanced AppleTalk Routing. August 1993
1503	Algorithms for Automating Administration in SNMPv2 Managers. August 1993
1502	X.400 Use of Extended Character Sets. August 1993
1501	OS/2 User Group. August 1993
1500	Internet Official Protocol Standards. (Obsoletes RFC 1410; Obsoleted RFC 1540) August 1993

RFC	Title
1499	Not yet issued
1498	On the Naming and Binding of Network Destinations. August 1993
1497	BOOTP Vendor Information Extensions. (Obsoletes RFC 1395; Obsoleted by RFC 1533; Updates RFC 951) August 1993
1496	Rules for Downgrading Messages from X.400/88 to X.400/84 When MIME Content-Types are Present in the Messages Updates RFC 1328. August 1993
1495	Mapping between X.400 and RFC-822 Message Bodies. (Obsoletes RFC 1327) August 1993
1494	Equivalences between 1988 X.400 and RFC-822 Message Bodies. August 1993
1493	Definitions of Managed Objects for Bridges. (Obsoletes RFC 1286) July 1993
1492	An Access Control Protocol. Sometimes Called TACACS. July 1993
1491	A Survey of Advanced Usages of X.500. July 1993
1490	Multiprotocol Interconnect over Frame Relay. (Obsoletes RFC 1294) July 1993
1489	Registration of a Cyrillic Character Set. July 1993
1488	The X.500 String Representation of Standard Attribute Syntaxes. July 1993
1487	X.500 Lightweight Directory Access Protocol. July 1993
1486	An Experiment in Remote Printing. (Obsoleted by RFC 1528, RFC 1529) July 1993
1485	A String Representation of Distinguished Names (OSI-DS 23 v5). July 1993
1484	Using the OSI Directory to achieve User Friendly Naming (OSI-DS 24 v1.2). July 1993
1483	Multiprotocol Encapsulation over ATM Adaptation Layer 5. July 1993
1482	Aggregation support in the NSFNET Policy-Based Routing Database. June 1993
1481	IAB Recommendation for an Intermediate Strategy to Address the Issue of Scaling. July 1993
1480	The US Domain. (Obsoletes RFC 1386) June 1993
1479	Inter-Domain Policy Routing Protocol Specification: Version 1. July 1993
1478	An Architecture for Inter-Domain Policy Routing. June 1993
1477	IDPR as a Proposed Standard. July 1993
1476	RAP: Internet Route Access Protocol. June 1993
1475	TP/IX: The Next Internet. June 1993

RFC	Title
1474	The Definitions of Managed Objects for the Bridge Network Control Protocol of the Point-to-Point Protocol. June 1993
1473	The Definitions of Managed Objects for the IP Network Control Protocol of the Point-to-Point Protocol. June 1993
1472	The Definitions of Managed Objects for the Security Protocols of the Point-to-Point Protocol. June 1993
1471	The Definitions of Managed Objects for the Link Control Protocol of the Point-to-Point Protocol. June 1993
1470	FYI on a Network Management Tool Catalog: Tools for Monitoring and Debugging TCP/IP Internets and Interconnected Devices. (Obsoletes RFC 1147) June 1993
1469	IP Multicast over Token-Ring Local Area Networks. June 1993
1468	Japanese Character Encoding for Internet Messages. June 1993
1467	Status of CIDR Deployment in the Internet. (Obsoletes RFC 1367) August 1993
1466	Guidelines for Management of IP Address Space. (Obsoletes 1366) May 1993
1465	Routing Coordination for X.400 MHS Service within a Multi Protocol/Multi Network Environment Table Format V3 for Static Routing. May 1993
1464	Using the Domain Name System To Store Arbitrary String Attributes. May 1993
1463	FYI on Introducing the Internet- A Short Bibliography of Introductory Internetworking Readings or the Network Novice (FYI 19). May 1993
1462	FYI on "What is the Internet?" (Also FYI 20) May 1993
1461	SNMP MIB extension for Multiprotocol Interconnect over X.25. May 1993
1460	Post Office Protocol V3. (Obsoletes RFC 1225) May 1993
1459	Internet Relay Chat Protocol. May 1993
1458	Requirements for Multicast Protocols. May 1993
1457	Security Label Framework for the Internet. May 1993
1456	Conventions for Encoding the Vietnamese Language VISCII: Vietnamese Standard code for Information Interchange VIQR: Vietnamese Quoted-Readable Specification Revision 1.1. May 1993
1455	Physical Link Security Type of Service. May 1993
1454	Comparison of Proposals for Next Version of IP. May 1993
1453	A Comment on Packet Video Remote Conferencing and the Transport/Network Layers. April 1993

RFC	Title
1452	Coexistence between v1 and v2 of the Internet-standard Network Management Framework 1993
1451	Manager-to-Manager Management Information Base. April 1993
1450	Management Information Base for v2 of the Simple Network Management protocol (SNMPv2). April 1993
1449	Transport Mappings for v2 of the Simple Network Management Protocol (SNMPv2). April 1993
1448	Protocol Operations for v2 of the Simple Network Management Protocol (SNMPv2). April 1993
1447	Party MIB for v2 of the Simple Network Management Protocol (SNMPv2). April 1993
1446	Security Protocols for v2 of the Simple Network Management Protocol (SNMPv2). April 1993
1445	Administrative Model for version 2 of the Simple Network Management Protocol SNMPv2). April 1993
1444	Conformance Statements for v2 of the Simple Network Management Protocol (SNMPv2). April 1993
1443	Textual Conventions for v2 of the Simple Network Management Protocol (SNMPv2). April 1993
1442	Structure of Management Information for v2 of the Simple Network Management Protocol (SNMPv2). April 1993
1441	Introduction to v2 of the Internet-standard Network Management Framework. April 1993
1440	SIFT/UFT: Sender-Initiated/Unsolicited File Transfer. July 1993
1439	The Uniqueness of Unique Identifiers. March 1993
1438	Internet Engineering Task Force Statements of Boredom (SOBs). April 1993
1437	The Extension of MIME Content-Types to a New Medium. April 1993
1436	The Internet Gopher Protocol (A distributed document search and retrieval protocol). March 1993
1435	IESG Advice from Experience with Path MTU Discovery. March 1993
1434	Data Link Switching: Switch-to-Switch Protocol. March 1993
1433	Directed ARP. March 1993
1432	Recent Internet Books. March 1993

RFC	Title
1431	DUA Metrics. February 1993
1430	A Strategic Plan for Deploying an Internet X.500 Directory Service. February 1993
1429	Listserv Distributed Protocol. February 1993
1428	Transition of InternetMail from Just-Send-8 to 8-bit SMTP/MIME. February 1993
1427	SMTP Service Extension for Message Size Declaration. February 1993
1426	SMTP Service Extension for 8-bit MIME transport. February 1993
1425	SMTP Service Extensions. February 1993
1424	Privacy Enhancement for Internet Electronic Mail: Part IV: Key Certification and Related Service. February 1993
1423	Privacy Enhancement for Internet Electronic Mail: Part III: Algorithms, Modes, and Identifiers. (Obsoletes RFC 1115) February 1993
1422	Privacy Enhancement for Internet Electronic Mail: Part II: Certificate-Based key Management. (Obsoletes RFC 1114) February 1993
1421	Privacy Enhancement for Internet Electronic Mail: Part I: Message Encryption and Authentication Procedures. (Obsoletes RFC 1113) February 1993
1420	SNMP over IPX. (Obsoletes 1298) March 1993
1419	SNMP over AppleTalk. March 1993
1418	SNMP over OSI. (Obsoletes 1161) March 1993
1417	The North American Directory Forum NADF Standing Documents: A Brief Overview. (Obsoletes RFC 1295, 1255, 1218) February 1993
1416	Telnet Authentication Option. (Obsoletes RFC 1409) February 1993
1415	FTP-FTAM Gateway Specification. January 1993
1414	Identification MIB. February 1993
1413	Identification Protocol. (Obsoletes RFC 931) February 1993
1412	Telnet Authentication: SPX. January 1993
1411	Telnet Authentication: Kerberos V4. January 1993
1410	IAB Official Protocol Standards. (Obsoletes RFC 1360, 1280, 1250, 1200, 1100, 1083, 1130, 1140; Obsoleted by RFC 1500) March 1993
1409	Telnet Authentication Option. (Obsoleted by RFC 1416) January 1993
1408	Telnet Environment Option. January 1993

RFC	Title
1407	Definitions of Managed Objects for the DS3/E3 Interface Type. (Obsoletes RFC 1233) January 1993
1406	Definitions of Managed Objects for the DS1 and E1 Interface Types. (Obsoletes RFC 1232) January 1993
1405	Mapping between X.400(1984/1988) and Mail-11 (DECnet mail). January 1993
1404	Model for Common Operational Statistics. January 1993
1403	BGP OSPF Interaction. (Obsoletes RFC 1364) January 1993
1402	There's Gold in them thar Networks! or Searching for Treasure in all the Wrong Places. (Obsoletes RFC 1290) January 1993
1401	Correspondence between the IAB and DISA on the use of DNS throughout the Internet. January 1993
1400	Transition and Modernization of the Internet Registration Service. March 1993
1399	Not yet issued.
1398	Definitions of Managed Objects for the Ethernet-like Interface Types. (Obsoletes RFC 1284) January 1993
1397	Default Route Advertisement In BGP2 And BGP3 Versions of The Border Gateway Protocol. January 1993
1396	The Process for Organization of Internet Standards—Working Group (POISED). January 1993
1395	BOOTP Vendor Information Extensions. (Obsoletes RFC 1084, 1048: Obsoleted by RFC 1497; Updates RFC 951) January 1993
1394	Relationship of Telex Answerback Codes to Internet Domains. January 1993
1393	Traceroute Using an IP Option. January 1993
1392	Internet Users' Glossary. (Also FYI 18). January 1993
1391	The Tao of IETF—A Guide for New Attendees of the Internet Engineering Task Force. (Obsoleted by RFC 1539) January 1993
1390	Transmission of IP and ARP over FDDI Networks. January 1993
1389	RIP Version 2 MIB Extension. January 1993
1388	RIP Version 2—Carrying Additional Information. (Updates RFC 1058) January 1993
1387	RIP Version 2 Protocol Analysis. January 1993
1386	The US Domain. (Obsoleted by RFC 1480) December 1992

RFC	Title
1385	EIP: The Extended Internet Protocol: A Framework for Maintaining Backward Compatibility. November 1992
1384	Naming Guidelines for Directory Pilots. January 1992
1383	An Experiment in DNS Based IP Routing. December 1992
1382	SNMP MIB Extension for the X.25 Packet Layer. November 1992
1381	SNMP MIB Extension for X.25 LAPB. November 1992
1380	IESG Deliberations on Routing and Addressing. November 1992
1379	Extending TCP for Transactions—Concepts. November 1992
1378	The PPP AppleTalk Control Protocol (ATCP). November 1992
1377	The PPP OSI Network Layer Control Protocol (OSINLCP). November 1992
1376	The PPP DECnet Phase IV Control Protocol (DNC). November 1992
1375	Suggestion for New Classes of IP Addresses. November 1992
1374	IP and ARP on HIPPI. October 1992
1373	PORTABLE DUAs. October 1992
1372	Telnet Remote Flow Control Option. (Obsoletes RFC 1080) October 1992
1371	Choosing a Common IGP for the IP Internet (The IESG's Recommendation to the IAB) October 1992
1370	Applicability Statement for OSPF. October 1992
1369	Implementation Notes and Experience for the Internet Ethernet MIB. October 1992
1368	Definitions of Managed Objects for IEEE 802.3 Repeater Devices. (Obsoleted by RFC 1516) October 1992
1367	Schedule for IP Address Space Management Guidelines. (Obsoleted by RFC 1467) October 1992
1366	Guidelines for Management of IP Address Space. (Obsoleted by RFC 1466) October 1992
1365	An Address Extension Proposal. September 1992
1364	BGP OSPF Interaction. (Obsoleted by RFC 1403) September 1992
1363	A Proposed Flow Specification. September 1992
1362	Novell IPX Over Various WAN Media (IPXWAN). (Obsoleted by RFC 1551) September 1992

RFC	Title
1361	Simple Network Time Protocol (SNTP). August 1992
1360	IAB Official Protocol Standards. (Obsoletes RFCs 1280, 1250, 1100, 1083, 1130, 1140,1200; Obsoleted by RFC 1410) September 1992
1359	Connecting to the Internet: What Connecting Institutions Should Anticipate. (Also FYI 16) August 1992
1358	Charter of the Internet Architecture Board (IAB). August 1992
1357	A Format for E-mailing Bibliographic Records. July 1992
1356	Multiprotocol interconnect on X.25 and ISDN in the Packet Mode. (Obsoletes RFC 877) August 1992
1355	Privacy and Accuracy Issues in Network Information Center Databases . August 1992
1354	IP Forwarding Table MIB. July 1992
1353	Definitions of Managed Objects for Administration of SNMP Parties. July 1992
1352	SNMP Security Protocols. July 1992
1351	SNMP Administrative Model. July 1992
1350	The TFTP Protocol (revision 2). (Obsoletes RFC 783) July 1992
1349	Type of Service in the Internet Protocol Suite. (Updates RFCs 1248, 1247, 1195, 1123, 1122, 1060, 791) July 1992
1348	DNS NSAP RRs. (Updates RFCs 1034, 1035) July 1992
1347	TCP and UDP with Bigger Addresses (TUBA); A Simple Proposal for Internet Addressing and Routing. June 1992
1346	Resource Allocation, Control, and Accounting for the Use of Network Resources. June 1992
1345	Character Mnemonics & Character Sets. June 1992
1344	Implications of MIME for Internet Mail Gateways. June 1992
1343	A User Agent Configuration Mechanism For Multimedia Mail Format Information. June 1992
1342	Representation of Non-ASCII Text in Internet Message Headers. (Obsoleted by 1522) June 1992
1341	MIME (Multipurpose Internet Mail Extensions) Mechanisms for Specifying and Describing the Format of Internet Message Bodies. (Obsoleted by RFC 1521) June 1992

RFC	Title
1340	Assigned Numbers. (Obsoletes RFCs 1060, 1010, 990, 960, 943, 923, 900, 870, 820, 790, 776, 770, 762, 758, 755, 750, 739, 604, 503, 433, 349, and IEN 127) July 1992.
1339	Remote Mail checking Protocol. June 1992
1338	Supernetting: an Address Assignment and Aggregation Strategy. (Obsoleted by RFC 1519) June 1992
1337	TIME-WAIT Assassination Hazards in TCP. May 1992
1336	Who's Who in the Internet: Biographies of IAB, IESG, and IRSG Members. (Obsoletes RFC 1251, FYI 9) May 1992
1335	A Two-Tier Address Structure for the Internet: A Solution to the problem of address Space Exhaustion. May 1992
1334	PPP Authentication Protocols. October 1992
1333	PPP Link Quality Monitoring. May 1992
1332	The PPP Internet Protocol Control Protocol (IPCP). (Obsoletes RFC 1172) May 1992
1331	The Point-to-Point Protocol (PPP) for the Transmission of Multi-protocol Datagrams over Point-to-Point Links. (Obsoletes RFC 1171, RFC 1172 Obsoleted by RFC 1548) May 1992
1330	Recommendations for the Phase I Deployment of OSI Directory Services (X.500) and OSI Message Handling Services (X.400) within the ESnet Community. May 1992
1329	Thoughts on Address Resolution for Dual MAC FDDI Networks. May 1992
1328	X.400 1988 to 1984 downgrading. (Updated by RFC 1496) May 1992
1327	Mapping between X.400(1988)/ISO 10021 and RFC 822. (Obsoletes RFC 987, 1026, 1138, 1148; Obsoleted by RFC 1495; Updates RFC 822) May 1992
1326	Mutual Encapsulation Considered Dangerous. May 1992
1325	FYI on Questions and Answers—Answers to Commonly asked New Internet-User Questions. (Obsoletes RFC 1206, FYI 4) May 1992
1324	A Discussion on Computer Network Conferencing. May 1992
1323	TCP Extensions for High Performance. (Obsoletes RFC 1072, 1185) May 1992
1322	A Unified Approach to Inter-Domain Routing. May 1992
1321	The MD5 Message-Digest Algorithm. April 1992
1320	The MD4 Message-Digest Algorithm. (Obsoletes RFC 1186) April 1992
1319	The MD2 Message-Digest Algorithm. April 1992

RFC	Title
1318	Definitions of Managed Objects for Parallel-printer-like Hardware Devices. April 1992
1317	Definitions of Managed Objects for RS-232-like Hardware Devices. April 1992
1316	Definitions of Managed Objects for Character Stream Devices. April 1992
1315	Management Information Base for Frame Relay DTEs. April 1992
1314	A File Format for the Exchange of Images in the Internet. April 1992
1313	Today's Programming for KRFC AM 1313 Internet Talk Radio. April 1992
1312	Message Send Protocol 2. (Obsoletes RFC 1159). April 1992
1311	Introduction to the STD Notes. March 1992
1310	The Internet Standards Process. March 1992
1309	Technical Overview of Directory Services Using the X.500 Protocol. (Also FYI 14) March 1992
1308	Executive Introduction to Directory Services Using the X.500 Protocol. (Also FYI 13) March 1992
1307	Dynamically Switched Link Control Protocol. March 1992
1306	Experiences Supporting By-Request Circuit-Switched T3 Networks. March 1992
1305	Network Time Protocol (Version 3) Specification, Implementation, and Analysis. (Obsoletes RFC 1119, 1059, 958) March 1992
1304	Definitions of Managed Objects for the SIP Interface Type. February 1992
1303	A Convention for Describing SNMP-based Agents. February 1992
1302	Building A Network Information Services Infrastructure. (Also FYI 12) February 1992
1301	Multicast Transport Protocol. February 1992
1300	Remembrances of Things Past. February 1992
1299	Not yet issued.
1298	SNMP over IPX. (Obsoleted by RFC 1420) February 1992
1297	NOC Internal Integrated Trouble Ticket System Functional Specification Wishlist (_NOC TT REQUIREMENTS_). January 1992
1296	Internet Growth (1981-1991). January 1992
1295	User Bill of Rights for entries and listings in the Public Directory. (Obsoleted by RFC 1417) January 1992

RFC	Title
1294	Multiprotocol Interconnect over Frame Relay. (Obsoleted by RFC 1490) January 1992
1293	Inverse Address Resolution Protocol. January 1992
1292	A Catalog of Available X.500 Implementations (Also FYI 11) January 1992
1291	Mid-Level Networks—Potential Technical Services. December 1991
1290	There's Gold in them Thar Networks! or Searching for Treasure in all the Wrong Places. (Obsoleted by RFC 1402) (Also FYI 10) December 1991
1289	DECnet Phase IV MIB Extensions. (Obsoleted by RFC 1559) December 1991
1288	The Finger User Information Protocol. (Obsoletes RFC 1196, 1194, 742) December 1991
1287	Towards the Future Internet Architecture. December 1991
1286	Definitions of Managed Objects for Bridges (Obsoleted by RFC 1493, 1525) December 1991
1285	FDDI Management Information Base. (Updated by RFC 1512) January 1991
1284	Definitions of Managed Objects for the Ethernet-like Interface Types. (Obsoleted by RFC 1398) December 1991
1283	SNMP over OSI. (Obsoletes RFC 1161; Obsoleted by RFC 1418) December 1991
1282	BSD Rlogin. (Obsoletes RFC 1258) December 1991
1281	Guidelines for the Secure Operation of the Internet. November 1991
1280	IAB Official Protocol Standards. (Obsoletes RFCs 1250, 1100, 1083, 1130, 1140, 1200; Obsoleted by RFC 1360) March 1991
1279	X.500 and Domains. November 1991
1278	A string encoding of Presentation Address. November 1991
1277	Encoding Network Addresses to support operation over non-OSI lower layers. November 1991
1276	Replication and Distributed Operations extensions to provide an Internet Directory using X.500. November 1991
1275	Replication Requirements to provide an Internet Directory using X.500. November 1991
1274	The COSINE and Internet X.500 Schema. November 1991
1273	A measurement study of changes in service-level reachability in the global TCP/IP Internet. November 1991
1272	Internet accounting: background. November 1991

RFC	Title
1271	Remote network monitoring Management Information Base. (Obsoleted by RFC 1513) November 1991
1270	SNMP communications services. October 1991
1269	Definitions of Managed Objects for the Border Gateway Protocol (version 3). October 1991
1268	Application of the Border Gateway Protocol in the Internet. (Obsoletes RFC 1164) October 1991
1267	A Border Gateway Protocol 3 (BGP-3). (Obsoletes RFC 1105, 1163) October 1991
1266	Experience with the BGP protocol. October 1991
1265	BGP protocol analysis. October 1991
1264	Internet routing protocol standardization criteria. October 1991
1263	TCP Extensions considered harmful. October 1991
1262	Guidelines for Internet measurement activities. October 1991
1261	Transition of NIC services. September 1991
1260	Not yet issued.
1259	Building the open road: The NREN as test-bed for the national public network. September 1991
1258	BSD Rlogin. (Obsoleted by RFC 1282) September 1991
1257	Isochronous applications do not require jitter-controlled networks. September 1991
1256	ICMP router discovery messages. September 1991
1255	Naming scheme for c=US. (Obsoletes RFC 1218; Obsoleted by RFC 1417) September 1991
1254	Gateway congestion control survey. August 1991
1253	OSPF version 2: Management Information Base. (Obsoletes RFC 1252) August 1991
1252	OSPF version 2: Management Information Base. (Obsoletes RFC 1248; Obsoleted by RFC 1253) August 1991
1251	Who's who in the Internet: Biographies of IAB, IESG, and IRSG members. (Obsoleted by RFC 1336) (Also FYI 9) August 1991
1250	IAB official protocol standards. (Obsoletes RFC 1200; Obsoleted by RFC 1360) August 1991
1249	DIXIE protocol specification. August 1991

RFC	Title
1248	OSPF version 2: Management Information Base. (Obsoleted by RFC 1252) Updated by RFC 1349. July 1991
1247	OSPF version 2. (Obsoletes RFC 1131; Updated by RFC 1349) July 1991
1246	Experience with the OSPF protocol. July 1991
1245	OSPF protocol analysis. July 1991
1244	Site Security Handbook. (Also FYI 8) July 1991
1243	AppleTalk Management Information Base. July 1991
1242	Benchmarking terminology for network interconnection devices. July 1991
1241	Scheme for an internet encapsulation protocol: Version 1. July 1991
1240	OSI connectionless transport services on top of UDP: Version 1. June 1991
1239	Reassignment of experimental MIBs to standard MIBs. (Updates RFC 1229, 1230, 1231, 1232, 1233) June 1991
1238	CLNS MIB for use with Connectionless Network Protocol (ISO 8473) and End System to Intermediate System (ISO 9542). (Obsoletes RFC 1162) June 1991
1237	Guideline for OSI NSAP allocation in the Internet July 1991
1236	IP to X.121 address mapping for DDN IP to X 121 address mapping for DDN. June 1991
1235	Coherent File Distribution Protocol. June 1991
1234	Tunneling IPX traffic through IP networks. June 1991
1233	Definitions of managed objects for the DS3 Interface type. (Obsoleted by RFC 1407; Updated by RFC 1239) May 1991
1232	Definitions of managed objects for the DS1 Interface type. (Obsoleted by RFC 1406; Updated by RFC 1239) May 1991
1231	IEEE 802.5 Token Ring MIB IEEE 802.5 Token Ring MIB. (Updated by RFC 1239) May 1991
1230	IEEE 802.4 Token Bus MIB IEEE 802.4 Token Bus MIB. (Updated by RFC 1239) May 1991
1229	Extensions to the generic-interface MIB. (Updated by RFC 1239) May 1991
1228	SNMP-DPI: Simple Network Management Protocol Distributed Program Interface. May 1991
1227	SNMP MUX protocol and MIB. May 1991
1226	Internet protocol encapsulation of X.25 frames Internet protocol encapsulation of X.25 frames. May 1991

RFC	Title
1225	Post Office Protocol: Version 3. (Obsoletes RFC 1081; Obsoleted by RFC 1460) May 1991
1224	Techniques for managing asynchronously generated alerts. May 1991
1223	OSI CLNS and LLC1 protocols on Network Systems HYPERchannel. May 1991
1222	Advancing the NSFNET routing architecture. May 1991
1221	Host Access Protocol (HAP) specification: Version 2. (Updates RFC 907) April 1991
1220	Point-to-Point Protocol extensions for bridging. April 1991
1219	On the assignment of subnet numbers. April 1991
1218	Naming scheme for c=US. (Obsoleted by RFC 1417) April 1991
1217	Memo from the Consortium for Slow Commotion Research (CSCR). April 1991
1216	Gigabit network economics and paradigm shifts. April 1991
1215	Convention for defining traps for use with the SNMP. March 1991
1214	OSI internet management: Management Information Base. April 1991
1213	Management Information Base for network management of TCP/IP-based internets: MIB-II. (Obsoletes RFC 1158) March 1991
1212	Concise MIB definitions. March 1991
1211	Problems with the maintenance of large mailing lists. March 1991
1210	Network and infrastructure user requirements for transatlantic research collaboration: Brussels, July 16-18, and Washington, July 24-25 1990. March 1991
1209	Transmission of IP datagrams over the SMDS Service. March 1991
1208	Glossary of networking terms. March 1991
1207	FYI on Questions and Answers: Answers to commonly asked experienced Internet user questions. February 1991
1206	FYI on Questions and Answers: answers to commonly asked new Internet user questions. (Obsoletes RFC 1177; Obsoleted by RFC 1325) February 1991
1205	5250 Telnet interface. February 1991
1204	Message Posting Protocol (MPP). February 1991
1203	Interactive Mail Access Protocol: Version 3. (Obsoletes RFC 1064) February 1991
1202	Directory Assistance service. February 1991
1201	Transmitting IP traffic over ARCNET networks (Obsoletes RFC 1051) February 1991

RFC	Title
1200	Defense Advanced Research Projects Agency, Internet Activities Board; DARPA IAB official protocol standards. (Obsoletes RFC 1104; Obsoleted by RFC 1360). April 1991
1199	RFC Numbers 1100-1199. December 1991
1198	FYI on the X window system. (Also FYI 6) January 1991
1197	Using ODA for translating multimedia information. December 1990
1196	Finger User Information Protocol. (Obsoletes RFC 1194; Obsoleted by RFC 1288) December 1990
1195	Use of OSI IS-IS for routing in TCP/IP and dual environments. (Updated by RFC 1349) December 1990
1194	Finger User Information Protocol. (Obsoletes RFC 742; Obsoleted by RFC 1288) November 1990
1193	Client Requirements for real-time communication services. November 1990
1192	Commercialization of the Internet summary report. November 1990
1191	Path MTU discovery. (Obsoletes RFC 1063) November 1990
1190	Experimental Internet Stream Protocol: Version 2 (ST-11). (Obsoletes IEN 119) October 1990
1189	Common Management Information Services and Protocols for the Internet (CMOT and CMIP). (Obsoletes RFC 1095) October 1990
1188	Proposed standard for the transmission of IP datagrams over FDDI networks. (Obsoletes RFC 1103) October 1990
1187	Bulk table retrieval with the SNMP. October 1990
1186	MD4 message digest algorithm. (Obsoleted by RFC 1320) October 1990
1185	TCP extension for high-speed paths. (Obsoleted by RFC 1323) October 1990
1184	Telnet Linemode option. (Obsoletes RFC 1116) October 1990
1183	New DNS RR definitions. (Updates RFC 1034, 1035) October 1990
1182	Not yet issued.
1181	RIPE terms of reference. September 1990
1180	TCP/IP tutorial. January 1991
1179	Line printer daemon protocol. August 1990
1178	Choosing a name for your computer. (Also FYI 5) August 1990
1177	FYI on Questions and Answers: Answers to commonly asked new internet user questions. (Obsoleted by RFC 1206) August 1990

RFC	Title
1176	Interactive mail Access Protocol: Version 2. (Obsoletes RFC 1064) August 1990
1175	FYI on where to start: A bibliography of internetworking information. (Also FYI 3) August 1990
1174	IAB recommended policy on distributing internet identifier assignment and IAB recommended policy change to internet connected status. August 1990
1173	Responsibilities of host and network managers: A summary of the oral tradition of the Internet. August 1990
1172	Point-to-Point Protocol (PPP) initial configuration options. (Obsoleted by RFC 1332) July 1990
1171	Point-to-Point Protocol for the transmission of multi-protocol datagrams over Point-to-Point links. (Obsoletes RFC 1134; Obsoleted by RFC 1331) July 1990
1170	Public key standards and licenses. January 1991
1169	Explaining the role of GOSIP. August 1990
1168	Intermail and Commercial mail Relay services. July 1990
1167	Thoughts on the National Research and Education Network July 1990
1166	Internet numbers. (Obsoletes RFC 1117, 1062, 1020) July 1990
1165	Network Time Protocol (NTP) over the OSI Remote Operations Service. June 1990
1164	Application of the Border Gateway Protocol in the Internet. (Obsoleted by RFC 1268) June 1990
1163	Border Gateway Protocol (BGP). (Obsoletes RFC 1105; Obsoleted by RFC 1267) June 1990
1162	Connectionless Network Protocol (ISO 8473) and End System to Intermediate System (ISO 9542) Management Information Base. (Obsoleted by RFC 1238) June 1990
1161	SNMP over OSI. (Obsoleted by RFC 1283) June 1990
1160	Internet Activities Board. (Obsoletes RFC 1120) May 1990
1159	Message Send Protocol. (Obsoleted by RFC 1312) June 1990
1158	Management Information Base for network management of TCP/IP-based internets: MIB-II. (Obsoleted by RFC 1213) May 1990
1157	Simple Network Management Protocol (SNMP). (Obsoletes RFC 1098) May 1990
1156	Management Information Base for network management of TCP/IP-based internets. (Obsoletes RFC 1066) May 1990
1155	Structure and identification of management information for TCP/IP-based internets. (Obsoletes RFC 1065) May 1990
1154	Encoding header field for internet messages. (Obsoleted by RFC 1505) April 1990

RFC	Title
1153	Digest message format. April 1990
1152	Workshop report: Internet research steering group workshop on very-high-speed networks. April 1990
1151	Version 2 of the Reliable Data Protocol (RDP). (Updates RFC 908)April 1990
1150	FYI on FYI: Introduction to the FYI notes. (Also FYI 1) March 1990
1149	Standard for the transmission of IP datagrams on avian carriers. April 1990
1148	Mapping between X.400(1988)/ISO 10021 and RFC 822. (Obsoleted by RFC 1327; Updates RFC 822, 987, 1026, 1138) March 1990
1147	FYI on a network management tool catalog: Tools for monitoring and debugging TCP/IP internets and inter connected devices. (Also FYI 2) (Obsoleted by RFC 1470) April 1990
1146	TCP alternate checksum options. (Obsoletes RFC 1145) March 1990
1145	TCP alternate checksum options. (Obsoleted by RFC 1146) February 1990
1144	Compressing TCP/IP headers for low-speed serial links. February 1990
1143	Q method of implementing Telnet option negotiation. February 1990
1142	OSI IS-IS Intra-domain Routing Protocol. February 1990
1141	Incremental updating of the Internet checksum. (Updates RFC 1071) January 1990
1140	DARPA IAB official protocol standards. (Obsoletes RFC 1130; Obsoleted by RFC 1360) May 1990
1139	Echo function for ISO 8473. January 1990
1138	Mapping between X.400(1988)/ISO 10021 and RFC 822. (Obsoleted by RFC 1327; Updates RFC 822, 987, 1026; Updated by RFC 1148) December 1989
1137	Mapping between full RFC 822 and RFC 822 with restricted encoding. (Updates RFC 976) December 1989
1136	Administrative Domains and Routing Domains: A model for routing in the Internet. December 1989
1135	Helminthiasis of the Internet. December 1989
1134	Point-to-Point Protocol: A proposal for multi-protocol transmission of data grams over Point-to-Point links. (Obsoleted by RFC 1171) November 1989
1133	Routing between the NSFNET and the DDN. November 1989
1132	Standard for the transmission of 802.2 packets over IPX networks. November 1989
1131	OSPF specification. (Obsoleted by RFC 1247) October 1989

RFC	Title
1130	Defense Advanced Research Projects Agency, Internet Activities Board: DARPA IAB IAB official protocol standards. (Obsoletes RFC 1100; Obsoleted by RFC 1360) October 1989
1129	Internet time synchronization: The Network Time Protocol 1989. October
1128	Measured performance of the Network Time Protocol in the Internet system. October 1989
1127	Perspective on the Host Requirements RFCs. October 1989
1126	Goals and functional requirements for inter-autonomous system routing. October 1989
1125	Policy requirements for inter Administrative Domain routing. November 1989
1124	Policy issues in interconnecting networks. September 1989
1123	Requirements for Internet hosts—application and support. (Updated by RFC 1349) October 1989
1122	Requirements for Internet hosts communication layers. (Updated by RFC 1349) October 1989
1121	Act One: The Poems. September 1989
1120	Internet Activities Board. (Obsoleted by RFC 1160) September 1989
1119	Network Time Protocol (version 2) specification and implementation. (Obsoletes RFC 1059, 958; Obsoleted by RFC 1305) September 1989
1118	Hitchhikers guide to the Internet. September 1989
1117	Internet numbers. (Obsoletes RFC 1062, 1020, 997; Obsoleted by RFC 1166) August 1989
1116	Telnet Linemode option. (Obsoleted by RFC 1184) August 1989
1115	Privacy enhancement for Internet electronic mail: Part III—Algorithms, modes, and identifiers [Draft]. (Obsoleted by RFC 1423) August 1989
1114	Privacy enhancement for Internet electronic mail: Part II—Certificate-based key management [Draft]. (Obsoleted by RFC 1422) August 1989
1113	Privacy enhancement for Internet electronic mail: Part I—Message encipherment and authentication procedures [Draft]. (Obsoletes RFC 989, 1040; Obsoleted by RFC 1421) August 1989
1112	Host extensions for IP multicasting. (Obsoletes RFC 988, 1054) August 1989
1111	Request for comments on Request for Comments: Instructions to RFC authors. (Obsoletes RFC 825; Obsoleted by RFC 1543) August 1989
1110	Problem with the TCP big window option. August 1989

RFC	Title
1109	Report of the second Ad Hoc Network Management Review Group. August 1989
1108	Security Option for the Internet Protocol. (Obsoletes RFC 1038) November 1991
1107	Plan for Internet directory services. July 1989
1106	TCP big window and NAK options. June 1989
1105	Border Gateway Protocol (BGP). (Obsoleted by RFC 1267) June 1989
1104	Models of policy based routing. June 1989
1103	Proposed standard for the transmission of IP datagrams over FDDI Networks. (Obsoleted by RFC 1188) June 1989
1102	Policy routing in Internet protocols. May 1989
1101	DNS encoding of network names and other types. (Updates RFC 1034, 1035) April 1989
1100	Defense Advanced Research Projects Agency, Internet Activities Board; DARPA IAB official protocol standards. (Obsoletes RFC 1083; Obsoleted by RFC 1360) April 1989
1099	Request for comments Summary RFC Numbers 1000-1099. December 1991
1098	Simple Network Management Protocol (SNMP). (Obsoletes RFC 1067; Obsoleted by RFC 1157) April 1989
1097	Telnet subliminal-message option. April 1989
1096	Telnet X display location option. March 1989
1095	Common Management Information Services and Protocol over TCP/IP (CMOT). (Obsoleted by RFC 1189) April 1989
1094	NFS: Network File System Protocol specification. March 1989
1093	NSFNET routing architecture. February 1989
1092	EGP and policy based routing in the new NSFNET backbone. February 1989
1091	Telnet terminal-type option. (Obsoletes RFC 930) February 1989
1090	SMTP on X.25 SMTP on X 25. February 1989
1089	SNMP over Ethernet. February 1989
1088	Standard for the Transmission of IP datagrams over NetBIOS networks. February 1989
1087	Defense Advanced Research Projects Agency, Internet Activities Board; DARPA IAB Ethics and the Internet. January 1989
1086	ISO-TP0 bridge between TCP and X.25. December 1988

RFC	Title
1085	ISO presentation services on top of TCP/IP based internets. December 1988
1084	BOOTP vendor information extensions. (Obsoletes RFC 1048; Obsoleted by RFC 1395) December 1988
1083	Defense Advanced Research Projects Agency, Internet Activities Board; DARPA IAB official protocol standards. (Obsoletes RFC 1011; Obsoleted by RFC 1360) December 1988
1082	Post Office Protocol: Version 3: Extended service offerings. December 1988
1081	Post Office Protocol: Version 3. (Obsoleted by RFC 1225) November 1988
1080	Telnet remote flow control option. (Obsoleted by RFC 1372) November 1988
1079	Telnet terminal speed option. December 1988
1078	TCP port service Multiplexer (TCPMUX). November 1988
1077	Critical issues in high bandwidth networking. November 1988
1076	HEMS monitoring and control language. (Obsoletes RFC 1023) November 1988
1075	Distance Vector Multicast Routing Protocol. November 1988
1074	NSFNET backbone SPF based Interior Gateway Protocol. October 1988
1073	Telnet window size option. October 1988
1072	TCP extensions for long-delay paths. (Obsoleted by RFC 1323) October 1988
1071	Computing the Internet checksum. (Updated by RFC 1141) September 1988
1070	Use of the Internet as a subnetwork for experimentation with the OSI network layer. February 1989
1069	Guidelines for the use of Internet-IP addresses in the ISO Connectionless-Mode Network Protocol. (Obsoletes RFC 986) February 1989
1068	Background File Transfer Program (BFTP). August 1988
1067	Simple Network Management Protocol. (Obsoleted by RFC 1098) August 1988
1066	Management Information Base for network management of TCP/IP-based internets. (Obsoleted by RFC 1156) August 1988
1065	Structure and identification of management information for TCP/IP-based internets. (Obsoleted by RFC 1155) August 1988
1064	Interactive Mail Access Protocol: Version 2. (Obsoleted by RFC 1176, 1203) July 1988
1063	IP MTU discovery options. (Obsoleted by RFC 1191) July 1988
1062	Internet numbers. (Obsoletes RFC 1020; Obsoleted by RFC 1117) August 1988

RFC	Title
1061	Not yet issued.
1060	Assigned numbers. (Obsoletes RFC 1010; Obsoleted by RFC 1340; Updated by RFC 1349) March 1990
1059	Network Time Protocol (version 1) specification and implementation. (Obsoleted by RFC 1305) July 1988
1058	Routing Information Protocol. (Updated by RFC 1388) June 1988
1057	RPC: Remote Procedure Call Protocol specification: Version 2.(Obsoletes RFC 1050) June 1988
1056	PCMAIL: A distributed mail system for personal computers. (Obsoletes RFC 993) June 1988
1055	Nonstandard for transmission of IP datagrams over serial lines: SLIP. June 1988
1054	Host extensions for IP multicasting. (Obsoletes RFC 988; Obsoleted by RFC 1112) May 1988
1053	Telnet X.3 PAD option Telnet X3 PAD option Telnet X 3 PAD option. May 1988
1052	IAB recommendation for the development of Internet network management standards. April 1988
1051	Standard for the transmission of IP datagrams and ARP packets over ARCNET networks. (Obsoleted by RFC 1201) March 1988
1050	RPC: Remote Procedure Call Protocol specification. (Obsoleted by RFC 1057) April 1988
1049	Content-type header field for Internet messages. April 1988
1048	BOOTP vendor information extensions. (Obsoleted by RFC 1395) February 1988
1047	Duplicate messages and SMTP. February 1988
1046	Queuing algorithm to provide type-of-service for IP links. February 1988
1045	VMTP: Versatile Message Transaction Protocol: Protocol specification. February 1988
1044	Internet Protocol on Network System's HYPERchannel: Protocol specification. February 1988
1043	Telnet Data Entry Terminal option: DODIIS implementation. (Updates RFC 732) February 1988
1042	Standard for the transmission of IP datagrams over IEEE 802 networks. (Obsoletes RFC 948) February 1988
1041	Telnet 3270 regime option. January 1988

RFC	Title
1040	Privacy enhancement for Internet electronic mail: Part I—Message encipherment and authentication procedures. (Obsoletes RFC 989; Obsoleted by RFC 1113) January 1988
1039	DoD statement on Open Systems Interconnection protocols. (Obsoletes RFC 945) January 1988
1038	Draft revised IP security option. (Obsoleted by RFC 1108) January 1988
1037	NFILE: A file access protocol. December 1987
1036	Standard for interchange of USENET messages. (Obsoletes RFC 850) 1987 December
1035	Domain names: Implementation and specification. (Obsoletes RFC 973, 882, 883; Updated by RFC 1348, 1183, 1101) November 1987
1034	Domain names: Concepts and facilities. (Obsoletes RFC 973, 882, 883; Updated by RFC 1348, 1183, 1101) November 1987
1033	Domain administrators operations guide. November 1987
1032	Domain administrators guide. November 1987
1031	MILNET name domain transition. November 1987
1030	On testing the NETBLT Protocol over diverse networks. November 1987
1029	More fault-tolerant approach to address resolution for a Multi-LAN system of Ethernets. May 1988
1028	Simple Gateway Monitoring Protocol. November 1987
1027	Using ARP to implement transparent subnet gateways. October 1987
1026	Addendum to RFC 987: (Mapping between X.400 and RFC-822). (Obsoleted by RFC 1327; Updates 987; Updated by 1138, 1148) September 1987
1025	TCP and IP bake-off. September 1987
1024	HEMS variable definitions. October 1987
1023	HEMS monitoring and control language. (Obsoleted by RFC 1076) October 1987
1022	High-level Entity Management Protocol (HEMP). October 1987
1021	High-level Entity Management System (HEMS). October 1987
1020	Internet numbers. (Obsoletes RFC 997; Obsoleted by RFC 1062, 1117) November 1987
1019	Report of the Workshop on Environments for Computational Mathematics. September 1987
1018	Some comments on SQuID. August 1987

RFC	Title
1017	Network requirements for scientific research: Internet task force on scientific computing. August 1987
1016	Something a host could do with source quench: The Source Quench Introduced Delay (SQuID). July 1987
1015	Implementation plan for interagency research Internet. July 1987
1014	XDR: External Data Representation standard. June 1987
1013	X Window System Protocol, version 11: Alpha update April 1987. June 1987
1012	Bibliography of Request For Comments 1 through 999. June 1987
1011	Official Internet protocols. (Obsoletes RFC 997; Obsoleted by RFC 1083) May 1987
1010	Assigned numbers. (Obsoletes RFC 990; Obsoleted by RFC 1340) May 1987
1009	Requirements for Internet gateways. (Obsoletes RFC 985) June 1987
1008	Implementation guide for the ISO Transport Protocol. June 1987
1007	Military supplement to the ISO Transport Protocol. June 1987
1006	ISO transport services on top of the TCP: Version 3. (Obsoletes RFC 983) May 1987
1005	ARPANET AHIP-E Host Access Protocol (enhanced AHIP) May 1987
1004	Distributed-protocol authentication scheme. April 1987
1003	Issues in defining an equations representation standard. March 1987
1002	DARPA IAB End-to-End Services Task Force, NetBIOS Working Group; Protocol standard for a NetBIOS service on a TCP/UDP transport: Detailed specifications. March 1987
1001	DARPA IAB End-to-End Services Task Force NetBIOS Working Group; Protocol standard for a NetBIOS service on a TCP/UDP transport: Concepts and methods. March 1987
1000	The request for comments reference guide. (Obsoletes RFC 999) August 1987
999	Request For comments summary notes: 900-999. (Obsoleted by RFC 1000) April 1987
998	NETBLT: A bulk data transfer protocol NETBLT a bulk data transfer protocol. (Obsoletes RFC 969) March 1987
997	Internet numbers. (Obsoleted by RFC 1020, 1117; Updates RFC 990) March 1987
996	Statistics server. February 1987
995	International Organization for Standardization; ISO End System to Intermediate System Routing Exchange Protocol for use in conjunction with ISO 8473. April 1986

RFC	Title
994	International Organization for Standardization; ISO Final text of DIS 8473, Protocol for Providing the Connectionless-mode Network Service. (Obsoletes RFC 926) March 1986
993	PCMAIL: A distributed mail system for personal computers. (Obsoletes RFC 984; Obsoleted by RFC 1056) December 1986
992	On communication support for fault tolerant process groups. November 1986
991	Official ARPA-Internet protocols. (Obsoletes RFC 961; Obsoleted by RFC 1011) November 1986
990	Assigned numbers. (Obsoletes RFC 960; Obsoleted by RFC 1340; Updated by RFC 997) November 1986
989	Privacy enhancement for Internet electronic mail: Part I—Message encipherment and authentication procedures. (Obsoleted by RFC 1040, 1113) February 1987
988	Host extensions for IP multicasting. (Obsoletes RFC 966; Obsoleted by RFC 1054, 1112) July 1986
987	Mapping between X.400 and RFC 822. (Obsoleted by RFC 1327; Updated by RFC 1026, 1138, 1148) June 1986
986	Guidelines for the use of Internet-IP addresses in the ISO Connectionless-Mode Network Protocol [Working draft]. (Obsoleted by RFC 1069) June 1986
985	National Science Foundation, Network Technical Advisory Group; NSF NTAG Requirements for Internet gateways—Draft Requirements for Internet gateways draft. (Obsoleted by RFC 1009) May 1986
984	PCMAIL: A Distributed Mail System for Personal Computers. (Obsoleted by RFC 993) May 1986
983	ISO transport arrives on top of the TCP. (Obsoleted by RFC 1006). April 1986
982	Guidelines for the specification of the structure of the Domain Specific Part (DSP) of the ISO standard NSAP address. April 1986
981	Experimental multiple-path routing algorithm. March 1986
980	Protocol document order information. March 1986
979	PSN End-to-End functional specification. March 1986
978	Voice File Interchange Protocol (VFIP). February 1986
977	Network News Transfer Protocol. February 1986
976	UUCP mail interchange format standard. (Updated by RFC 1137) February 1986
975	Autonomous confederations. February 1986
974	Mail routing and the domain system. January 1986

RFC	Title
973	Domain system changes and observations. (Obsoleted by RFC 1034, 1035; Updates RFC 882, 883) January 1986
972	Password Generator Protocol. January 1986
971	Survey of data representation standards. January 1986
970	On packet switches with infinite storage. December 1985
969	NETBLT: A Bulk Data Transfer Protocol. (Obsoleted by RFC 998) December 1985
968	Twas the night before start-up. December 1985
967	All victims together. December 1985
966	Host Groups: A Multicast Extension to the Internet Protocol. (Obsoleted by RFC 988) December 1985
965	Format for a graphical communication protocol. December 1985
964	Some problems with the specification of the Military Standard Transmission Control Protocol. November 1985
963	Some problems with the specification of the Military Standard Internet Protocol. November 1985
962	TCP-4 prime. November 1985
961	Official ARPA-Internet protocols. (Obsoletes RFC 944; Obsoleted by RFC 991) December 1985
960	Assigned numbers. (Obsoletes RFC 943; Obsoleted by RFC 1340) December 1985
959	File Transfer Protocol. (Obsoletes RFC 765 [IEN 149]) October 1985
958	Network Time Protocol (NTP). (Obsoleted by RFC 1305) September 1985
957	Experiments in network clock synchronization. September 1985
956	Algorithms for synchronizing network clocks. September 1985
955	Towards a transport service for transaction processing applications. September 1985
954	NICNAME/WHOIS. (Obsoletes RFC 812) October 1985
953	Hostname Server. (Obsoletes RFC 811) October 1985
952	DoD Internet host table specification. (Obsoletes RFC 810) October 1985
951	Bootstrap Protocol. (Updated by RFC 1497, 1395, 1532, 1542) September 1985
950	Internet standard subnetting procedure. (Updates RFC 792) August 1985
949	FTP unique-named store command. July 1985

RFC	Title
948	Two methods for the transmission of IP datagrams over IEEE 802.3 networks. (Obsoleted by RFC 1042) June 1985
947	Multi-network broadcasting within the Internet. June 1985
946	Telnet terminal location number option. May 1985
945	DoD statement on the NRC report. (Obsoleted by RFC 1039) May 1985
944	Official ARPA-Internet protocols. (Obsoleted by RFC 924; Obsoleted by RFC 961) April 1985
943	Assigned numbers. (Obsoletes 923; Obsoleted by 1340) April 1985
942	National Research Council; NRC Transport protocols for Department of Defense data networks. February 1985
941	International Organization for Standardization; ISO Addendum to the network service definition covering network layer addressing. April 1985
940	Gateway Algorithms and Data Structures Task Force; GADS Toward an Internet standard scheme for subnetting. April 1985
939	National Research Council; NRC Executive summary of the NRC report on transport protocols for Department of Defense data networks. February 1985
938	Internet Reliable Transaction Protocol functional and interface specification. February 1985
937	Post Office Protocol: Version 2. (Obsoletes RFC 918) February 1985
936	Another Internet subnet addressing scheme. February 1985
935	Reliable link layer protocols. January 1985
934	Proposed standard for message encapsulation. January 1985
933	Output marking Telnet option. January 1985
932	Subnetwork addressing scheme. January 1985
931	Authentication server. (Obsoletes RFC 912; Obsoleted by RFC 1413) January 1985
930	Telnet terminal type option. (Obsoletes RFC 884; Obsoleted by RFC 1091) January 1985
929	Proposed Host-Front End Protocol. December 1984
928	Introduction to proposed DoD standard H-FP. December 1984
927	TACACS user identification Telnet option. December 1984
926	International Organization for Standardization; ISO Protocol for providing the connectionless mode network services. (Obsoleted by RFC 994) December 1984

RFC	Title
925	Multi-LAN address resolution. October 1984
924	Official ARPA-Internet protocols for connecting personal computers to the Internet. (Obsoletes RFC 901; Obsoleted by RFC 944) October 1984
923	Assigned numbers. (Obsoletes RFC 900; Obsoleted by RFC 1340) October 1984
922	Broadcasting Internet datagrams in the presence of subnets. October 1984
921	Domain name system implementation schedule, revised. (Updates RFC 897) October 1984
920	Domain requirements. October 1984
919	Broadcasting Internet datagrams. October 1984
918	Post Office Protocol. (Obsoleted by RFC 937) October 1984
917	Internet subnets. October 1984
916	Reliable Asynchronous Transfer Protocol (RATP). October 1984
915	Network mail path service. December 1984
914	Thinwire protocol for connecting personal computers to the Internet. September 1984
913	Simple File Transfer Protocol. September 1984
912	Authentication service. (Obsoleted by RFC 931) September 1984
911	EGP Gateway under Berkeley UNIX 4.2. August 1984
910	Multimedia mail meeting notes. August 1984
909	Loader Debugger Protocol. July 1984
908	Reliable Data Protocol. (Updated by RFC 1151) July 1984
907	Bolt Beranek and Newman, Inc; BBN Host Access Protocol specification. (Updated by RFC 1221) July 1984
906	Bootstrap loading using TFTP. June 1984
905	ISO Transport Protocol specification ISO DP 8073. (Obsoletes RFC 892) April 1984
904	Exterior Gateway Protocol formal specification. (Updates RFC 827, 888) April 1984
903	Reverse Address Resolution Protocol. June 1984
902	ARPA Internet Protocol policy. July 1984
901	Official ARPA-Internet protocols. (Obsoletes RFC 880; Obsoleted by RFC 924) June 1984
900	Assigned Numbers. (Obsoletes RFC 870; Obsoleted by RFC 1340) June 1984

RFC	Title
899	Requests for Comments Summary Notes: 800-899. May 1984
898	Gateway special interest group meeting notes. April 1984
897	Domain name system implementation schedule. (Updates RFC 881; Updated by RFC 921) February 1984
896	Congestion control in IP/TCP internetworks. January 1984
895	Standard for the transmission of IP datagrams over experimental Ethernet networks. April 1984
894	Standard for the transmission of IP datagrams over Ethernet networks. April 1984
893	Trailer encapsulations. April 1984
892	International Organization for Standardization; ISO Transport Protocol specification [Draft]. (Obsoleted by RFC 905) December 1983
891	DCN local-network protocols. December 1983
890	Exterior Gateway Protocol implementation schedule. February 1983
889	Internet delay experiments. December 1983
888	"STUB" Exterior Gateway Protocol STUB Exterior Gateway Protocol. (Updated by RFC 904). January 1983
887	Resource Location Protocol. December 1983
886	Proposed standard for message header munging. December 1983
885	Telnet end of record option. December 1983
884	Telnet terminal type option. (Obsoleted by RFC 930) December 1983
883	Domain names: Implementation specification. (Obsoleted by RFC 1034, 1035; Updated by RFC 973) November 1983
882	Domain names: Concepts and facilities. (Obsoleted by RFC 1034, 1035; Updated by RFC 973) November 1983
881	Domain names plan and schedule. (Updated by RFC 897) November 1983
880	Official protocols. (Obsoletes RFC 840; Obsoleted by RFC 901) October 1983
879	TCP maximum segment size and related topics. November 1983
878	ARPANET 1822L Host Access Protocol. (Obsoletes RFC 851) December 1983
877	Standard for the transmission of IP datagrams over public data networks. (Obsoleted by RFC 1356) September 1983
876	Survey of SMTP implementations. September 1983

RFC	Title
875	Gateways, architectures, and heffalumps. September 1982
874	A Critique of X.25. September 1982
873	Illusion of vendor support. September 1982
872	TCP-on-a-LAN. September 1982
871	Perspective on the ARPANET reference model. September 1982
870	Assigned numbers. (Obsoletes RFC 820; Obsoleted by RFC 1340) October 1983
869	Host Monitoring Protocol. December 1983
868	Time Protocol. May 1983
867	Daytime Protocol. May 1983
866	Active users. May 1983
865	Quote of the Day Protocol. May 1983
864	Character Generator Protocol. May 1983
863	Discard Protocol. May 1983
862	Echo Protocol. May 1983
861	Telnet extended options: List option. (Obsoletes NIC 16239) May 1983
860	Telnet timing mark option. (Obsoletes NIC 16238) May 1983
859	Telnet status option. (Obsoletes RFC 651) May 1983
858	Telnet Suppress Go Ahead option. (Obsoletes NIC 15392) May 1983
857	Telnet echo option. (Obsoletes NIC 15390) May 1983
856	Telnet binary transmission. (Obsoletes NIC 15389) May 1983
855	Telnet option specifications. (Obsoletes NIC 18640) May 1983
854	Telnet Protocol specification. (Obsoletes RFC 765, NIC 18639) May 1983
853	Not issued.
852	ARPANET short blocking feature. April 1983
851	ARPANET 1822L Host Access Protocol. (Obsoletes RFC 802; Obsoleted by RFC 878) April 1983
850	Standard for interchange of USENET messages. (Obsoleted by RFC 1036) June 1983
849	Suggestions for improved host table distribution. May 1983
848	Who provides the "little" TCP services? March 1983

RFC	Title
847	Summary of Smallberg surveys. (Obsoletes RFC 846) February 1983
846	Who talks TCP? Survey of 22 February 1983. (Obsoletes RFC 845; Obsoleted by RFC 847). February 1983
845	Who talks TCP? Survey of 15 February 1983 (Obsoletes RFC 843; Obsoleted by RFC 846) February 1983
844	Who talks ICMP, Too? Survey of 18 February 1983. (Updates RFC 843) February 1983
843	Who talks TCP? Survey of 8 February 83. (Obsoletes RFC 842; Obsoleted by RFC 845; Updated by RFC 844) February 1983
842	Who talks TCP? Survey of 1 February 83. (Obsoletes RFC 839; Obsoleted by RFC 843) February 1983
841	National Bureau of Standards; NBS Specification for message format for Computer Based Message Systems. January 1983
840	Official protocols. (Obsoleted by RFC 880) April 1983
839	Who talks TCP? (Obsoletes RFC 838; Obsoleted by RFC 842) January 1983
838	Who talks TCP? (Obsoletes RFC 837; Obsoleted by RFC 839) January 1983
837	Who talks TCP? (Obsoletes RFC 836; Obsoleted by RFC 838) January 1983
836	Who talks TCP? (Obsoletes RFC 835; Obsoleted by RFC 837) January 1983
835	Who talks TCP? (Obsoletes RFC 834; Obsoleted by RFC 836) December 1982
834	Who talks TCP? (Obsoletes RFC 833; Obsoleted by RFC 835) December 1982
833	Who talks TCP? (Obsoletes RFC 832; Obsoleted by RFC 834) December 1982
832	Who talks TCP? (Obsoleted by RFC 833) December 1982
831	Backup access to the European side of SATNET. December 1982
830	Distributed system for Internet name service. October 1982
829	Packet satellite technology reference sources. November 1982
828	Data communications: IFIP's international "network" of experts. August 1982
827	Exterior Gateway Protocol (EGP). (Updated by RFC 904) October 1982
826	Ethernet Address Resolution Protocol: Or converting network protocol addresses to 48.bit Ethernet address for transmission on Ethernet hardware. November 1982
825	Request for comments on Requests For Comments. (Obsoleted by RFC 1111) November 1982
824	CRONUS Virtual Local Network. August 1982

RFC	Title
823	DARPA Internet gateway. (Updates IEN 109, IEN 30) September 1982
822	Standard for the format of ARPA Internet Text messages. (Obsoletes RFC 733; Updated by RFC 1327, 1148, 1138) August 1982
821	Simple Mail Transfer Protocol. (Obsoletes RFC 788) August 1982.
820	Assigned numbers. (Obsoletes RFC 790; Obsoleted by RFC 1340) August 1982
819	Domain naming convention for Internet user applications. August 1982
818	Remote User Telnet service. November 1982
817	Modularity and efficiency in protocol implementation. July 1982
816	Fault isolation and recovery. July 1982
815	IP datagram reassembly algorithms July 1982
814	Name, addresses, ports, and routes. July 1982
813	Window and acknowledgment strategy in TCP. July 1982
812	NICNAME/WHOIS. (Obsoleted by RFC 954) March 1982
811	Hostnames Server. (Obsoleted by RFC 953) March 1982
810	DoD Internet host table specification. (Obsoletes RFC 608; Obsoleted by RFC 852) March 1982
809	UCL facsimile system. February 1982
808	Summary of compute mail services meeting held at BBN on 10 January 1979. March 1982
807	Multimedia mail meeting notes. February 1982
806	National Bureau of Standards; NBS Proposed Federal Information Processing Stand: Specification for message format for computer-based message systems. (Obsoleted by RFC 841) September 1981
805	Computer mail meeting notes. February 1982
804	International Telecommunication Union, International Telegraph and Telephone Consultative Committee; ITU CCITT CCITT draft recommendation T.4 [Standardization of Group 3 facsimile apparatus for document transmission]. 1981
803	Dacom 450/500 facsimile data transcoding. November 1981
802	ARPANET 1822L Host Access Protocol. (Obsoleted by RFC 851) November 1981
801	NCP/TCP transition plan. November 1981
800	Requests For Comments Summary Notes 700-799. November 1981

RFC	Title
799	Internet name domains. September 1981
798	Decoding facsimile data from the Rapicom 450. September 1981
797	Format for bitmap files. September 1981
796	Address mappings. (Obsoletes IEN 115) September 1981
795	Service mappings. September 1981
794	Preemption. (Updates IEN 125) September 1981
793	Transmission Control Protocol. September 1981
792	Internet Control Message Protocol. (Obsoletes RFC 777; Updated by RFC 950) September 1981
791	Internet Protocol. (Obsoletes RFC 760; Updated by RFC 1349) September 1981
790	Assigned numbers. (Obsoletes RFC 776; Obsoleted by RFC 1340) September 1981
789	Vulnerabilities of network control protocols: An example. July 1981
788	Simple Mail Transfer Protocol. (Obsoletes RFC 780; Obsoleted by RFC 821) November 1981
787	Connectionless data transmission survey/tutorial. July 1981
786	Mail Transfer Protocol: ISI TOPS20 MTP-NIMAIL interface. July 1981
785	Mail Transfer Protocol: ISI TOPS20 file definitions. July 1981
784	Mail Transfer Protocol: ISI TOPS20 implementation. July 1981
783	TFTP Protocol (revision 2). (Obsoletes IEN 133; Obsoleted by RFC 1350) June 1981
782	Virtual Terminal management model. May 1981
781	Specification of the Internet Protocol (IP) timestamp option. May 1981
780	Mail Transfer Protocol. (Obsoletes RFC 772; Obsoleted by RFC 788) May 1981
779	Telnet send-location option. April 1981
778	DCNET Internet Clock Service. April 1981
777	Internet Control Message Protocol. (Obsoletes RFC 760; Obsoleted by RFC 792) April 1981
776	Assigned numbers. (Obsoletes RFC 770; Obsoleted by RFC 1340) January 1981
775	Directory-oriented FTP commands. December 1980
774	Internet Protocol Handbook. (Obsoletes RFC 766) October 1980

RFC	Title
773	Comments on NCP/TCP mail service transition strategy. October 1980
772	Mail Transfer Protocol. (Obsoleted by RFC 780) September 1980
771	Mail transition plan. September 1980
770	Assigned numbers. (Obsoletes RFC 762; Obsoleted by RFC 1340) September 1980
769	Rapicom 450 facsimile file format. September 1980
768	User Datagram Protocol. August 1980
767	Structured format for transmission of multi-media documents. August 1980
766	Internet Protocol Handbook: Table of contents. (Obsoleted by RFC 774) July 1980
765	File Transfer Protocol specification. (Obsoletes RFC 542; Obsoleted by RFC 959) June 1980
764	Telnet Protocol specification. (Obsoleted by RFC 854) June 1980
763	Role mailboxes May 1980
762	Assigned numbers. (Obsoletes RFC 758; Obsoleted by RFC 1340) January 1980
761	DoD standard Transmission Control Protocol. January 1980
760	DoD standard Internet Protocol. (Obsoletes IEN 123; Obsoleted by RFC 791, 777) January 1980
759	Internet Message Protocol. August 1980
758	Assigned numbers. (Obsoletes RFC 755; Obsoleted by RFC 1340) August 1980
757	Suggested solution to the naming, addressing, and delivery problem for ARPANET message systems. September 1979
756	NIC name server: A datagram-based information utility. July 1979
755	Assigned numbers. (Obsoletes RFC 750; Obsoleted by RFC 1340) May 1979
754	Out-of-net host addresses for mail. April 1979
753	Internet Message Protocol. May 1979
752	Universal host table. January 1979
751	Survey of FTP mail and MLFL. December 1978
750	Assigned numbers. (Obsoletes RFC 739; Obsoleted by RFC 1340) September 1978
749	Telnet SUPDUP-Output option. September 1978
748	Telnet randomly-lose option. April 1978
747	Recent extensions to the SUPDUP Protocol. March 1978

RFC	Title
746	SUPDUP graphics extension. 1978 March
745	JANUS interface specifications. March 1978
744	A Message Archiving & Retrieval Service. January 1978
743	FTP extension: XRSQ/XRCP. December 1977
742	NAME/FINGER Protocol. (Obsoleted by RFC 1288) December 1977
741	Specifications for the Network Voice Protocol (NVP). November 1977
740	NETRJS Protocol. (Obsoletes RFC 599) November 1977
739	Assigned numbers. (Obsoletes RFC 604, 503; Obsoleted by RFC 1340) November 1977
738	Time server. October 1977
737	FTP extension: XSEN October 1977
736	Telnet SUPDUP option. October 1977
735	Revised Telnet byte macro option. (Obsoletes RFC 729) November 1977
734	SUPDUP Protocol. October 1977
733	Standard for the format of ARPA network text messages. (Obsoletes RFC 724; Obsoleted by RFC 822) November 1977
732	Telnet Data Entry Terminal option. (Obsoletes RFC 731; Updated by RFC 1043) September 1977
731	Telnet Data entry Terminal option. (Obsoleted by RFC 732) June 1977
730	Extensible field addressing. May 1977
729	Telnet byte macro option. (Obsoleted by RFC 735) May 1977
728	Minor pitfall in the Telnet Protocol. April 1977
727	Telnet logout option. April 1977
726	Remote Controlled Transmission and Echoing Telnet option. March 1977
725	RJE protocol for a resource sharing network. March 1977
724	Proposed official standard for the format of ARPA Network messages. (Obsoleted by RFC 733) May 1977
723	Not issued.
722	Thoughts on interactions in distributed services. September 1976
721	Out-of-band control signals in a Host-to-Host Protocol. September 1976
720	Address specification syntax for network mail. August 1976

RFC	Title
719	Discussion on RCTE. July 1076
718	Comments on RCTE from the Tenex implementation experience. June 1976
717	Assigned network numbers. July 1976
716	Interim revision to Appendix F of BBN 1822. May 1976
715	Not issued.
714	Host-Host Protocol for an ARPANET-type network. (NOL) April 1976
713	MSDTP-Message Services Data Transmission Protocol. April 1976
712	Distributed Capability Computing System (DCCS). (NOL) February 1976
711	Not issued.
710	Not issued.
709	Not issued.
708	Elements of a distributed programming system. January 1976
707	High-level framework for network-based resource sharing. December 1975
706	On the junk mail problem. November 1975
705	Front-end Protocol B6700 version. November 1975
704	IMP/Host and Host/IMP Protocol change. (Obsoletes RFC 687) September 1975
703	July, 1975 survey of New-Protocol Telnet Servers. (NOL) July 1975
702	September, 1974 survey of New-Protocol Telnet servers. (NOL) July 1974
701	August, 1974 survey of New-Protocol Telnet servers. August 1974
700	Protocol experiment. August 1974
699	Request For Comments Summary Notes: 600-699. November 1982
698	Telnet extended ASCII option. July 1975
697	CWD command of FTP. (NOL) July 1975
696	Comments on the IMP/Host and Host/IMP Protocol changes. (NOL) July 1975
695	Official change in Host-Host Protocol. July 1975
694	Protocol information. (NOL) June 1975
693	Not issued.
692	Comments on IMP/Host Protocol changes (RFCs 687 and 690) June 1975

RFC	Title
691	One more try on the FTP. May 1975
690	Comments on the proposed Host/IMP Protocol changes. (NOL) (Updates RFC 687; Updated by RFC 692) June 1975
689	Tenex NCP finite state machine for corrections. May 1975
688	Tentative schedule for the new Telnet implementation for the TIP. (NOL) June 1975
687	IMP/Host and Host/IMP Protocol changes. (Obsoleted by RFC 704: Updated by RFC 690) June 1975
686	Leaving well enough alone. (NOL) May 1975
685	Response time in cross network debugging. April 1975
684	Commentary on procedure calling as a network protocol. April 1975
683	FTPSRV-Tenex extension for paged files. April 1975
682	Not issued.
681	Network UNIX. March 1975
680	Message Transmission Protocol. (NOL) April 1975
679	February, 1975 survey of New-Protocol Telnet servers. (NOL) February 1975
678	Standard file formats. December 1974
677	Maintenance of duplicate databases. (NOL) January 1975
676	Not issued.
675	Specification on Internet Transmission Control Program. (NOL) December 1974
674	Procedure call documents: Version2. December 1974
673	Not issued.
672	Multi-site data collection facility. December 1974
671	Note on Reconnection Protocol. (NOL) December 1974
670	Not issued.
669	November, 1974 survey of New-Protocol Telnet servers. (NOL) December 1974
668	Not issued.
667	BBN host ports. (NOL) December 1974
666	Specification of the Unified User-Level Protocol. (NOL) November 1974

RFC	Title
665	Not issued.
664	Not issued.
663	Lost message detection and recovery protocol. November 1974
662	Performance improvement in ARPANET file transfers from Multics. November 1974
661	Protocol information. (NOL) November 1974
660	Some changes to the IMP and the IMP/Host interface. October 1974
659	Announcing additional Telnet options. (NOL) October 1974
658	Telnet output linefeed disposition. October 1974
657	Telnet output vertical tab disposition option. October 1974
656	Telnet output vertical tabstops option. October 1974
655	Telnet output formfeed disposition option. October 1974
654	Telnet output horizontal tab disposition option. October 1974
653	Telnet output horizontal tabstops option. October 1974
652	Telnet output carriage-return disposition option. October 1974
651	Revised Telnet status option. (Obsoleted by RFC 859) October 1974
650	Not issued.
649	Not issued.
648	Not issued
647	Proposed protocol for connecting host computers to ARPA-like networks via front end processors. (NOL) November 1974
646	Not issued.
645	Network Standard Data Specification syntax. (NOL) June 1974
644	On the problem of signature authentication for network mail. July 1974
643	Network Debugging Protocol. July 1974
642	Ready line philosophy and implementation. (NOL) July 1974
641	Not issued.
640	Revised FTP reply codes. June 1974
639	Not issued.
638	IMP/TIP preventive maintenance schedule. (NOL) April 1974

RFC	Title
637	Change of network address for SU-DSL. (NOL) April 1974
636	TIP/Tenex reliability improvements. June 1974
635	Assessment of ARPANET protocols. (NOL) April 1974
634	Change in network address for Haskins Lab. (NOL) April 1974
633	IMP/TIP preventive maintenance schedule. (NOL) April 1974
632	Throughput degradations for single packet messages. (NOL) May 1974
631	International meeting on minicomputers and data communication: Call for papers. (NOL) April 1974
630	FTP error code usage for more reliable mail service. (NOL) April 1974
629	Scenario for using the Network Journal. (NOL) March 1974
628	Status of RFC numbers and a note on preassigned journal numbers. (NOL) March 1974
627	ASCII text file of hostnames. (NOL) March 1974
626	On a possible lockup condition in IMP subnet due to message sequencing. March 1974
625	On-line hostnames service. (NOL) March 1974
624	Comments on the File Transfer Protocol. (Obsoletes RFC 607) February 1974
623	Comments on on-line host name service. (NOL) February 1974
622	Scheduling IMP/TIP down time. (NOL) February 1974
621	NIC user directories at SRI ARC. (NOL) March 1974
620	Request for Monitor Host Table Updates. March 1974
619	Mean round-trip times in the ARPANET. (NOL) March 1974
618	Few observations on NCP statistics. February 1974
617	Note on socket member assignment. February 1974
616	Latest network maps. February 1973
615	Proposed Network Standard Data Pathname syntax. March 1974
614	Response to RFC 607: "Comments on the File Transfer Protocol." (Updates RFC 607) January 1974
613	Network connectivity: A response to RFC 603. (NOL) (Updates RFC 603) January 1974
612	Traffic statistics (December 1973). (NOL) January 1974

RFC	Title
611	Two changes to the IMP/Host Protocol to improve user/network communications. (NOL) February 1974
610	Further datalanguage design concepts. (NOL) December 1973
609	Statement of upcoming move on NIC/NLS service. (NOL) January 1974
608	Host names on-line. (NOL) (Obsoleted by RFC 810) January 1974
607	Comments on the File Transfer Protocol. (Obsoleted by RFC 624; Updated by RFC 614) January 1974
606	Host names on-line. December 1973
605	Not issued.
604	Assigned link numbers. (NOL) (Obsoletes RFC 317; Obsoleted by RFC 1340) December 1973
603	Response to RFC 597: Host status. (NOL) (Updates RFC 597; Updated by RFC 613) December 1973
602	"The stockings were hung by the chimney with care." December 1973
601	Traffic statistics (November 1973). (NOL) December 1973
600	Interfacing an Illinois plasma terminal to the ARPANET. (NOL) November 1973
599	Update on NetRJS. (Obsoletes RFC 189; Obsoleted by RFC 740) December 1973
598	Network Information Center; SRI NIC RFC index, December 5, 1973. (NOL) December 1973
597	Host status. (NOL) (Updated by RFC 603) December 1973
596	Second thoughts on Telnet Go-Ahead. (NOL) December 1973
595	Second thoughts in defense on the Telnet Go-Ahead. (NOL) December 1973
594	Speedup of Host-IMP interface. (NOL) December 1973
593	Telnet and FTP implementation schedule change. (NOL) November 1973
592	Some thoughts on system design to facilitate resource sharing. (NOL) November 1973
591	Addition to the Very Distant Host specifications. (NOL) November 1973
590	MULTICS address change. (NOL) November 1973
589	CCN NETRJS server messages to remote user. (NOL) November 1973
588	London node is now up. (NOL) October 1973
587	Announcing new Telnet options. (NOL) October 1973

RFC	Title
586	Traffic statistics (October 1973). (NOL) November 1973
585	ARPANET users interest working group meeting. (NOL) November 1973
584	Charter for ARPANET Users Interest Working Group. (NOL) November 1973
583	Not issued.
582	Comments on RFC 580: Machine readable protocols. (NOL) (Updates RFC 580) November 1973
581	Corrections to RFC 560: Remote Controlled Transmission and Echoing Telnet option. (NOL) November 1973
580	Note to protocol designers and implementers. (Updated by RFC 582) October 1973
579	Traffic statistics. (September 1973) November 1973
578	Using MIT-Mathlab MACSYMA from MIT-DMS Muddle. (NOL) October 1973
577	Mail priority. (NOL) October 1973
576	Proposal for modifying linking. (NOL) September 1973
575	Not issued.
574	Announcement of a mail facility at UCSB. (NOL) September 1973
573	Data and file transfer: Some measurement results. (NOL) September 1973
572	Not issued.
571	Tenex FTP problem. (NOL) November 1973
570	Experimental input mapping between NVT ASCII and UCSB on-line systems. (NOL) October 1973
569	NETED: A common editor for the ARPA network. October 1973
568	Response to RFC 567: Cross-country network bandwidth. (NOL) (Updated by RFC 568) September 1973
567	Cross-country network bandwidth. (Updated by RFC 568) September 1973
566	Traffic statistics (August 1973). (NOL) September 1973
565	Storing network survey data at the datacomputer. (NOL) August 1973
564	Not issued.
563	Comments on the RCTE Telnet option. (NOL) August 1973
562	Modifications to the Telnet specification. (NOL) August 1973
561	Standardizing Network mail headers. (Updated by RFC 680) September 1973

RFC	Title
560	Remote Controlled Transmission and Echoing Telnet option. (NOL) August 1973
559	Comments on the new Telnet Protocol and its implementation. (NOL) August 1973
558	Not issued.
557	Revelations in network host measurements. (NOL) August 1973
556	Traffic statistics (July 1973). (NOL) August 1973
555	Responses to critiques of the proposed mail protocol. (NOL) July 1973
554	Not issued.
553	Draft design for a text/graphics protocol. (NOL) July 1973
552	Single access to standard protocols. (NOL) July 1973
551	[Letter from Feinroth re: NYU, ANL, and LBL entering the net, and FTP protocol]. (NOL) August 1973.
550	NIC NCP experiment. (NOL) August 1973
549	Minutes of Network Graphics Group meeting, 15-17 July 1973. (NOL) July 1973
548	Hosts using the IMP Going Down message. (NOL) August 1973
547	Change to the Very Distant Host specification. (NOL) August 1973
546	Tenex load averages for July 1973. (NOL) August 1973
545	Of what quality be the UCSB resources evaluators? (NOL) July 1973
544	Locating on-line documentation at SRI-ARC. (NOL) July 1973
543	Network journal submission and delivery. (NOL) July 1973
542	File Transfer Protocol. (Obsoletes RFC 354; Obsoleted by RFC 765) July 1973
541	Not issued.
540	Not issued.
539	Thoughts on the mail protocol proposed in RFC 524. (NOL) July 1973
538	Traffic statistics (June 1973). (NOL) July 1973
537	Announcement of NGG meeting July 16-17. (NOL) June 1973
536	Not issued.
535	Comments on File Access Protocol. (NOL) July 1973
534	Lost message detection. (NOL) July 1973
533	Message-ID numbers. (NOL) July 1973

RFC	Title
532	UCSD-CC Server-FTP facility. (NOL) July 1973
531	Feast or famine? A response to two recent RFC's about network information. (NOL) June 1973
530	Report on the Survey project. (NOL) June 1973
529	Note on protocol synch sequences. (NOL) June 1973
528	Software checksumming in the IMP and network reliability. (NOL) June 1973
527	ARPAWOCKY. May 1973
526	Technical meeting: Digital image processing software systems. (NOL) June 1973
525	MIT-MATHLAB meets UCSB-OLS—an example of resource sharing. (NOL) June 1973
524	Proposed Mail Protocol. (NOL) June 1973
523	SURVEY is in operation again (NOL) June 1973
522	Traffic statistics (May 1973). (NOL) June 1973
521	Restricted use of IMP DDT. (NOL) May 1973
520	Memo to FTP group: Proposal for File Access Protocol. (NOL) June 1973
519	Resource evaluation. (NOL) June 1973
518	ARPANET accounts. (NOL) June 1973
517	Not issued.
516	Lost message detection. (NOL) May 1973
515	Specifications for datalanguage: Version 0/9. (NOL) June 1973
514	Network make-work. (NOL) June 1973
513	Comments on the new Telnet specifications. (NOL) May 1973
512	More on lost message detection. (NOL) May 1973
511	Enterprise phone service to NIC from ARPANET sites. (NOL) May 1973
510	Request for network mailbox addresses. (NOL) May 1973
509	Traffic statistics (April 1973). (NOL) May 1973
508	Real-time data transmission on the ARPANET. (NOL) May 1973
507	Not issued.
506	FTP command naming problem. (NOL) June 1973

RFC	Title
505	Two solutions to a file transfer access problem. (NOL) June 1973
504	Distributed resources workshop. (NOL) April 1973
503	Socket number list. (NOL) (Obsoletes RFC 433; Obsoleted by RFC 1340) April 1973
502	Not issued.
501	Un-muddling "free file transfer." (NOL) May 1973
500	Integration of data management systems on a computer network. (NOL) April 1973
499	Harvard's network RJE. (NOL) April 1973
498	On mail service to CCN. (NOL) April 1973
497	Traffic statistics (March 1973). (NOL) April 1973
496	TNLS quick reference card is available. (NOL) April 1973
495	Telnet Protocol specifications. (NOL) (Obsoletes RFC 158) May 1973
494	Availability of MIX and MIXAL in the Network. (NOL) April 1973
493	Graphics Protocol. (NOL) April 1973
492	Response to RFC 467. (NOL) (Updates RFC 467) April 1973
491	What is "free"? (NOL) April 1973
490	Surrogate RJS for UCLA-CCN. (NOL) March 1973.
489	Comment on resynchronization of connection status proposal. (NOL) March 1973
488	NLS classes at network sites. (NOL) March 1973
487	Free file transfer. (NOL) April 1973
486	Data transfer revisited. (NOL) March 1973
485	MIX and MIXAL at UCSB. (NOL) March 1973
484	Not isssued.
483	Cancellation of the resource notebook framework meeting. (NOL) March 1973
482	Traffic statistics (February 1973). (NOL) March 1973
481	Not issued.
480	Host-dependent FTP parameters. (NOL) March 1973
479	Use of FTP by the NIC Journal. (NOL) March 1973
478	FTP server-server interaction, II. (NOL) March 1973

RFC	Title
477	Remote Job Service at UCSB. (NOL) May 1973
476	IMP/TIP memory retrofit schedule (rev. 2). (NOL) (Obsoletes RFC 447) March 1973
475	FTP and network mail system. (NOL) March 1973
474	Announcement of NGWG meeting: Call for papers. (NOL) March 1973
473	MIX and MIXAL? (NOL) February 1973
472	Illinois' reply to Maxwell's request for graphics information (NIC 14925). (NOL) March 1973
471	Workshop on multi-site executive programs. (NOL) March 1973
470	Change in socket for TIP news facility. (NOL) March 1973
469	Network mail meeting summary. (NOL) March 1973
468	FTP data compression. (NOL) March 1973
467	Proposed change to Host-Host Protocol: Resynchronization of connection status. (NOL) (Updated by RFC 492) February 1973
466	Telnet logger/server for host LL-67. (NOL) February 1973
465	Not issued.
464	Resource notebook framework. (NOL) February 1973
463	FTP comments and response to RFC 430. (NOL) February 1973
462	Responding to user needs. (NOL) February 1973
461	Telnet Protocol meeting announcement. (NOL) February 1973
460	NCP survey. (NOL) February 1973
459	Network questionnaires. (NOL) February 1973
458	Mail retrieval via FTP. (NOL) February 1973
457	TIPUG. (NOL) February 1973
456	Memorandum: Date change of mail meeting. (NOL) February 1973
455	Traffic statistics for January 1973. (NOL) February 1973
454	File Transfer Protocol—meeting announcement and a new proposed document. (NOL) February 1973
453	Meeting announcement to discuss a network mail system. (NOL) February 1973
452	Not issued.

RFC	Title
451	Tentative proposal for a Unified User Level Protocol. (NOL) February 1973
450	MULTICS sampling timeout change. (NOL) February 1973
449	Current flow-control scheme for IMPSYS. (NOL) (Updates RFC 442) January 1973
448	Print files in FTP. (NOL) February 1973
447	IMP/TIP memory retrofit schedule. (NOL) (Obsoletes RFC 434; Obsoleted by RFC 476) January 1973
446	Proposal to consider a network program resource notebook. (NOL) January 1973
445	IMP/TIP preventive maintenance schedule. (NOL) January 1973
444	Not issued.
443	Traffic statistics. (December 1972). (NOL) January 1973
442	Current flow-control scheme for IMPSYS. (NOL) (Updated by RFC 449) January 1973
441	Inter-Entity Communication—An experiment. (NOL) January 1973
440	Scheduled network software maintenance. (NOL) January 1973
439	PARRY encounters the DOCTOR. (NOL) January 1973
438	FTP server-server interaction. (NOL) January 1973
437	Data Reconfiguration Service at UCSB. (NOL) January 1973
436	Announcement of RJS at UCSB. (NOL) January 1973
435	Telnet issues. (NOL) (Updates RFC 318) January 1973
434	IMP/TIP memory retrofit schedule. (NOL) January 1973
433	Socket number list. (NOL) (Obsoletes RFC 349; Obsoleted by RFC 1340) December 1972
432	Network logical map. (NOL) December 1972
431	Update on SMFS login and logout. (NOL) (Obsoletes RFC 399) December 1972
430	Comments on File Transfer Protocol. (NOL) February 1972
429	Character generator process. (NOL) December 1972
428	Not issued.
427	Not issued.
426	Reconnection Protocol. (NOL) 1973 January
425	But my NCP costs $500 a day.... 1972 December

RFC	Title
424	Not issued.
423	UCLA Campus Computing Network liaison staff for ARPANET. (NOL) (Obsoletes RFC 389) December 1972
422	Traffic statistics (November 1972). (NOL) December 1972
421	Software consulting service for network users. (NOL) November 1972
420	CCA ICCC weather demo. (NOL) January 1973
419	To: Network liaisons and station agents. (NOL) December 1972
418	Server file transfer under TSS/360 at NASA Ames. (NOL) November 1972
417	Link usage violation. (NOL) December 1972
416	ARC system will be unavailable for use during Thanksgiving week. (NOL) November 1972
415	Tenex bandwidth. (NOL) November 1972
414	File Transfer Protocol (FTP) status and further comments. (NOL) (Updates RFC 385) December 1972
413	Traffic statistics (October 1972). (NOL) November 1972
412	User FTP documentation. (NOL) November 1972
411	New MULTICS network software features. (NOL) November 1972
410	Removal of the 30-second delay when hosts come up. (NOL) November 1972
409	Tenex interface to UCSB's Simple-Minded File System. (NOL) December 1972
408	NETBANK. (NOL) October 1972
407	Remote Job Entry Protocol. (Obsoletes RFC 360) October 1972
406	Scheduled IMP software releases. (NOL) October 1972
405	Correction to RFC 404. (NOL) (Obsoletes RFC 404) October 1972
404	Host address changes involving Rand and ISI. (NOL) (Obsoleted by RFC 405) October 1972
403	Desirability of a network 1108 service. (NOL) January 1973
402	ARPA Network mailing lists. (NOL) (Obsoletes RFC 363) October 1972
401	Conversion of NGP-0 coordinates to device-specific coordinates. (NOL) October 1972
400	Traffic statistics (September 1972). (NOL) October 1972
399	SMFS login and logout. (NOL) (Obsoleted by RFC 431; Updates RFC 122) September 1972

RFC	Title
398	ICP sockets. (NOL) September 1972
397	Not issued.
396	Network Graphics Working Group meeting—second iteration. (NOL) November 1972
395	Switch settings on IMPs and TIPs. (NOL) October 1972
394	Two proposed changes to the IMP-Host Protocol. (NOL) September 1972
393	Comments on Telnet Protocol changes. (NOL) October 1972
392	Measurement of host costs for transmitting network data. (NOL) September 1972
391	Traffic statistics (August 1972). (NOL) (Obsoletes RFC 378) September 1972
390	TSO scenario. (NOL) September 1972
389	UCLA Campus Computing Network liaison staff for ARPA Network. (NOL) (Obsoleted by RFC 423) August 1972.
388	NCP statistics. (NOL) (Updates RFC323) August 1972
387	Some experiences in implementing Network Graphics Protocol Level 0. (NOL) August 1972
386	Letter to TIP users-2. (NOL) August 1972
385	Comments on the File Transfer Protocol. (NOL) (Updates RFC 354; Updated by RFC 414) August 1972
384	Official site idents for organizations in the ARPA Network. (NOL) (Obsoletes RFC 289) August 1972
383	Not issued.
382	Mathematical software on the ARPA Network. (NOL) August 1972
381	Three aids to improved network operation. (NOL) July 1972
380	Not issued.
379	Using TSO at CCN. (NOL) August 1972
378	Traffic statistics (July 1972). (NOL) (Obsoleted by RFC 391) August 1972
377	Using TSO via ARPA Network Virtual Terminal. (NOL) August 1972
376	Network host status. (NOL) (Obsoletes RFC 370) August 1972
375	Not issued.
374	IMP system announcement. (NOL) July 1972
373	Arbitrary character sets. (NOL) July 1972

RFC	Title
372	Notes on a conversation with Bob Kahn on the ICCC. (NOL) July 1972
371	Demonstration at International Computer Communications Conference. (NOL) July 1972
370	Network host status. (NOL) (Obsoletes RFC 367; Obsoleted by RFC 376) July 1972
369	Evaluation of ARPANET services January-March, 1972. (NOL) July 1972
368	Comments on "Proposed Remote Job Entry Protocol." (NOL) July 1972
367	Network host status. (NOL) (Obsoletes RFC 366; Obsoleted by RFC 370) July 1972
366	Network host status. (NOL) (Obsoletes RFC 362; Obsoleted by RFC 367) July 1972
365	Letter to all TIP users. (NOL) July 1972
364	Serving remote users on the ARPANET. (NOL) July 1972
363	Stanford Research Inst., Network Information Center; SRI NIC ARPA Network mailing lists. (NOL) (Obsoletes RFC 329; Obsoleted by RFC 402) August 1972
362	Network host status. (NOL) (Obsoletes RFC 353; Obsoleted by RFC 366) June 1972
361	Daemon processes on host 106. (NOL) July 1972
360	Proposed Remote Job Entry Protocol. (NOL) (Obsoleted by RFC 407) June 1972
359	Status of the release of the new IMP System. (3600) (NOL) (Obsoletes RFC 343) June 1972
358	Not issued.
357	Echoing strategy for satellite links. (NOL) June 1972
356	ARPA Network Control Center. (NOL) June 1972
355	Response to NWG/RFC 346. (NOL) June 1972
354	File Transfer Protocol. (NOL) (Obsoletes RFC 264, RFC 265; Obsoleted by RFC 542; Updated by RFC 385) July 1972
353	Network host status. (NOL) (Obsoletes RFC 344; Obsoleted by RFC 362) June 1972
352	TIP site information form. (NOL) June 1972
351	Graphics information form for the ARPANET graphics resources notebook. (NOL) June 1972
350	User accounts for UCSB On-Line System. (NOL) May 1972
349	Proposed standard socket numbers. (NOL) (Obsoleted by RFC 1340) May 1972
348	Discard process. (NOL) May 1972

RFC	Title
347	Echo process. (NOL) May 1972
346	Satellite considerations. (NOL) May 1972
345	Interest in mixed integer programming. (MPSX on NIC 360/91 at CCN) (NOL) May 1972
344	Network host status. (NOL) (Obsoletes RFC 342; Obsoleted by RFC 353) May 1972
343	IMP System change notification. (NOL) (Obsoletes RFC 331; Obsoleted by RFC 359) May 1972
342	Network host status. (NOL) (Obsoletes RFC 332; Obsoleted by RFC 344) May 1972
341	Not issued.
340	Proposed Telnet changes. (NOL) May 1972
339	MLTNET: A "Multi Telnet" subsystem for Tenex. (NOL) May 1972
338	EBCDIC/ASCII mapping for network RJE. (NOL) May 1972
337	Not issued.
336	Level 0 Graphic Input Protocol. (NOL) May 1972
335	New interface—IMP/360. (NOL) May 1972.
334	Network use on May 8. (NOL) May 1972
333	Proposed experiment with a Message Switching Protocol. (NOL) May 1972
332	Network host status. (NOL) (Obsoletes RFC 330; Obsoleted by RFC 342) April 1972
331	IMP System change notification. (NOL) (Obsoleted by RFC 343) April 1972
330	Network host status. (NOL) (Obsoletes RFC 326; Obsoleted by RFC 332) April 1972
329	Stanford Research Inst., Network Information Center; SRI NIC ARPA Network mailing lists. (NOL) (Obsoletes RFC 303; Obsoleted by RFC 363) May 1972
328	Suggested Telnet Protocol changes. (NOL) April 1972
327	Data and File Transfer workshop notes. (NOL) April 1972
326	Network host status. (NOL) (Obsoletes RFC 319; Obsoleted by RFC 330) April 1972
325	Network Remote Job Entry program—NETRJS. (NOL) April 1972
324	RJE Protocol meeting. (NOL) April 1972
323	Formation of Network Measurement Group (NMG). (NOL) (Updated by RFC 388) March 1972
322	Well-known socket numbers. (NOL) March 1972

RFC	Title
321	CBI networking activity at MITRE. (NOL) March 1972
320	Workshop on hard copy line printers. (NOL) March 1972
319	Network host status. (NOL) (Obsoletes RFC 315; Obsoleted by RFC 326) March 1972
318	[Ad hoc Telnet Protocol]. (NOL) (Updates RFC 158; Updated by RFC 435) April 1972
317	Official Host-Host Protocol modification: Assigned link numbers. (NOL) (Obsoleted by RFC 604) March 1972
316	ARPA Network Data Management Working Group. (NOL) February 1972
315	Network host status. (NOL) (Obsoletes RFC 306; Obsoleted by RFC 319) March 1972
314	Network Graphics Working Group meeting. (NOL) March 1972
313	Computer based instruction. (NOL) March 1972
312	Proposed change in IMP-to-Host Protocol. (NOL) March 1972
311	New console attachments to the USCB host. (NOL) February 1972
310	Another look at Data and File Transfer Protocols. (NOL) April 1972
309	Data and File Transfer workshop announcement. (NOL) March 1972
308	ARPANET host availability data. (NOL) March 1972
307	Using network Remote Job Entry. (NOL) February 1972
306	Network host status. (NOL) (Obsoletes RFC 298; Obsoleted by RFC 315) February 1972
305	Unknown host numbers. (NOL) February 1972
304	Data management system proposal for the ARPA network. (NOL) February 1972
303	Stanford Research Institute, Network Information Center; SRI NIC ARPA Network mailing lists. (NOL) (Obsoletes RFC 300; Obsoleted by RFC 329) February 1972
302	Exercising the ARPANET. (NOL) February 1972
301	BBN IMP (#5) and NCC schedule March 4, 1971. (NOL) February 1972
300	ARPA Network mailing lists. (NOL) (Obsoletes RFC 211; Obsoleted by RFC 303) January 1972
299	Information management system. (NOL) February 1972
298	Network host status. (NOL) (Obsoletes RFC 293; Obsoleted by RFC 306) February 1972
297	TIP message buffers. (NOL) January 1972

RFC	Title
296	DS-1 display system. (NOL) January 1972
295	Report of the Protocol Workshop. (NOL) January 1972
294	On the use of "set data type" transaction in File Transfer Protocol. (NOL) (Updates RFC 265) January 1972
293	Network host status. (NOL) (Obsoletes RFC 288; Obsoleted by RFC 298) January 1972
292	Graphics Protocol: Level 0 only. (NOL) January 1972
291	Data management meeting announcement. (NOL) January 1972
290	Computer networks and data sharing: A bibliography. (NOL) (Obsoletes RFC 243) January 1972
289	What we hope is an official list of host names. (NOL) (Obsoleted by RFC 384) December 1971
288	Network host status. (NOL) (Obsoletes RFC 287; Obsoleted by RFC 293) January 1972
287	Status of network hosts. (NOL) (Obsoletes RFC 267; Obsoleted by RFC 288) December 1971
286	Network library information system. (NOL) December 1971
285	Network graphics. (NOL) December 1971
284	Not issued.
283	NETRJT: Remote Job Service Protocol for TIPS. (NOL) (Updates RFC 189) December 1971
282	Graphics meeting report. (NOL) December 1971
281	Suggested addition to File Transfer Protocol. (NOL) December 1971
280	Draft of host names. (NOL) November 1971
279	Not issued.
278	Revision of the Mail Box Protocol. (NOL) (Obsoletes RFC 221) November 1971
277	Not issued.
276	NIC course. (NOL) November 1971
275	Not issued.
274	Establishing a local guide for network usage. (NOL) November 1971
273	More on standard host names. (NOL) (Obsoletes RFC 237) October 1971

RFC	Title
272	Not issued.
271	IMP System change notifications. (NOL) January 1972
270	Correction to BBN Report No. 1822 (NIC NO 7958). (NOL) (Updates NIC 7959) January 1972
269	Some experience with file transfer. (NOL) (Updates RFC 122, 238) December 1971
268	Graphics facilities information. (NOL) November 1971
267	Network host status. (NOL) (Obsoletes RFC 266; Obsoleted by RFC 287) November 1971
266	Network host status. (NOL) (Obsoletes RFC 255; Obsoleted by RFC 267) November 1971
265	File Transfer Protocol. (NOL) (Obsoletes RFC 172; Obsoleted by RFC 354; Updated by RFC 294) November 1971
264	Data Transfer Protocol. (NOL) (Obsoletes RFC 171; Obsoleted by RFC 354) December 1971
263	"Very distant" host interface. (NOL) December 1971
262	Not issued.
261	Not issued.
260	Not issued.
259	Not issued.
258	Not issued.
257	Not issued.
256	IMPSYS change notification. (NOL) November 1971
255	Status of network hosts. (NOL) (Obsoletes RFC 252; Obsoleted by RFC 266) October 1971
254	Scenarios for using ARPANET computers. (NOL) October 1971
253	Second Network Graphics meeting details. (NOL) October 1971
252	Network host status. (NOL) (Obsoletes RFC 240; Obsoleted by RFC 255) October 1971
251	Weather data. (NOL) October 1971
250	Some thoughts on file transfer. (NOL) October 1971
249	Coordination of equipment and supplies purchase. (NOL) October 1971

RFC	Title
248	Not issued.
247	Proffered set of standard host names. (NOL) (Obsoletes RFC 226) October 1971
246	Network Graphics meeting. (NOL) October 1971
245	Reservations for Network Group meeting. (NOL) October 1971
244	Not issued.
243	Network and data sharing bibliography. (NOL) (Obsoleted by RFC 290) October 1971
242	Data descriptive language for shared data. (NOL) July 1971
241	Connecting computers to MLC ports. (NOL) September 1971
240	Site status. (NOL) (Obsoletes RFC 235; Obsoleted by RFC 252) September 1971
239	Host mnemonics proposed in RFC 226. (NIC 7625) (NOL) September 1971
238	Comments on DTP and FTP proposals. (NOL) (Updates RFC 171, 172; Updated by RFC 269) September 1971
237	NIC view of standard host names. (NOL) (Obsoleted by RFC 273) September 1971
236	Standard host names. (NOL) (Obsoleted by RFC 240) September 1971
235	Site status. (NOL) (Obsoleted by RFC 240) September 1971
234	Network Working Group meeting schedule. (NOL) (Updates RFC 222, 204) October 1971
233	Standardization of host call letters. (NOL) September 1971
232	Postponement of network graphics meeting. (NOL) September 1971
231	Service center standards for remote usage: A user's view. (NOL) September 1971
230	Toward reliable operation of minicomputer-based terminals on a TIP. (NOL) September 1971
229	Standard host names. (NOL) (Updates RFC 70) September 1971
228	Clarification. (NOL) (Updates RFC 70) September 1971
227	Data transfer rates (Rand/UCLA). (NOL) (Updates RFC 113) September 1971
226	Standardization of host mnemonics. (NOL) (Obsoleted by RFC 247) September 1971
225	Rand/UCSB network graphics experiment. (NOL) (Updates RFC 74) September 1971
224	Comments on Mailbox Protocol. (NOL) September 1971

RFC	Title
223	Network Information Center schedule for network users. (NOL) September 1971
222	Subject: System programmer's workshop. (NOL) (Updates RFC 212; Updated by RFC 234) September 1971
221	Mail Box Protocol: Version 2. (NOL) (Obsoletes RFC 196; Obsoleted by RFC 278) August 1971
220	Not issued.
219	User's view of the datacomputer. (NOL) September 1971
218	Changing the IMP status reporting facility. (NOL) September 1971
217	Specifications changes for OLS, RJE/RJOR, and SMFS. (NOL) (Updates RFC 74, 105, 122) September 1971
216	Telnet access to UCSB's On-Line System. (NOL) September 1971
215	NCP, ICP, and Telnet: The Terminal IMP implementation. (NOL) August 1971
214	Network checkpoint. (NOL) (Obsoletes RFC 198) August 1971
213	IMP System change notification. (NOL) August 1971
212	University of Southern California, Information Sciences Institute; USC ISI NWG meeting on network usage. (NOL) (Obsoletes RFC 207; Updated by RFC 222) August 1971
211	ARPA network mailing lists. (NOL) (Obsoletes RFC 168; Obsoleted by RFC 300) August 1971
210	Improvement of flow control. (NOL) August 1971
209	Host/IMP interface documentation. (NOL) August 1971
208	Address tables. (NOL) August 1971
207	September Network Working Group meeting. (NOL) (Obsoleted by RFC 212) August 1971
206	User Telnet—description of an initial implementation. (NOL) August 1971
205	NETCRT—a character display protocol. (NOL) August 1971
204	Sockets in use. (NOL) (Updated by RFC 234) August 1971
203	Achieving reliable communication. (NOL) August 1971
202	Possible deadlock in ICP. (NOL) July 1971
201	Not issued.
200	RFC list by number. (NOL) (Obsoletes RFC 170, 160; Obsoleted by MIC 7724) August 1971

RFC	Title
199	Suggestions for a network data-tablet graphics protocol. (NOL) July 1971
198	Site certification—Lincoln Labs 360/67. (NOL) (Obsoletes RFC 193; Obsoleted by RFC 214) July 1971
197	Initial Connection Protocol—Reviewed. (NOL) July 1971
196	Mail Box Protocol. (NOL) (Obsoleted by RFC 221) July 1971
195	Data computers-data descriptions and access language. (NOL) July 1971
194	Data Reconfiguration Service—compiler/interpreter implementation notes. (NOL) July 1971
193	Network checkout. (NOL) (Obsoleted by RFC 198) July 1971
192	Some factors which a Network Graphics Protocol must consider. (NOL) July 1971
191	Graphics implementation and conceptualization at Augmentation Research Center. (NOL) July 1971
190	DEC PDP-10-IMLAC communication system. (NOL) July 1971
189	Interim NETRJS specifications. (Obsoletes RFC 88; Obsoleted by RFC 599; Updated by RFC 283) July 1971
188	Data management meeting announcement. (NOL) January 1971
187	Network/440 protocol concept. (NOL) July 1971
186	Network graphics loader. (NOL) July 1971
185	NIC distribution of manuals and handbooks. (NOL) July 1971
184	Proposed graphic display modes. (NOL) July 1971
183	EBCDIC codes and their mapping to ASCII. (NOL) July 1971
182	Compilation of list of relevant site reports. (NOL) June 1971
181	Modifications to RFC 177. (NOL) (Updates RFC 177) July 1971
180	File system questionnaire. (NOL) June 1971
179	Link number assignments. (Updates RFC 107) June 1971
178	Network graphic attention handling. (NOL) June 1971
177	Device independent graphical display description. (NOL) (Updates RFC 125; Updated by RFC 181) June 1971
176	Comments on "Byte size for connections." (NOL) June 1971
175	Socket conventions reconsidered. (NOL) June 1971

RFC	Title
174	UCLA Computer science graphics overview. (NOL) June 1971
173	Network data management committee meeting announcement. (NOL) June 1971
172	File Transfer Protocol. (NOL) (Obsoleted by RFC 265; Updates RFC 114; Updated by RFC 238) June 1971
171	Data Transfer Protocol. (NOL) (Obsoleted by RFC 264; Updates RFC 114; Updated by RFC 238) June 1971
170	Stanford Research Institute, Network Information Center; SRI NIC. RFC list by number. (NOL) (Obsoleted by RFC 200) June 1971
169	Computer networks. (NOL) May 1971
168	ARPA Network mailing lists. (NOL) (Obsoletes RFC 155; Obsoleted by RFC 211) May 1971
167	Socket conventions reconsidered. (NOL) May 1971
166	Data Reconfiguration Service: An implementation specification. (NOL) May 1971
165	Preferred official Initial Connection Protocol. (NOL) (Obsoletes RFC 145, 143, 123; Updated by NIC 7101) May 1971
164	Minutes of Network Working group meeting, 5/16 through 5/19/71. (NOL) May 1971
163	Data transfer protocols. (NOL) May 1971
162	NETBUGGERS3. (NOL) May 1971
161	Solution to the race condition in the ICP. (NOL) May 1971
160	Stanford Research Inst., Network Information Center; SRI NIC RFC brief. (NOL) (Obsoleted by RFC200; Updates NIC 6716) May 1971
159	Not issued.
158	Telnet Protocol: A proposed document. (NOL) (Obsoleted by RFC 495; Updates RFC 139; Updated by RFC 318) May 1971
157	Invitation to the Second Symposium on Problems in the Optimization of Data communications Systems. (NOL) May 1971
156	Status of the Illinois site: Response to RFC 116. (NOL) (Updates RFC 116) April 1971
155	ARPA Network mailing lists. (NOL) (Obsoletes RFC 95; Obsoleted by RFC 168) May 1971
154	Exposition style. (NOL) May 1971
153	SRI ARC-NIC status. (NOL) May 1971

RFC	Title
152	SRI Artificial Intelligence status report. (NOL) May 1971
151	Comments on a proffered official ICP: RFCs 123, 127. (NOL) (Updates RFC 127) May 1971
150	Use of IPC facilities: A working paper. (NOL) May 1971
149	Best laid plans. (Updates RFC 140) May 1971
148	Comments on RFC 123. (NOL) (Updates RFC 123) May 1971
147	Definition of a socket. (NOL) (Updates RFC 129) May 1971
146	Views on issues relevant to data sharing on computer networks. (NOL) May 1971
145	Initial Connection Protocol control commands. (NOL) (Obsoletes RFC 127; Obsoleted by RFC 165) May 1971
144	Data sharing on computer networks. (NOL) April 1971
143	Regarding proffered official ICP. (NOL) (Obsoleted by RFC 165) May 1971
142	Time-out mechanism in the Host-Host Protocol. (NOL) May 1971
141	Comments on RFC 114: A File Transfer Protocol. (NOL) (Updates RFC 114) April 1971
140	Agenda for the May NWG meeting. (NOL) (Updated by RFC 149) May 1971
139	Discussion of Telnet Protocol. (NOL) (Updates RFC 137; Updated by RFC 158) May 1971
138	Status report on proposed Data Reconfiguration Service. (NOL) April 1971
137	Telnet Protocol—a proposed document (NOL) (Updated by RFC 139) April 1971
136	Host accounting and administrative procedures. (NOL) April 1971
135	Response to NWG/RFC 110. (NOL) (Updates RFC 110) April 1971
134	Network Graphics meeting. (NOL) April 1971
133	File transfer and recovery. (NOL) April 1971
132	Typographical error in RFC 107. (NOL) (Obsoleted by RFC 154; Updates RFC 107) April 1971
131	Response to RFC 116: May NWG meeting. (NOL) (Updates RFC 116) April 1971
130	Response to RFC 111: Pressure from the chairman. (NOL) (Updates RFC 111) April 1971
129	Request for comments on socket name structure. (NOL) (Updated by RFC 147) April 1971

RFC	Title
128	Bytes. (NOL) April 1971
127	Comments on RFC 123. (NOL) (Obsoleted by RFC 145; Updates RFC 123; Updated by RFC 151) April 1971
126	Graphics facilities at Ames Research Center. (NOL) April 1971
125	Response to RFC 86: Proposal for network standard format for a graphics data stream. (NOL) (Updates RFC 86; Updated by RFC 177) April 1971
124	Typographical error in RFC 107. (NOL) (Updates RFC 107) April 1971
123	Proffered official ICP. (NOL) (Obsoletes 66, 80; Obsoleted by RFC 165; Updates RFC 98, 101; Updated by RFC 127, RFC 148) April 1971
122	Network specifications for UCSB's Simple-Minded File System. (NOL) (Updated by RFC 217, 269, 399) April 1971
121	Network on-line operators. (NOL) April 1971
120	Network PL1 subprograms. (NOL) April 1971
119	Network Fortran subprograms. (NOL) April 1971
118	Recommendations for facility documentation. (NOL) April 1971
117	Some comments on the official protocol. (NOL) April 1971
116	Structure of the May NWG Meeting. (NOL) (Updates RFC 99; Updated by RFC 131, 156) April 1971
115	Some Network Information Center policies on handling documents. (NOL) April 1971
114	File Transfer Protocol. (NOL) (Updated by RFC 141, 172, 171) April 1971
113	Network activity report: UCSB Rand. (NOL) (Updated by RFC 227) April 1971
112	User/Server Site Protocol: Network host questionnaire responses (NOL) April 1971
111	Pressure from the chairman. (NOL) (Updates RFC 107; Updated by RFC 130) March 1971
110	Conventions for using an IBM 2741 terminal as a user console for access to network server hosts. (NOL) (Updated by RFC 135) March 1971
109	Level III Server Protocol for the Lincoln Laboratory NIC 360/67 Host. (NOL) March 1971
108	Attendance list at the Urbana NWG meeting, February 17-19, 1971. (NOL) (Updates RFC 101) March 1971
107	Output of the Host-Host Protocol glitch cleaning committee. (NOL) (Updates RFC 102; Updated by RFC 179, RFC 132, RFC 124, RFC 111, NIC 7147) March 1971

RFC	Title
106	User/Server Site Protocol network host questionnaire. (NOL) March 1971
105	Network specifications for Remote Job Entry and Remote Job Output Retrieval at UCSB. (NOL) (Updated by RFC 217) March 1971
104	Link 191. (NOL) February 1971
103	Implementation of interrupt keys. (NOL) February 1971
102	Output of the Host-Host Protocol glitch cleaning committee. (NOL) (Updated by RFC 107) February 1971
101	Notes on the Network Working Group meeting, Urbana, Illinois, February 17, 1971. (NOL) (Updated by RFC 108, RFC 123) February 1971
100	Categorization and guide to NWG/RFCs. (NOL) February 1971
99	Network meeting. (NOL) (Updated by RFC 116) February 1971
98	Logger Protocol proposal. (NOL) (Updated by RFC 123) 1971 February
97	First cut at a proposed Telnet Protocol. (NOL) February 1971
96	Interactive network experiment to study modes of access to the Network Information Center. (NOL) February 1971
95	Distribution of NWG/RFC's through the NIC. (NOL) (Obsoleted by RFC 155) February 1971
94	Some thoughts on network graphics. (NOL) February 1971
93	Initial Connection Protocol. (NOL) January 1971
92	Not issued.
91	Proposed User-User Protocol. (NOL) December 1970
90	CCN as a network service center. (NOL) January 1971
89	Some historic moments in networking. (NOL) January 1971
88	NETRJS: A third level protocol for Remote Job Entry. (NOL) (Obsoleted by RFC 189) January 1971
87	Topic for discussion at the next Network Working Group meeting. (NOL) January 1971
86	Proposal: Network standard format for data stream control graphics display. (NOL) (Updated by RFC 125) January 1971
85	Network Working Group meeting. (NOL) December 1970
84	List of NWG/RFC's 1-80. (NOL) December 1970
83	Language-machine for data reconfiguration. (NOL) December 1970

RFC	Title
82	Network meeting notes. (NOL) December 1970
81	Request for reference information. (NOL) December 1970
80	Protocols and data formats. (NOL) (Obsoleted by RFC 123) December 1970
79	Logger Protocol error. (NOL) November 1970
78	NCP status report: USCB/Rand. (NOL) October 1970
77	Network meeting report. (NOL) November 1970
76	Connection by name: User oriented protocol. (NOL) October 1970
76A	Syntax and semantics for the terminal user control language for the proposed PDP-11 ARPA Network terminal system. (NOL) October 1970
75	Network meeting. (NOL) October 1970
74	Specifications for network use of the UCSB On-Line System. (NOL) (Updated by RFC 217, RFC 225) October 1970
73	Response to NWG/RFC 67. (NOL) September 1970
72	Proposed moratorium on changes to network protocol. (NOL) September 1970
71	Reallocation in case of input error. (NOL) September 1970
70	Note on padding. (NOL) (Updated by RFC 228) October 1970
69	Distribution list change for MIT. (NOL) (Updates RFC 52) September 1970
68	Comments on memory allocation control commands: CEASE, ALL, GVB, RET, and RFNM. (NOL) August 1970
67	Proposed change to Host/IMP spec to eliminate marking. (NOL) 1970
66	NIC—third level ideas and other noise. (NOL) (Obsoleted by RFC 123) August 1970
65	Comments on Host/Host Protocol document #1. (NOL) August 1970
64	Getting rid of marking. (NOL) July 1970
63	Belated network meeting report. (NOL) July 1970
62	Systems for interprocess communication in a resource-sharing computer network. (NOL) (Obsoletes RFC 61) August 1970
61	Note on interprocess communication in a resource-sharing computer network. (NOL) (Obsoleted by RFC 62) July 1970
60	Simplified NCP Protocol. (NOL) 1970 July
59	Flow control—fixed versus demand allocation Flow control fixed versus demand allocation. (NOL) June 1970

RFC	Title
58	Logical message synchronization. (NOL) 1970 June
57	Thoughts and reflections on NWG/RFC 54. (NOL) (Updates RFC 54) June 1970
56	Third level protocol: Logger Protocol. (NOL) June 1970
55	Prototypical implementation of the NCP. (NOL) June 1970
54	Official protocol proffering. (NOL) (Updated by RFC 57) June 1970
53	Official protocol mechanism. (NOL) June 1970
52	Updated distribution list. (NOL) (Updated by RFC 69) July 1970
51	Proposal: A Network Interchange Language. (NOL) May 1970
50	Comments on the Meyer proposal. (NOL) April 1970
49	Conversations with S. Crocker (UCLA). (NOL) April 1970
48	Possible protocol plateau. (NOL) April 1970
47	BBN's comments on NWG/RFC #33. (NOL) (Updates RFC 33) April 1970
46	ARPA Network protocol notes. (NOL) April 1970
45	New protocol is coming. (NOL) April 1970
44	Comments: NWG/RFC 33 & 36. (NOL) (Updates #36) April 1970
43	Proposed meeting. (NOL) April 1970
42	Message data types. (NOL) March 1970
41	IMP-IMP teletype communication. (NOL) March 1970
40	More comments on the forthcoming protocol. (NOL) March 1970
39	Comments on protocol re: NWG/RFC #36. (NOL) (Updates RFC 36) March 1970
38	Comments on network protocol from NWG/RFC #36. (NOL) March 1970
37	Network meeting epilogue, etc. (NOL) March 1970
36	Protocol notes. (NOL) (Updates RFC 33; Updated by RFC 39, RFC 44) March 1970
35	Network meeting. (NOL) March 1970
34	Some brief preliminary notes on the Augmentation Research Center clock. (NOL) February 1970
33	New Host-Host Protocol. (NOL) (Obsoletes RFC 11; Updated by RFC 36, RFC 47) February 1970
32	Connecting M.I.T. computers to the ARPA computer-to-computer communication network. (NOL) January 1969

RFC	Title
31	Binary message forms in computer. (NOL) February 1968
30	Documentation conventions. (Obsoletes RFC 27) February 1970
29	Response to RFC 28. January 1970
28	Time standards. January 1970
27	Documentation conventions. (Obsoletes RFC 24; Obsoleted by RFC 30) December 1969
26	Not issued.
25	No high link numbers. October 1969
24	Documentation conventions. (Obsoletes RFC 16; Obsoleted by RFC 27) November 1969
23	Transmission of multiple control messages. October 1969
22	Host-host control message formats. (NOL) October 1969
21	Network meeting. October 1969
20	ASCII format for network interchange. (NOL) October 1969
19	Two protocol suggestions to reduce congestion at swap bound nodes. October 1969.
18	[Link assignments] September 1969.
17	Some questions re: Host-IMP Protocol. August 1969.
16	M.I.T. (Obsoletes RFC 10; Obsoleted by RFC 24) September 1969
15	Network subsystem for time sharing hosts. (NOL) September 1969
14	Not issued.
13	[Referring to NWG/RFC 11]. (NOL) August 1969
12	IMP-Host interface flow diagrams. (NOL) August 1969
11	Implementation of the Host-Host software procedures in GORDO. (NOL) (Obsoleted by RFC 33) August 1969
10	Documentation conventions. (Obsoletes RFC 3; Obsoleted by RFC 16) July 1969
9	Host software. (NOL) May 1969
8	Functional specifications for ARPA Network. May 1969
7	Host-IMP interface. (NOL) May 1969
6	Conversation with Bob Kahn. April 1969
5	Decode Encode Language. June 1969

RFC	Title
4	Network timetable. (NOL) March 1969
3	Documentation conventions. (Obsoleted by RFC 10) April 1969
2	Host software. (NOL) April 1969
1	Host software. (NOL) April 1969

BIBLIOGRAPHY

Abbatiello, Judy and Ray Sarch, editors. *Telecommunications and Data Communications Factbook*. Ramsey, NJ: CCMI/McGraw-Hill, Inc., 1987.

Apple Computer, Inc. *Planning and Managing AppleTalk Networks*. Menlo Park, CA: Addison-Wesley Publishing Company, 1991.

Apple Computer, Inc. *Technical Introduction to the Macintosh Family Second Edition*. Menlo Park, CA: Addison-Wesley Publishing Company, 1992.

Ashley, Ruth, and Judi N. Fernandez. *Job Control Language*. New York: John Wiley & Sons, Inc., 1984.

Aspray, William. *John Von Neumann and The Origins of Modern Computing*. Cambridge, MA: MIT Press, 1990.

ATM Forum. *ATM User-Network Interface Specification*. Englewood Cliffs, NJ: Prentice-Hall, Inc., 1993.

Bach, Maurice J. *The Design of The UNIX Operating System*. Englewood Cliffs, NJ: Prentice-Hall, Inc., 1986.

Baggott, Jim. *The Meaning of Quantum Theory*, New York: Oxford University Press, 1992.

Bashe, Charles J., Lyle R. Johnson, John H. Palmer, and Emerson W. Pugh. *IBM's Early Computers*. Cambridge, MA: MIT Press, 1986.

Berson, Alex. *APPC Introduction to LU6.2*. New York: McGraw-Hill, Inc., 1990.

Black, Ulysses. *The V Series Recommendations Protocols for Data Communications Over the Telephone Network*. New York: McGraw-Hill, Inc., 1991.

———. *Data Networks Concepts, Theory, and Practice*. Englewood Cliffs, NJ: Prentice-Hall, Inc., 1989.

———. *The X Series Recommendations Protocols for Data Communications Networks*. New York: McGraw-Hill, Inc., 1991.

———. *TCP/IP and Related Protocols*. New York: McGraw-Hill, Inc., 1992.

Blyth, W. John and Mary M. *Telecommunications: Concepts, Development, and Management*. Mission Hills, CA: Glencoe/McGraw-Hill, Inc., 1990.

Bohl, Marilyn. *Information Processing, Third Edition*. Chicago: Science Research Associates, Inc., 1971.

Bradbeer, Robin, Peter De Bono, and Peter Laurie. *The Beginner's Guide to Computers*. Menlo Park, CA: Addison-Wesley Publishing Company, 1982.

Brookshear, J. Glenn. *Computer Science An Overview*. Menlo Park, CA: The Benjamin Cummings Publishing Company, Inc., 1988.

Bryant, David. *Physics*. London: Hodder and Stoughton Ltd., 1971.

Campbell, Joe. *The RS-232 Solution*. Alameda, CA: Sybex, 1984.

———. *C Programmer's Guide to Serial Communications*. Carmel, IN: Howard W. Sams & Company, 1987.

Chorafas, Dimitris N. *Local Area Network Reference*. New York: McGraw-Hill, Inc., 1989.

Comer, Douglas E. *Internetworking with TCP/IP*, 2 vols. Englewood Cliffs, NJ: Prentice-Hall, Inc., 1991.

Dayton, Robert L. *Telecommunications: The Transmission of Information*. New York: McGraw-Hill, Inc., 1991.

Dern, Daniel P. *The Internet Guide for New Users*. New York: McGraw-Hill, Inc., 1994.

Digital Equipment Corporation. *OpenVMS Software Overview*. AA-PVXHA-TE, Maynard, MA: Digital Equipment Corporation, 1993.

———. *DECnet Digital Network Architecture (Phase V): Network Routing Layer Functional Specification*. EK-DNA03-FS-001, Maynard, MA: Digital Equipment Corporation, 1991.

———. *DECnet/OSI for OpenVMS: Introduction and Planning*. AA-PNHTB-TE, Maynard, MA: Digital Equipment Corporation, 1993.

———. *OpenVMS DCL Dictionary*, 2 vols. AA-PV5LA-TK, Maynard, MA: Digital Equipment Corporation, 1993.

———. *OpenVMS Glossary*. AA-PV5UA-TK, Maynard, MA: Digital Equipment Corporation, 1993.

Edmunds, John J. *SAA/LU 6.2 Distributed Networks and Applications*. New York: McGraw-Hill, Inc., 1992.

Feit, Sidnie. *TCP/IP Architecture, Protocols, and Implementation*. New York: McGraw-Hill, Inc., 1993.

Forney, James S. *MS-DOS Beyond 640K: Working with Extended and Expanded Memory*. Blue Ridge Summit, PA: Windcrest Books, 1993.

Fortier, Paul J. *Handbook of LAN Technology*. New York: Intertext Publications/Multiscience Press, 1989.

Gasman, Lawrence. *Broadband Networking*. New York: Van Nostrand Reinhold, 1994.

Graubart-Cervone, H. Frank. *VSE/ESA JCL Utilities, Power, and VSAM*. New York: McGraw-Hill, Inc., 1994.

Groff, James R., and Paul N. Weinbert. *Understanding UNIX: A Conceptual Guide.* Carmel, IN: Que Corporation, 1983.

Hecht, Jeff. *Understanding Fiber Optics.* Carmel, IN: Howard W. Sams & Company, 1990.

Hewlett Packard Company. *Using the X Window System.* B1171-90037, Ft. Collins, CO: Hewlett Packard Company, 1991.

———. *HP OpenView SNMP Agent Administrator's Reference.* J2322-90002, Ft. Collins, CO: Hewlett Packard Company, 1992.

———. *HP OpenView SNMP Management Platform Administrator's Reference.* J2313-90001, Ft. Collins, CO: Hewlett Packard Company, 1992.

———. *HP OpenView Windows User's Guide.* J2316-90000, Ft. Collins, CO: Hewlett Packard Company, 1992.

———. *Using HP-UX: HP 9000 Workstations.* B2910-90001, Ft. Collins, CO: Hewlett Packard Company, 1992.

IBM Corporation. *IBM Virtual Machine Facility: Terminal User's Guide.* GC20-1810-9, Poughkeepsie, NY: IBM Corporation, 1980.

———. *IBM System/370 Extended Architecture: Principles of Operation.* SA22-7085-0, Research Triangle Park, NC: IBM Corporation, 1983.

———. *IBM 3270 Information Display System: 3274 Control Unit Description and Programmer's Guide.* GA23-0061-2, Research Triangle Park, NC: IBM Corporation, 1985.

———. *JES3 Introduction.* GC23-0039-2, Poughkeepsie, NY: IBM Corporation, 1986.

———. *IBM System/370: Principles of Operation.* GA22-7000-10, Poughkeepsie, NY: IBM Corporation, 1987.

———. *3270 Information Display System: Introduction.* GA27-2739-22, Research Triangle Park, NC: IBM Corporation, 1988.

———. *IBM Enterprise Systems Architecture/370: Principles of Operation.* SA22-7200-0, Poughkeepsie, NY: IBM Corporation, 1988.

———. *MVS/ESA Operations: System Commands Reference Summary.* GX22-0013-1, Poughkeepsie, NY: IBM Corporation, 1989.

———. *Enterprise System/9000 Models 120, 130, 150, and 170: Introducing the System.* GA24-4186-00, Endicott, NY: IBM Corporation, 1990.

———. *Enterprise Systems Architecture/390: Principles of Operation.* SA22-7201-00, Poughkeepsie, NY: IBM Corporation, 1990.

———. *IBM 3172 Interconnect Controller: Presentation Guide.* White Plains, NY: IBM Corporation, 1990.

————. *IBM VSE/ESA: System Control Statements.* SC33-6513-00, Mechanics-burg, PA: IBM Corporation, 1990.

————. *IBM VSE/POWER: Networking.* SC33-6573-00, Mechanicsburg, PA: IBM Corporation, 1990.

————. *MVS/ESA SP Version 4 Technical Presentation Guide.* GG24-3594-00, Poughkeepsie, NY: IBM Corporation, 1990.

————. *Virtual Machine/Enterprise Systems Architecture.* GC24-5441, Endi-cott, NY: IBM Corporation, 1990.

————. *VM/ESA and Related Products: Overview.* GG24-3610-00, Poughkeepsie, NY: IBM Corporation, 1990.

————. *3174 Establishment Controller: Functional Description.* GA23-0218-08, Research Triangle Park, NC: IBM Corporation, 1991.

————. *Dictionary of Computing.* SC20-1699-8, Poughkeepsie, NY: IBM Corporation, 1991.

————. *Enterprise Systems Architecture/390 ESCON I/O Interface: Physical Layer.* SA23-0394-00, Kingston, NY: IBM Corporation, 1991.

————. *Enterprise Systems Connection.* GA23-0383-01, Kingston, NY: IBM Corporation, 1991.

————. *Enterprise Systems Connection: ESCON I/O Interface.* SA22-7202-01, Poughkeepsie, NY: IBM Corporation, 1991.

————. *Enterprise Systems Connection Manager.* GC23-0422-01, Kingston, NY: IBM Corporation, 1991.

————. *Installation Guidelines for the IBM Token-Ring Network Products.* GG24-3291-02, Research Triangle Park, NC: IBM Corporation, 1991.

————. *NetView: NetView Graphic Monitor Facility Operation.* SC31-6099-1, Research Triangle Park, NC: IBM Corporation, 1991.

————. *Systems Network Architecture: Concepts and Products.* GC30-3072-4, Research Triangle Park, NC: IBM Corporation, 1991.

————. *Systems Network Architecture: Technical Overview.* GC30-3073-3, Research Triangle Park, NC: IBM Corporation, 1991.

————. *Systems Network Architecture: Type 2.1 Node Reference, Version 1.* SC20-3422-2, Research Triangle Park, NC: IBM Corporation, 1991.

————. *Virtual Machine/Enterprise System Architecture: General Information.* GC24-5550-02, Endicott, NY: IBM Corporation, 1991.

————. *3172 Interconnect Controller: Operator's Guide.* GA27-3970-00, Research Triangle Park, NC: IBM Corporation, 1992.

———. *3172 Interconnect Controller: Planning Guide.* GA27-3867-05, Research Triangle Park, NC: IBM Corporation, 1992.

———. *3270 Information Display System: Data Stream Programmer's Reference.* GA23-0059-07, Research Triangle Park, NC: IBM Corporation, 1992.

———. *APPN Architecture and Product Implementations Tutorial.* GG24-3669-01, Research Triangle Park, NC: IBM Corporation, 1992.

———. *ES/9000 Multi-Image Processing, Volume 1: Presentation and Solutions Guidelines.* GG24-3920-00, Poughkeepsie, NY: IBM Corporation, 1992.

———. *High-Speed Networking Technology: An Introductory Survey.* GG24-3816-00, Raleigh, NC: IBM Corporation, 1992.

———. *IBM Networking Systems: Planning and Reference.* SC31-6191-00, Research Triangle Park, NC: IBM Corporation, 1992.

———. *MVS/ESA and Data in Memory: Performance Studies.* GG24-3698-00, Poughkeepsie, NY: IBM Corporation, 1992.

———. *MVS/ESA: General Information for MVS/ESA System Product, Version 4.* GC28-1600-04, Poughkeepsie, NY: IBM Corporation, 1992.

———. *Sockets Interface for CICS—Using TCP/IP Version 2 Release 2 for MVS: User's Guide.* GC31-7015-00, Research Triangle Park, NC: IBM Corporation, 1992.

———. *Synchronous Data Link Control: Concepts* GA27-3093-04, Research Triangle Park, NC: IBM Corporation, 1992.

———. *TCP/IP Version 2 Release 2.1 for MVS: Offload of TCP/IP Processing.* SA31-7033-00, Research Triangle Park, NC: IBM Corporation, 1992.

———. *TCP/IP Version 2 Release 2.1 for MVS: Planning and Customization.* SC31-6085-02, Research Triangle Park, NC: IBM Corporation, 1992.

———. *The IBM 6611 Network Processor.* GG24-3870-00, Raleigh, NC: IBM Corporation, 1992.

———. *VM/ESA: CMS Primer.* SC24-5458-02, Endicott, NY: IBM Corporation, 1992.

———. *VM/ESA Release 2 Overview.* GG24-3860-00, Poughkeepsie, NY: IBM Corporation, 1992.

———. *VSE/ESA Version 1.3: An Introduction Presentation Foil Master.* GG24-4008-00, Raleigh, NC: IBM Corporation, 1992.

———. *IBM Network Products Implementation Guide.* GG24-3649-01, Raleigh NC: IBM Corporation, 1993.

———. *IBM VSE/Interactive Computing and Control Facility: Primer.* SC33-6561-01, Charlotte, NC: IBM Corporation, 1993.

————. *LAN File Services/ESA: MVS Guide and Reference.* SH24-5265-00, Endicott, NY: IBM Corporation, 1993.

————. *LAN File Services/ESA: VM Guide and Reference.* SH24-5264-00, Endicott, NY: IBM Corporation, 1993.

————. *LAN Resource Extension and Services/VM: Guide and Reference.* SC24-5622-01, Endicott, NY: IBM Corporation, 1993.

————. *MVS/ESA JES2 Commands.* GC23-0084-04, Poughkeepsie, NY: IBM Corporation, 1993.

————. *MVS/ESA: System Commands.* GC28-1626-05, Poughkeepsie, NY: IBM Corporation, 1993.

————. *NetView: Command Quick Reference.* SX75-0090-00, Research Triangle Park, NC: IBM Corporation, 1993.

————. *NetView: Installation and Administration.* SC31-7084-00, Research Triangle Park, NC: IBM Corporation, 1993.

————. *System Information Architecture: Formats.* GA27-3136, Research Triangle Park, NC: IBM Corporation, 1993.

————. *System Network Architecture: Architecture Reference, Version 2.* SC30-3422-03, Research Triangle Park, NC: IBM Corporation, 1993.

————. *The Host as a Data Server Using LANRES and Novell NetWare.* GG24-4069-00, Poughkeepsie, NY: IBM Corporation, 1993.

————. *Virtual Machine/Enterprise Systems Architecture.* SC24-5460-03, Endicott, NY: IBM Corporation, 1993.

————. *VM/ESA: CMS Command Reference.* SC24-5461-03, Endicott, NY: IBM Corporation, 1993.

————. *VM/ESA: CP Command and Utility Reference.* SC24-5519-03, Endicott, NY: IBM Corporation, 1993.

————. *VTAM: Operation.* SC31-6420-00, Research Triangle Park, NC: IBM Corporation, 1993.

————. *VTAM: Resource Definition Reference Version 4 Release 1 for MVS/ESA.* SC31-6427-00, Research Triangle Park, NC: IBM Corporation, 1993.

————. *LAN Resource Extension and Services/MVS: General Information.* GC24-5625-03, Endicott, NY: IBM Corporation, 1994

————. *LAN Resource Extension and Services/MVS: Guide and Reference.* SC24-5623-02, Endicott, NY: IBM Corporation, 1994.

————. *LAN Resource Extension and Services/VM: General Information.* GC24-5618-03, Endicott, NY: IBM Corporation, 1994.

———. *MVS/ESA: JES2 Command Reference Summary.* GX22-0017-03, Poughkeepsie, NY: IBM Corporation. 1993.

Jain, Bijendra N., and Ashok K. Agrawala. *Open Systems Interconnection.* New York: McGraw-Hill, Inc., 1993.

Kessler, Gary C. *ISDN.* New York: McGraw-Hill, Inc., 1990.

Kessler, Gary C., and David A. Train. *Metropolitan Area Networks: Concepts, Standards, and Services.* New York: McGraw-Hill, Inc., 1992.

Killen, Michael. *SAA and UNIX: IBM's Open Systems Strategy.* New York: McGraw-Hill, Inc., 1992.

———. *SAA: Managing Distributed Data.* New York: McGraw-Hill, Inc., 1992.

Kochan, Stephen G, and Patrick H. Wood. *Exploring the UNIX System.* Indianapolis, IN: Hayden Books, 1984.

Madron, Thomas W. *Local Area Networks: The Next Generation.* New York: John Wiley & Sons, Inc., 1988.

Martin, James. *Local Area Networks Architectures and Implementations.* Englewood Cliffs, New Jersey: Prentice-Hall, Inc., 1989.

McClain, Gary R. *Open Systems Interconnection Handbook.* New York: Inter-text Publications/Multiscience Press, 1991.

Meijer, Anton. *Systems Network Architecture: A Tutorial.* London: Pitman Press, 1987.

Merrow, Bill. *VSE/ESA Performance Management and Fine Tuning.* New York: McGraw-Hill, Inc., 1993.

———. *VSE/ESA Concepts and Facilities.* New York: McGraw-Hill, Inc., 1994.

Nash, Stephen G., editor. *A History of Scientific Computing.* New York: ACM Press, 1990.

Naugle, Matthew G. *Local Area Networking.* New York: McGraw-Hill, Inc., 1991.

———. *Network Protocol Handbook.* New York McGraw-Hill, Inc., 1994.

Nemzow, Martin A. W. *The Ethernet Management Guide: Keeping the Link, Second Edition.* New York: McGraw-Hill, Inc., 1992.

O'Dell, Peter. *The Computer Networking Book.* Chapel Hill, NC: Ventana Press, Inc., 1989.

Parker, Sybil P. *McGraw-Hill Dictionary of Science and Engineering.* New York: McGraw-Hill, Inc., 1984.

Pugh, Emerson W. *Memories that Shaped an Industry.* Cambridge, MA: MIT Press, 1984.

Pugh, Emerson W., Lyle R. Johnson, and John H. Palmer. *IBM's 360 and Early 370 Systems.* Cambridge, MA: MIT Press.

Ranade, Jay, and George C. Sackett. *Introduction to SNA Networking Using VTAM/NCP.* New York: McGraw-Hill, Inc., 1989.

Rose, Marshall T. *The Open Book: A Practical Perspective on OSI.* Englewood Cliffs, NJ: Prentice Hall, Inc., 1990.

———. *The Simple Book: An Introduction to Management of TCP/IP-Based Internets.* Englewood Cliffs, NJ: Prentice-Hall, Inc., 1991.

Samson, Stephen L. *MVS Performance Management.* New York: McGraw-Hill, Inc., 1990.

Savit, Jeffrey. *VM/CMS Concepts and Facilities.* New York: McGraw-Hill, Inc., 1993.

Schatt, Stan. *Understanding Local Area Networks, Second Edition.* Carmel, IN: Howard W. Sams & Company, 1990.

Schlar, Sherman K. *Inside X.25: A Manager's Guide.* New York: McGraw-Hill, Inc., 1990.

Seyer, Martin D. *RS-232 Made Easy: Connecting Computers, Printers, Terminals, and Modems.* Englewood Cliffs, NJ: Prentice-Hall, Inc., 1991.

Sidhu, Gursharan S., Richard F. Andrews, and Alan B. Oppenheimer. *Inside AppleTalk, Second Edition,* Menlo Park, CA: Addison-Wesley Publishing Company, 1990.

Spohn, Darren L. *Data Network Design.* New York: McGraw-Hill, Inc., 1993.

Stallings, William. *Handbook of Computer Communications Standards,* 3 vols. New York: Macmillan Publishing Company, 1987.

Stamper, David A. *Business Data Communications,* Menlo Park, CA: The Benjamin Cummings Publishing Company, 1986,

Tang, Adrian, and Sophia Scoggins. *Open Networking with OSI.* Englewood Cliffs, NJ: Prentice-Hall, Inc., 1992.

Umar, Amjad. *Distributed Computing: A Practical Synthesis.* Englewood Cliffs, NJ: Prentice-Hall, Inc., 1993.

White, Gene. *Internetworking and Addressing,* New York: McGraw-Hill, Inc., 1992.

Zwass, Vladimir. *Introduction to Computer Science.* New York: Barnes & Noble Books, 1981.

INDEX

SOFTWARE AND INFORMATION LICENSE

The software and information on this diskette (collectively referred to as the "Product") are the property of The McGraw-Hill Companies, Inc. ("McGraw-Hill") and are protected by both United States copyright law and international copyright treaty provision. You must treat this Product just like a book, except that you may copy it into a computer to be used and you may make archival copies of the Products for the sole purpose of backing up our software and protecting your investment from loss.

By saying "just like a book," McGraw-Hill means, for example, that the Product may be used by any number of people and may be freely moved from one computer location to another, so long as there is no possibility of the Product (or any part of the Product) being used at one location or on one computer while it is being used at another. Just as a book cannot be read by two different people in two different places at the same time, neither can the Product be used by two different people in two different places at the same time (unless, of course, McGraw-Hill's rights are being violated).

McGraw-Hill reserves the right to alter or modify the contents of the Product at any time.

This agreement is effective until terminated. The Agreement will terminate automatically without notice if you fail to comply with any provisions of this Agreement. In the event of termination by reason of your breach, you will destroy or erase all copies of the Product installed on any computer system or made for backup purposes and shall expunge the Product from your data storage facilities.

LIMITED WARRANTY

McGraw-Hill warrants the physical diskette(s) enclosed herein to be free of defects in materials and workmanship for a period of sixty days from the purchase date. If McGraw-Hill receives written notification within the warranty period of defects in materials or workmanship, and such notification is determined by McGraw-Hill to be correct, McGraw-Hill will replace the defective diskette(s). Send request to:

Customer Service
McGraw-Hill
Gahanna Industrial Park
860 Taylor Station Road
Blacklick, OH 43004-9615

The entire and exclusive liability and remedy for breach of this Limited Warranty shall be limited to replacement of defective diskette(s) and shall not include or extend to any claim for or right to cover any other damages, including but not limited to, loss of profit, data, or use of the software, or special, incidental, or consequential damages or other similar claims, even if McGraw-Hill has been specifically advised as to the possibility of such damages. In no event will McGraw-Hill's liability for any damages to you or any other person ever exceed the lower of suggested list price or actual price paid for the license to use the Product, regardless of any form of the claim.

THE McGRAW-HILL COMPANIES, INC. SPECIFICALLY DISCLAIMS ALL OTHER WARRANTIES, EXPRESS OR IMPLIED, INCLUDING BUT NOT LIMITED TO, ANY IMPLIED WARRANTY OF MERCHANTABILITY OR FITNESS FOR A PARTICULAR PURPOSE. Specifically, McGraw-Hill makes no representation or warranty that the Product is fit for any particular purpose and any implied warranty of merchantability is limited to the sixty day duration of the Limited Warranty covering the physical diskette(s) only (and not the software or in-formation) and is otherwise expressly and specifically disclaimed.

This Limited Warranty gives you specific legal rights; you may have others which may vary from state to state. Some states do not allow the exclusion of incidental or consequential damages, or the limitation on how long an implied warranty lasts, so some of the above may not apply to you.

This Agreement constitutes the entire agreement between the parties relating to use of the Product. The terms of any purchase order shall have no effect on the terms of this Agreement. Failure of McGraw-Hill to insist at any time on strict compliance with this Agreement shall not constitute a waiver of any rights under this Agreement. This Agreement shall be construed and governed in accordance with the laws of New York. If any provision of this Agreement is held to be contrary to law, that provision will be enforced to the maximum extent permissible and the remaining provisions will remain in force and effect.

About the Author

Ed Taylor is the founder of Information World, Inc. He is a former network architect for IBM, where he designed large heterogeneous networks. He has served as a consultant to such firms as NEC, Chrysler, BASF, Hewlett-Packard, Dow Jones, Ore-Ida Foods, Mutual of New York, IBM, Orange County, California, Harris, and others.

Mr. Taylor has been in the networking and Internet industry for twelve years. He is the author of several definitive books in the industry. For example, he authored *The Network Architecture Design Handbook*, and other books in the Taylor Networking Series published by McGraw-Hill. His book *The McGraw-Hill Internetworking Handbook* is now in its second edition, and is also published in Japanese. Mr. Taylor authored the *Multiplatform Network Management* book which is being translated into Traditional Character Long Form Chinese in Taiwan. Some of Mr. Taylor's work has been presented at the Pentagon, and is also translated into French.

Ed has also been a featured speaker at major industry trade shows such as Comnet, UNIX Expo, Enterprise Expo, and Client/Server World to name a few. He has toured the United States, and spoken as well in Canada and France. Some of his professional affiliations are IEEE, NASPA, ACM, SPIE, and the Lakewood Training and Research Group.